Advances in
PARASITOLOGY
The Evolution of Parasitism —
a Phylogenetic Perspective

VOLUME 54

Editorial Board

M. Coluzzi, Director, Istituto de Parassitologia, Università Degli Studi di Roma 'La Sapienza', P. le A. Moro 5, 00185 Roma, Italy

C. Combes, Laboratoire de Biologie Animale, Université de Perpignan, Centre de Biologie et d'Ecologie Tropicale et Méditerranéenne, Avenue de Villeneuve, 66860 Perpignan Cedex, France

D.D. Despommier, Division of Tropical Medicine and Environmental Sciences, Department of Microbiology, Columbia University, 630 West 168th Street, New York, NY 10032, USA

J.J. Shaw, Instituto de Ciências Biomédicas, Universidade de São Paulo, av. Prof Lineu Prestes 1374, 05508-900 Cidade Universitária, São Paulo, SP, Brazil

K. Tanabe, Laboratory of Biology, Osaka Institute of Technology, 5-16-1 Ohmiya Asahi-Ku, Osaka 535, Japan

P. Wenk, Falkenweg 69, D-72076 Tübingen, Germany

Advances in PARASITOLOGY

The Evolution of Parasitism — a Phylogenetic Perspective

Series editors

J.R. BAKER, R. MULLER and D. ROLLINSON

Guest Editor

D.T.J. LITTLEWOOD

*The Natural History Museum,
London, England*

VOLUME 54

ELSEVIER
ACADEMIC
PRESS

Amsterdam Boston Heidelberg London New York Oxford Paris
San Diego San Francisco Singapore Sydney Tokyo

ELSEVIER B.V.	ELSEVIER Inc.	ELSEVIER Ltd	ELSEVIER Ltd
Sara Burgerhartstraat 25	525 B Street, Suite 1900	The Boulevard, Langford Lane	84 Theobalds Road
P.O. Box 211, 1000 AE Amsterdam	San Diego, CA 92101-4495	Kidlington, Oxford OX5 1GB	London WC1X 8RR
The Netherlands	USA	UK	UK

© 2003 Elsevier Ltd. All rights reserved.

This work is protected under copyright by Elsevier Ltd., and the following terms and conditions apply to its use:

Photocopying

Single photocopies of single chapters may be made for personal use as allowed by national copyright laws. Permission of the Publisher and payment of a fee is required for all other photocopying, including multiple or systematic copying, copying for advertising or promotional purposes, resale, and all forms of document delivery. Special rates are available for educational institutions that wish to make photocopies for non-profit educational classroom use.

Permissions may be sought directly from Elsevier's Rights Department in Oxford, UK: phone: (+44) 1865 843830, fax: (+44) 1865 853333, e-mail: permissions@elsevier.com. Requests may also be completed on-line via the Elsevier homepage (http://www.elsevier.com/locate/permissions).

In the USA, users may clear permissions and make payments through the Copyright Clearance Center, Inc., 222 Rosewood Drive, Danvers, MA 01923, USA; phone: (+1) (978) 7508400, fax: (+1) (978) 7504744, and in the UK through the Copyright Licensing Agency Rapid Clearance Service (CLARCS), 90 Tottenham Court Road, London W1P 0LP, UK; phone: (+44) 207 631 5555; fax: (+44) 207 631 5500. Other countries may have a local reprographic rights agency for payments.

Derivative Works

Tables of contents may be reproduced for internal circulation, but permission of the Publisher is required for external resale or distribution of such material. Permission of the Publisher is required for all other derivative works, including compilations and translations.

Electronic Storage or Usage

Permission of the Publisher is required to store or use electronically any material contained in this work, including any chapter or part of a chapter.

Except as outlined above, no part of this work may be reproduced, stored in a retrieval system or transmitted in any form or by any means, electronic, mechanical, photocopying, recording or otherwise, without prior written permission of the Publisher.

Address permissions requests to: Elsevier's Rights Department, at the fax and e-mail addresses noted above.

Notice

No responsibility is assumed by the Publisher for any injury and/or damage to persons or property as a matter of products liability, negligence or otherwise, or from any use or operation of any methods, products, instructions or ideas contained in the material herein. Because of rapid advances in the medical sciences, in particular, independent verification of diagnoses and drug dosages should be made.

First edition 2003

A catalogue record for this is available from the British Library.

ISBN: 0-12-031754-0

Typeset by Keyword Publishing Services, Barking, Essex
Printed and bound in Great Britain by MPG Books Ltd, Bodmin, Cornwall

∞ The paper used in this publication meets the requirements of ANSI/NISO Z39.48-1992 (Permanence of Paper).

CONTRIBUTORS TO VOLUME 54

F. J. AYALA, *Department of Ecology and Evolutionary Biology, University of California at Irvine, 321 Steinhaus Hall, Irvine, CA 92697-2525, USA*

M. L. BLAXTER, *Institute of Cell, Animal and Population Biology, University of Edinburgh, King's Buildings, Edinburgh EH9 3JT, UK*

R. A. BRAY, *Parasitic Worms Division, Department of Zoology, The Natural History Museum, Cromwell Road, London SW7 5BD, UK*

M. A. CHARLESTON, *Department of Zoology, University of Oxford, South Parks Road, Oxford OX1 3PS, UK*

C. J. CREEVEY, *Bioinformatics and Pharmacogenomics Laboratory, Department of Biology, National University of Ireland, Maynooth, Co. Kildare, Ireland*

T. H. CRIBB, *Department of Microbiology & Parasitology and Centre for Marine Studies, The University of Queensland, Brisbane, Australia 4072*

A. DRUMMOND, *Department of Zoology and Department of Statistics University of Oxford, South Parks Road, Oxford OX1 3PS, UK*

P. J. KEELING, *Canadian Institute for Advanced Research, Department of Botany, University of British Columbia, 3529-6270 University Boulevard, Vancouver, BC V6T 1Z4, Canada*

D. T. J. LITTLEWOOD, *Parasitic Worms Division, Department of Zoology, The Natural History Museum, Cromwell Road, London SW7 5BD, UK*

J. O. MCINERNEY, *Bioinformatics and Pharmacogenomics Laboratory, Department of Biology, National University of Ireland, Maynooth, Co. Kildare, Ireland*

S. MORAND, *Centre de Biologie et de Gestion des Populations, Campus International de Baillarguet, CS 30 016, 34988 Montferrier sur Lez, France*

P. D. OLSON, *Parasitic Worms Division, Department of Zoology, The Natural History Museum, Cromwell Road, London SW7 5BD, UK*

R. POULIN, *Department of Zoology, University of Otago, P.O. Box 56, Dunedin, New Zealand*

O. G. PYBUS, *Department of Zoology, University of Oxford, South Parks Road, Oxford OX1 3PS, UK*

A. RAMBAUT, *Department of Zoology, University of Oxford, South Parks Road, Oxford OX1 3PS, UK*

S. M. RICH, *Division of Infectious Disease, Tufts University School of Veterinary Medicine, 200 Westboro Road, North Grafton, MA 01536, USA*

J. B. WHITFIELD, *Department of Entomology, 320 Morrill Hall, 505 S. Goodwin Avenue, University of Illinois, Urbana, IL 61801, USA*

B. A. P. WILLIAMS, *Canadian Institute for Advanced Research, Department of Botany, University of British Columbia, 3529-6270 University Boulevard, Vancouver, BC V6T 1Z4, Canada*

PREFACE

We are extremely fortunate to have Tim Littlewood as the guest editor of this, our third, special volume of *Advances in Parasitology*. He has assembled an impressive list of authors with expertise in diverse fields of parasitology, systematics and evolutionary biology to produce a volume that provides an authoritative overview of our current understanding of the evolution of parasitism. Parasites do not have extensive fossil records, so to trace their evolutionary past it is necessary to delve into their molecular and morphological make-up and to look closely at relationships between existing taxa. Fortunately, the huge growth in acquisition of molecular data and the concurrent development of phylogenetics have enabled major new insights into the evolution of many of the parasite groups represented in this volume.

We sincerely thank the guest editor and all the authors who have contributed to the planning and writing of this special volume.

John Baker
Ralph Muller
David Rollinson

CONTENTS

CONTRIBUTORS TO VOLUME 54 v
PREFACE vii

Introduction — Phylogenies, Phylogenetics, Parasites and the Evolution of Parasitism

D. Timothy J. Littlewood

Cryptic Organelles in Parasitic Protists and Fungi

Bryony A. P. Williams and Patrick J. Keeling

    Abstract 10
1. Introduction 10
2. The Origin of Mitochondria and Plastids by Endosymbiosis 12
3. Cryptic Organelles and How to Find Them 21
4. Case Histories – Mitochondria 24
5. Case Histories – Plastids 38
6. Future Directions 49
    Acknowledgements 53
    References 54

Phylogenetic Insights into the Evolution of Parasitism in Hymenoptera

James B. Whitfield

    Abstract 69
1. Introduction 70
2. Some Questions about Hymenopteran Parasitoid Evolution Addressed using Phylogeny 74
3. Evolution from Parasitism to Other Lifestyles 84

4.	The Comparative Method and Parasitoids: Future Prospects	89
5.	Conclusion	91
	Acknowledgements	91
	References	91

Nematoda: Genes, Genomes and the Evolution of Parasitism

Mark L. Blaxter

	Abstract	102
1.	Nematode Genomes and the Evolution of Parasitism	103
2.	Nematode Parasitism	108
3.	Nematode Genomes and Parasitism	126
4.	Summary	165
	Acknowledgements	167
	References	167

Life Cycle Evolution in the Digenea: a New Perspective from Phylogeny

Thomas H. Cribb, Rodney A. Bray, Peter D. Olson, D. Timothy J. Littlewood

	Abstract	198
1.	Introduction	198
2.	Methods	198
3.	Background to the Digenea	203
4.	Mapping and Interpreting Life Cycle Traits	209
5.	Problems	240
	Appendix	244
	Acknowledgements	249
	References	249

Progress in Malaria Research: the Case for Phylogenetics

Stephen M. Rich and Francisco J. Ayala

	Abstract	255
1.	The Malaria Phylum: Apicomplexa	256

2. Morphology, Phylogenetics and *Plasmodium* Systematics 258
3. Evolution and Extant Distribution of Malignant Human
   Malaria: *P. falciparum* 266
4. Concluding Remarks 275
   References 275

Phylogenies, the Comparative Method and Parasite Evolutionary Ecology

Serge Morand and Robert Poulin

	Abstract	282
1.	Introduction	282
2.	Phylogenetic Effects and Constraints, and the Need for Phylogenies	283
3.	The Phylogenetically Independent Contrasts Method	284
4.	Diversity and Diversification	287
5.	The Phylogenetic Eigenvector Method	290
6.	The Study of Host–Parasite Co-adaptation Using the Independent Contrasts Method	292
7.	The Study of Host–Parasite Co-adaptation Using PER	293
8.	Scepticism about Comparative Methods: Why Bother with Phylogeny?	295
9.	Phylogenetically Structured Environmental Variation	296
10.	Conclusions	299
	Acknowledgements	299
	References	299

Recent Results in Cophylogeny Mapping

Michael A. Charleston

	Abstract	303
1.	Introduction	304
2.	Cophylogenetic Events	306
3.	Cophylogeny Mapping	308
4.	Complexity	312
5.	Modelling Cophylogeny	316

6. Tests of Significance 322
7. Confounding Cophylogeny 324
8. Discussion 326
   Acknowledgements 328
   References 328

Inference of Viral Evolutionary Rates from Molecular Sequences

Alexei Drummond, Oliver G. Pybus and Andrew Rambaut

   Abstract 332
1. Introduction 332
2. General Linear Regression and Other
   Distance-based Methods 337
3. Maximum Likelihood Estimation 343
4. Bayesian Inference of Evolutionary Rates 347
5. Discussion 350
   Acknowledgements 354
   References 355

Detecting Adaptive Molecular Evolution: Additional Tools for the Parasitologist

James O. McInerney, D. Timothy J. Littlewood and Christopher J. Creevey

   Abstract 360
1. What is Adaptive Molecular Evolution? 360
2. Methodological Advances 362
3. Example of Adaptive Evolution
   in the Malaria Rifin Proteins 366
4. Prospects 373
   Acknowledgements 376
   References 377

   INDEX 381
   CONTENTS OF VOLUMES IN THIS SERIES 399

Introduction – phylogenies, phylogenetics, parasites and the evolution of parasitism

D. Timothy J. Littlewood

*Department of Zoology, The Natural History Museum,
Cromwell Road, London SW7 5BD, UK*

"Nothing in biology makes sense except in the light of evolution" (Dobzhansky, 1973). This popular quote not only illustrates how biological disciplines are united, but also acts as a guiding principle in the pursuit of biological knowledge. Phylogenetics, the science of estimating and analysing evolutionary relationships between biological entities sharing common ancestors, has become a standard and increasingly powerful tool in evolutionary biology. Through a greater understanding of molecular evolution and the relative ease with which molecular data can be collected, and as a result of complex algorithms being developed and implemented using ever increasing computing power, phylogenetics has now impinged upon almost every aspect of biological research.

An evolutionary framework not only answers questions directly by providing genealogical links between taxa or other biological entities, whether extinct or extant, but also provides the foundation for testing hypotheses and interpreting comparative data. At its most fundamental level, in taxonomy, the resolved genealogy is an ideal foundation for nomenclature to reflect relatedness. However, phylogenies provide a veritable springboard for biological research that goes way beyond taxonomy and pushes the boundaries of systematics. At the heart of a phylogeny is the recognition of homology, in the strict evolutionary sense of the word (Patterson, 1982; Wagner, 1989), and the subsequent statements we can make concerning homology (Hall, 1994), and its counterpart homoplasy (Sanderson and Hufford, 1996), inferred from the phylogeny.

Parasites, by definition, require hosts and parasites necessarily become adapted to surviving and thriving on or in a host for some time, however fleeting, in their life history. They are differentiated from commensals or symbionts by the detrimental effects (usually measured in terms of reduced fitness) they impose upon their host(s) and the obligate relationship they must engage in with their host(s). Thus, parasites are united primarily by their life-style, however diverse, and not their phylogeny. Whether a parasitologist addresses evolutionary questions directly, or addresses parasitological questions in an evolutionary context, the consequences of historical and on-going natural selection cannot be ignored. Selection affects the parasite, the host and the interaction between them. Parasites use a wide variety of habitats where ecological constraints, imposed by biotic and abiotic factors, affect them as they would any free-living species. However, there is the added constraint that their hosts provide specialised, often hostile, habitats that parasites must first enter, actively or passively. Thereafter the parasites must grow and/or develop, reproduce and move to the next habitat, be it another host or as a free-living stage or gamete, in order to complete the life cycle. When more than one host is involved, the completion of a life cycle can appear to be a formidable task, and the means by which parasites achieve this task rank among some of the most spectacular feats in natural history.

The life of a parasite is thus inextricably linked with that of its host(s), and speciation within and between hosts, as lineages of hosts and parasites evolve, further qualifies parasitism as a unique way of life. Neither parasites nor parasitism can be fully understood without reference to the host(s). Once parasitic, it seems that this distinctive but highly variable life-style cannot revert to a host-free (i.e. free-living) state. Thus, parasites are locked, to various degrees, in evolutionary 'arms-races' with their host(s), and selection proceeds at the genomic, organismal and population level for both the parasite and the host(s) involved. Multiple hosts and complicated life history strategies often compound the difficulty in disentangling host-parasite interactions, but phylogenetics offers a testable framework within which to proceed, and a foundation upon which refutable hypotheses may be built.

Within the comprehensive, and still developing, phylogenetic toolkit are interpretative and diagnostic tools for identifying organisms and their genes, and methods for revealing and testing evolutionary patterns and processes. Some of these are exclusive to, or exemplified by, parasites. This volume is a collection of original articles including reviews and new data about parasite taxa or other biological units common to parasites (e.g. genes), sharing an evolutionary history. Contributors show how evolutionary relationships within and between taxa, their genomes and their genes reveal the nature of parasitism from the biochemical to the ecological. The interface between phylogenetics and parasitology allows us to ask how did parasitism begin, how did it proceed, what does it entail and, perhaps, where is it going?

INTRODUCTION

In 1996 Harvey *et al.* published a book entitled *New Uses for New Phylogenies* (Harvey *et al.*, 1996) in which authors gave examples to illustrate the title. Contributors demonstrated how advances in phylogenetics and the increasing numbers of organismal and gene phylogenies have impinged upon population genetics, developmental genetics, biogeography, conservation, and epidemiology. The utility of combining multiple independent data sets estimating evolutionary rates, measuring cospeciation and resolving a variety of old and new macroevolutionary problems with new techniques were highlighted. This volume aims to take a similar approach with an emphasis on new uses for new phylogenies of parasites and new uses for new phylogenetic methods applied to parasites and the study of parasitism.

Parasitic taxa offer some of the most bizarre tales that natural history can tell and few if any phyla can claim to be free from parasite infection. However, rather than focus solely on a diversity of taxa that might entice a diversity of taxon-focused parasitologists, this volume aims to display the diversity of the phylogeneticist's toolkit as used variously by parasitologists. Moreover, although many groups of obligate parasites are well recognised as monophyletic groups, as Brooks and McLennan point out, there is no one single characteristic that delineates parasitism (Brooks and McLennan, 1993). Indeed, who knows to what extent the diversity of parasitism shapes the diversity of organisms?

Chapters present exemplar approaches to particular parasitological groups. However, topic selection, instead of being dictated by parasite taxonomy, was guided by the need to detail an extensive range of phylogeny-based applications. Necessarily, this editorial approach results in themes applicable to many organisms, but also illustrates how the contributors' own unique questions and needs have benefited from a phylogenetic approach. Nevertheless, the tree of life is visited from roots to tips.

The first chapter shows how parasitism is integral to eukaryote structure and function. Bryony Williams and Patrick Keeling of the University of British Columbia, take on the task of discovering the origins of cryptic organelles in parasitic protists and fungi. Beginning with the endosymbiosis of eubacterial species that became mitochondria and plastids, these organelles have subsequently evolved into a variety of entities that may or may not have left cytological evidence for their presence. The origins and history of the mitochondria and plastids are revealed by phylogenetic analysis of genes (including Cpn60, HSP70 and FabI), establishing orthology and resolving interrelationships, and by employing molecular markers (transit peptides) to localise the organelle-like structures. That the authors have worked on cryptic organelles in parasitic taxa further underscores the importance of their work since the organelles, of prokaryotic origin and therefore with distinctive biochemistries, in turn offer therapeutic targets for the control of their pathogenic eukaryotic hosts. From an evolutionary point of view, the chapter also offers a useful starting point as we are taken to some of the earliest and deepest branches of the tree of life, and follow evolutionary histories that define eukaryotes and include some of the most medically important parasites.

Jim Whitfield of the University of Illinois concentrates on a very different branch of the tree of life. Not only have Hymenoptera fascinated entomologists and parasitologists alike but they also provide a magnificent opportunity to understand the evolution of parasitism in a successful and species-rich clade. Using phylogenies to resolve the interrelationships of parasitoid wasps, in particular, multiple suites of data are discussed to illustrate what parasitism has meant for the radiation and success of the group. Although the nature of parasitism within the group has been well studied, giving rise to specific terminology that describes hymenopteran life history strategies, it has only been through phylogenetic analysis that the success of parasitism within the group can be measured. Cophylogeny between parasitoids and their hosts and parasitoids and their own viral parasites are also considered.

Mark Blaxter of the University of Edinburgh provides a tour de force on the Nematoda that should at worst gratify and at best irritate the nematologists. Few authors would be able to take on the evolution of parasitism within this beguiling but ubiquitous and difficult phylum. Fewer still would be able to move so swiftly between genes, genomes, proteomes, species and beyond. Trees are used to demonstrate how nematode phylogenetics has informed parasitologists. Trees demonstrate the origins of parasitism within the phylum, life cycle evolution in the Strongyloidea, nematode–bacterial symbiosis (emphasising the associations with *Wolbachia*), the diversity of genomics programmes, and the evolution of nematode HOX genes and mitochondrial genomes. Additionally, with the first metazoan genome project of *Caenorhabditis elegans* complete, a mass of comparative data is brought to order by techniques including phylogenetics, a bonus to those trying to understand diseases caused by roundworm parasitism.

Tom Cribb of the University of Queensland, Rod Bray and Peter Olson of The Natural History Museum in London, and I concentrate on a very different group of helminths, the Digenea. As obligate flatworm parasites, the flukes utilise a wide variety of molluscan first hosts, variously and usually at least one or more intermediate host and a fantastic variety of vertebrate definitive hosts. Using a new molecular phylogeny of the Digenea, the authors explore known host associations and see how far phylogenetics can be used to infer the evolution of their complex life cycles. Without a fossil record for the group (typical for many parasites), only host associations, life history strategies, morphology and behavior from extant taxa can be relied upon to illuminate past events and the nature of ancestral or historical life cycles. The challenge has been to exhume a comprehensive history whilst not deviating from a tree-based approach. Much is gained by this attempt at objectivity including, inevitably, the formulation of many new questions.

Stephen Rich of Tufts University and Francisco Ayala of the University of California review the progress in malaria research and assess the extent to which phylogenetics has become an integral component of perhaps the biggest and most important parasite research programme. Plastid evolution, life cycle evolution, host association and the origins and evolutionary radiation of human malaria are discussed in the light of phylogenetics. Each area provides further insights into how the parasite may be better understood as well as tackled.

Serge Morand of the Campus International de Baillarguet (Montpellier) and Robert Poulin of the University of Otago demonstrate the utility of phylogenetics as a tool particularly useful to parasite evolutionary ecologists. They do so by reviewing the comparative approach in its diverse guises and through various methodological implementations. Through worked examples, they make a plea for parasite evolutionary ecologists to place their comparative data in a phylogenetic framework. Although on balance the authors prefer the eigenvector method of comparative analysis, they consider its limitations and introduce a new method for variation partitioning in a phylogenetic context.

Michael Charleston, recently of the University of Oxford, reviews advances in what must be the one area of parasitology where phylogenetics needs little or no introduction. Cophylogeny mapping allows the evolutionary radiation of host and parasites to be tracked together. Although there are compelling reasons for following host and parasite radiations concomitantly, any deviation from strict codivergence, as is found usually in nature, brings with it a need to explain and optimise the likelihood of historical scenarios that best reflects the host-parasite association through time. Cophylogeny mapping provides methods by which we can assess association by descent (cospeciation) and association by colonisation (host–switching) whilst considering lineage splitting (speciation) and extinction. This chapter investigates significance testing and the theoretical and practical problems encountered in unravelling cophylogeny in order to assess the degree to which lineages have coevolved and codiverged. Such tangled trees are increasingly easier to disentangle but there are still knots that may confound or lead to misinterpretation.

Alexei Drummond, Oliver Pybus and Andrew Rambaut of the University of Oxford consider methods of inferring rates of molecular change in the evolutionary divergence of viruses. Viruses offer a fast-evolving example of parasites that respond rapidly to selection and indeed usually evolve much faster than their hosts. The authors illustrate the utility of molecular substitution rate estimation as estimated under the broad categories of linear regression (distance-based), maximum likelihood and Bayesian inference methods. Worked examples employing these techniques on such viruses as Dengue and HIV-1, include the dating of events, the estimate of demographic parameters and an investigation of the effects of natural selection on molecular evolution.

Finally, James McInerney and Christopher Creevey of the National University of Ireland and I consider recent applications of phylogenetics to distinguish adaptive molecular evolution. Such a technique may prove to be a useful tool for the parasitologist attempting to identify genes undergoing positive selection. Using an example from the recently sequenced genome of *Plasmodium falciparum*, phylogeny-based methods are reviewed and employed in characterising adaptive evolution in a family of surface-exposed proteins (the Rifins). The authors conclude that recognising adaptive molecular evolution in lineages offers greater understanding of the molecular basis for the host–parasite response. In turn, such methods may assist in the identification of vaccine targets.

It remains for me to thank all the contributors of this volume for providing a collection of articles with which I am delighted, notwithstanding my direct involvement with two. Those authors who have had to put up with me during the development of this volume, should at least spare a charitable thought for my colleagues who also tolerated me as a co-author. I am indebted to the many reviewers who gave up their precious time and shared their expert knowledge to referee the chapters critically and swiftly. Editing a thematic volume such as this is a purely selfish exercise but one that I was lucky enough to have had endorsed by the series editors and the support of the authors. I thank them for their indulgence, support, and patience.

REFERENCES

Brooks, D.R. and McLennan, D.A. (1993). Parascript: parasites and the language of evolution. Washington: Smithsonian Institution Press.
Dobzhansky, T. (1973). Nothing in biology makes sense except in the light of evolution. *American Biology Teacher* **35**, 125–129.
Hall, B.K. (1994). Homology: the hierarchical basis of comparative biology. London: Academic Press.
Harvey, P.H., Leigh Brown, A.J., Maynard Smith, J. and Nee, S. (1996). New uses for new phylogenies. pp. 349, Oxford: Oxford University Press.

Patterson, C. (1982). Morphological characters and homology. In: *Problems of phylogenetic reconstruction* (K.A. Joysey and A.E. Friday, eds). pp. 21–74. New York: Academic Press.

Sanderson, M.J. and Hufford, L. (1996). Homoplasy: the recurrence of similarity in evolution. London: Academic Press.

Wagner, G.P. (1989). The origin of morphological characters and the biological basis of homology. *Evolution* **43**, 1157–1171.

Cryptic Organelles in Parasitic Protists and Fungi

Bryony A. P. Williams and Patrick J. Keeling

Canadian Institute for Advanced Research, Department of Botany, University of British Columbia, 3529-6270 University Boulevard, Vancouver, BC V6T 1Z4, Canada

Abstract	10
1. Introduction	10
2. The Origin of Mitochondria and Plastids by Endosymbiosis	12
2.1. Mitochondria	12
2.2. Plastids	16
3. Cryptic Organelles and How to Find Them	21
4. Case Histories – Mitochondria	24
4.1. Hydrogenosomes	24
4.1.1. Ciliate hydrogenosomes	24
4.1.2. Chytrid hydrogenosomes	25
4.1.3. Parabasalian hydrogenosomes	26
4.2. Entamoebid Mitosomes	30
4.3. Microsporidian Mitosomes	31
4.4. Mitochondria in Diplomonads?	36
5. Case Histories – Plastids	38
5.1. Apicomplexa	38
5.2. Green Algal Parasites *Helicosporidium* and *Prototheca*	46
5.3. Plastids in Parasitic Plants	47
6. Future Directions	49
Acknowledgement	53
References	54

ABSTRACT

A number of parasitic protists and fungi have adopted extremely specialised characteristics of morphology, biochemistry, and molecular biology, sometimes making it difficult to discern their evolutionary origins. One aspect of several parasitic groups that reflects this is their metabolic organelles, mitochondria and plastids. These organelles are derived from endosymbiosis with an alpha-proteobacterium and a cyanobacterium respectively, and are home to a variety of core metabolic processes. As parasites adapted, new demands, or perhaps a relaxation of demands, frequently led to significant changes in these organelles. At the extreme, the organelles are degenerated and transformed beyond recognition, and are referred to as "cryptic". Generally, there is no prior cytological evidence for a cryptic organelle, and its presence is only discovered through phylogenetic analysis of molecular relicts followed by their localisation to organelle-like structures. Since the organelles are derived from eubacteria, the genes for proteins and RNAs associated with them are generally easily recognisable, and since the metabolic activities retained in these organelles are prokaryotic, or at least very unusual, they often serve as an important target for therapeutics. Cryptic mitochondria are now known in several protist and fungal parasites. In some cases (e.g., *Trichomonas*), well characterised but evolutionarily enigmatic organelles called hydrogenosomes were shown to be derived from mitochondria. In other cases (e.g., *Entamoeba* and microsporidia), "amitochondriate" parasites have been shown to harbour a previously undetected mitochondrial organelle. Typically, little is known about the functions of these newly discovered organelles, but recent progress in several groups has revealed a number of potential functions. Cryptic plastids have now been found in a small number of parasites that were not previously suspected to have algal ancestors. One recent case is the discovery that helicosporidian parasites are really highly adapted green alga, but the most spectacular case is the discovery of a plastid in the Apicomplexa. Apicomplexa are very well-studied parasites that include the malaria parasite, *Plasmodium*, so the discovery of a cryptic plastid in Apicomplexa came as quite a surprise. The apicomplexan plastid is now very well characterised and has been shown to function in the biosynthesis of fatty acids, isopentenyl diphosphate and heme, activities also found in photosynthetic plastids.

1. INTRODUCTION

The fundamental difference between prokaryotes and eukaryotes is the compartmentalisation of the cell. The primary defining feature of eukaryotic cells is the enclosure of the bulk of their DNA within the nucleus, and eukaryotes are further characterised by the compartmentalisation of numerous cellular processes within membrane-bounded structures, or organelles. These include the Golgi apparatus, endoplasmic reticulum, peroxisomes, lysosomes, endosomes, mitochondria and plastids. The origins of these structures have been the subject of considerable debate and speculation for which there are two generally contrasting possibilities: an autogenous origin (i.e., derived from within the cell) versus an endosymbiotic origin (i.e., from a merger of two cells). The endomembrane system – composed of the endoplasmic reticulum, Golgi apparatus, endosomes, lysosomes, and the nucleus – is a continuous system and is widely held to have originated autogenously by invagination of vesicles of the plasma membrane. The nucleus has, from time to time, been suggested to be derived from an endosymbiotic bacterium based on a variety of rationales (e.g., Lake and Rivera, 1994; Lopez-Garcia and Moreira, 1999). However, the fact that the nucleus is continuous with the endomembrane system cannot easily be explained by phagocytosis of a bacterial cell, and endosymbiotic models for nuclear origins often mistakenly claim it to be a double membrane-bounded organelle, but it is really a single membrane folded on itself. Therefore, the more widely held view is that the nucleus resulted from the infolding of a primitive endoplasmic reticulum (Cavalier-Smith, 1991b). Similarly, the peroxisome has been proposed to be derived from a bacterial endosymbiont, partly based on the observation that they appear to grow and divide by fission (Lazarow and Fujiki, 1985). However, peroxisomes have no genome with which to trace a possible exogenous origin and, furthermore, it has been shown that their replication is dependent on the endoplasmic reticulum – suggesting a possible relationship to the endomembrane system (Titorenko et al., 1997). The current view thus appears to be that peroxisomes are also autogenously derived organelles (Martin, 1999; Cavalier-Smith, 2002). In contrast, it is now abundantly clear, based on a variety of evidence from morphology, biochemistry, and molecular phylogeny, that mitochondria and plastids are derived exogenously from endosymbiosis events involving eubacteria.

Mitochondria and plastids are structurally very distinctive and are almost universally recognisable by their characteristic features. The mitochondrion is often considered one of the defining structures of the eukaryotic cell and is thought to be virtually ubiquitous in this group. The plastid is similarly considered a defining feature of plants and algae. However, striking variations on the basic structure of these organelles do exist and, not surprisingly, this is most often seen when organisms make extreme adaptive changes that create new demands on the metabolic pathways that these organelles harbour. One common circumstance that appears to drive such change is the adaptation to a parasitic way of life.

In many groups of parasitic eukaryotes, the mitochondria and plastids have been substantially altered in morphology and biochemistry. Whilst these peculiar organelles are interesting from an evolutionary perspective, a thorough understanding of them also offers practical gains: they frequently contain unique biochemical pathways not found in other eukaryotes, and these differences provide new potential drug targets (McFadden and Roos, 1999). At its most extreme, parasitism leads to such a significant alteration in the organelle's ultrastructural features and biochemical pathways that they are no longer recognisable, and become "cryptic" organelles. Fortunately, in the case of mitochondria and plastids, their relict bacterial genome or other molecular clues can be used to find these organelles and determine their evolutionary origin, even when all other archetypal features have been lost. Here we review a number of instances in which mitochondria and plastids in parasitic protists and fungi have been modified to the extent that they are difficult to recognise. In each case, reconstructing molecular phylogenies has been instrumental in identifying the evolutionary origin of these organelles, in determining their function by identifying which enzymes and biochemical pathways are localised to the organelle, or even in hinting at the very existence of the organelles themselves.

2. THE ORIGIN OF MITOCHONDRIA AND PLASTIDS BY ENDOSYMBIOSIS

2.1. Mitochondria

As previously stated – mitochondria are unquestionably derived from endosymbiosis. Mitochondria were first recognised under the light microscope in the 1850s and approximately 100 years later they were recognised as housing the components of fatty acid oxidation, respiration, and oxidative phosphorylation, leading to the label, the 'powerhouse of the cell' (Kennedy and Lehninger, 1949). Fifty years on, mitochondria have been shown to have far wider roles than solely ATP production, and it is now known that their other essential functions include iron sulphur cluster assembly and export, and programmed cell death, amongst others (Kroemer *et al.*, 1995; Lill *et al.*, 1999). Surprisingly, as early as 1890 mitochondria were suggested to be autonomous structures derived from symbiotic bacteria (Altmann, 1890). This hypothesis was resurrected several times over the next century, but was not widely accepted until its popularisation during the 70s and 80s.

The debate over the endosymbiotic origin of mitochondria cannot be separated from the parallel debate over the origins of plastids, although the evidence for the endosymbiotic origin of mitochondria lagged behind the much clearer case for plastids (see below). By 1982 a long list of characteristics suggested very strongly that plastids had an endosymbiotic origin (Gray and Doolittle, 1982). These included the sensitivity of chloroplast protein synthesis to anti-bacterial antibiotics, biochemical and ultrastructural similarities to cyanobacteria, ribosomes with a eubacterial sedimentation coefficient, and, most importantly, the presence of a small genome that retained several eubacterial features. Several such characteristics were also found for mitochondria, but the endosymbiotic origin of the mitochondrion from a prokaryotic cell was unequivocally established by the discovery and phylogenetic analysis of its genome. The mitochondrial genome is generally, but not always, a single circular chromosome, and is highly reduced compared with the genomes of its autonomous eubacterial relatives (Gray and Spencer, 1996; Gray et al., 2001) (Table 1). Nevertheless, it contains a variety of genes now known to be eubacterial in nature. This was first revealed through phylogenetic analysis based on cytochrome C protein sequences and small subunit ribosomal RNA (SSU rRNA) (Schwartz and Dayhoff, 1978; Yang et al., 1985). SSU rRNA phylogeny clearly placed mitochondria closer to prokaryotes than to eukaryotes, and

Table 1 Size and characteristics for organellar genomes.

Organism	Organelle	Genome
Homo (animal)	mitochondrion	16.5 kbp, circular
Saccharomyces (fungus)	mitochondrion	85.8 kbp, circular
Arabidopsis (plant)	mitochondrion	367 kbp, linear (?)
Porphyra (red alga)	mitochondrion	36.8 kbp, circular
Plasmodium (apicomplexan)	mitochondrion	6 kbp, linear
Reclinomonas (jakobid)	mitochondrion	69 kbp, circular
Entamoeba (archamoeba)	mitosome	Uncertain
Encephalitozoon (microsporidian)	mitosome	None
Trichomonas (parabasalian)	hydrogenosome	None
Neocallimastix (chytrid fungus)	hydrogenosome	None
Nycotherus (*ciliate*)	hydrogenosome	present, but uncharacterised
Arabidopsis (plant)	plastid	154.5 kbp, circular
Epifagus (parasitic plant)	plastid	70 kbp, circular
Porphyra (red alga)	plastid	191 kbp, circular
Plasmodium (apicomplexan)	plastid (apicoplast)	35 kbp, circular
Prototheca (green alga)	plastid	54 kbp, circular
Helicosporidium (green alga)	plastid	present, but uncharacterised

also allowed the identification of the closest ancestral lineage of the mitochondrion, the alpha subdivision of the proteobacteria (Yang *et al.*, 1985). Interestingly, this ancestor was previously proposed on the basis of biochemical data (Whatley, 1981), and has now been widely supported by the analysis of key genomic sequences (Kurland and Andersson, 2000).

The origin of the mitochondrion is thought to have occurred something like the process shown in Figure 1. In this simplified scheme, an early eukaryote, or perhaps a protoeukaryote (it is not certain that this organism possessed a nucleus at this point) engulfed an alpha-proteobacterial cell. This suggests that the host must have at least developed a system for phagocytosis, and therefore the eukaryotic cytoskeleton was likely a key innovation that allowed the evolution of organelles by endosymbiosis

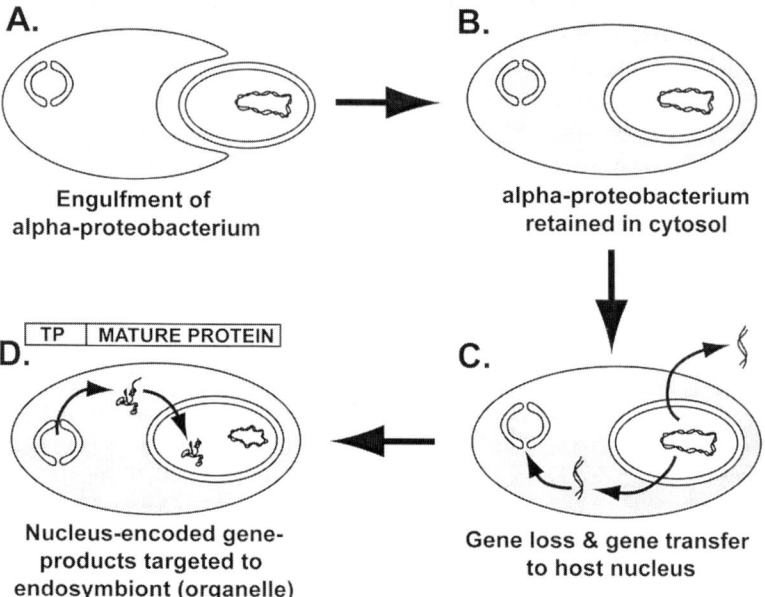

Figure 1 Schematic of the endosymbiotic origin of mitochondria. (A) A heterotrophic eukaryote (or some proto-eukaryote) engulfed an alpha-proteobacterium and (B) retained it within the cytoplasm, losing the membrane of the phagocytic vacuole. As the endosymbiotic relationship progressed (C), the endosymbiont lost genes no longer necessary and transferred others to the nuclear genome of the host. A system was established to target proteins to the organelle using an N-terminal leader sequence called a transit peptide (TP), so that protein products of transferred genes are localised to the organelle. A schematic of one such protein is shown in (D). All canonical mitochondria possess a small relict genome, but several mitochondrion-derived organelles have lost the genome altogether.

(Stanier, 1970). In any case, rather than digesting this alpha-proteobacterium for nutrients, it was retained in the cytosol for reasons that are not presently established. Various theories have been put forward to explain this association, which have been discussed elsewhere (e.g., Martin and Müller, 1998; Gupta, 1999; Lopez-Garcia and Moreira, 1999; Kurland and Andersson, 2000). The endosymbiont and host then became increasingly mutually dependent and integrated. Part of this process was a large-scale reduction of the endosymbiont at many levels. Many structural and biochemical characteristics of the bacterium were discarded, including the peptidoglycan wall, flagella, as well as most of its biochemical versatility. At the molecular level, this reduction meant a significant decrease in size of the endosymbiont genome (Gray et al., 1999; Kurland and Andersson, 2000). Much of this reduction reflects the loss of a large number of genes as their products became unnecessary, but the mitochondrial genome is still far too small to encode all of the genes required for those structural, biochemical, and housekeeping functions retained by the organelle. Most of these genes were transferred to the nucleus of the host cell, and their protein products are translated by cytosolic ribosomes and post-translationally targeted to the mitochondrion. This post-translational targeting system is generally mediated by distinctive N-terminal leader sequences called transit peptides on proteins bound for the mitochondrion. These are recognised by protein complexes in the outer and inner membranes of the mitochondrial envelope (Pfanner and Geissler, 2001). Many of the proteins in these membrane complexes are referred to as TIMs and TOMs (Translocation Inner Mitochondrial membrane and Translocation Outer Mitochondrial membrane, respectively), and these are assisted by several molecular chaperones that refold the proteins as they cross the membranes.

The near-ubiquitous distribution of mitochondria across the eukaryotic tree implies that the endosymbiotic origin of mitochondria occurred at an early stage of eukaryotic evolution, earlier than the endosymbiosis that gave rise to plastids. This suggests that eukaryotes that appear to lack mitochondria may represent extremely ancient eukaryotic lineages that diverged from a pre-mitochondrial ancestor. This idea was formalised in the Archezoa hypothesis, in which organisms that were thought to be primitively amitochondriate were placed within a paraphyletic group at the base of the eukaryotic tree (Cavalier-Smith, 1983, 1987). This group included the microsporidia (e.g., *Encephalitozoon*), the parabasalids (e.g., *Trichomonas*), the diplomonads (e.g., *Giardia*), retortomonads and oxymonads, and the Archamoebae (e.g., *Entamoeba*). Phylogenetic analysis of the first molecular data from these organisms appeared to provide stunning support for this theory: the SSU rRNA from microsporidia, Parabasalia, and diplomonads all clustered at the base of the eukaryotes with strong support (Vossbrinck et al., 1987; Sogin et al., 1989; Sogin, 1989). Initial analyses of protein-coding genes also supported this view (Hashimoto et al., 1994; Shirakura et al., 1994; Kamaishi et al., 1996a,b) and the phylogenetic debate was centred largely on which archezoan was the most primitive of all (Leipe et al., 1993). In the end, phylogenetic analyses of other genes demonstrated that none of the Archezoa were primitively amitochondriate (discussed below), and there is currently no direct evidence that any primitively amitochondriate eukaryote exists today.

2.2. Plastids

Plastids are the cellular compartments of plants and algae where photosynthesis takes place. The term plastid is a general term for any photosynthetic organelle and all their non-photosynthetic derivatives. Other terms refer to specific functional or phylogenetic subsets of plastids. For example, chloroplasts are photosynthetic plastids of plants, green algae and their derivatives (e.g., those of *Chlorarachnion* and *Euglena* – see below), while cyanelles are the plastids of glaucophyte algae. Some examples of functionally defined subsets of plastids include a variety of non-photosynthetic organelles such as leucoplasts, amyloplasts, and eliaoplasts.

As with mitochondria, there is now an abundance of evidence from morphology, biochemistry, cell biology, and molecular sequences supporting an endosymbiotic origin of plastids from a cyanobacterium (McFadden, 2001). Indeed, the first musings of a possible endosymbiotic origin of any organelle came from observing plastids under the light microscope (Schimper, 1883; Mereschkowsky, 1905). This idea received little attention for some time, but was rekindled in the mid-20th century for a number of the same reasons discussed for mitochondria, most importantly the discovery that the organelle contained DNA (Stocking and Gifford, 1959). Plastid genomes are, like most mitochondrial genomes, composed of a single circular chromosome that is highly reduced, but generally not so reduced as mitochondrial genomes (see Table 1).

It has been clear for some time that all plastids are derived from cyanobacteria. In addition to a strong relationship between nearly all plastid genes and their cyanobacterial homologues, plastids and cyanobacteria share a large number of extremely detailed and unique biochemical characteristics relating to photosynthesis. There are four bacterial groups where photosynthesis takes place, but the cyanobacteria and plastids are unique in possessing both photosystems I and II, and by generating molecular oxygen by splitting water molecules (Blankenship, 1994). While the cyanobacterial origin of plastids is not disputed, the huge diversity of plastid types, together with the even greater diversity of host cells in which plastids are found led to a great deal of controversy over the origin of plastids and their subsequent evolutionary history. Part of this controversy resulted from the fact that plastids are scattered all over the tree of eukaryotes: a number of algae are clearly more closely related to non-photosynthetic protists than to other groups of algae (e.g., the photosynthetic *Euglena* and the non-photosynthetic *Trypanosoma*). Such a distribution is consistent with multiple independent plastid origins, but the truth turned out to be much stranger. Unlike mitochondria, where the eubacterium was engulfed, retained and inherited faithfully thereafter, plastids have continued to spread from eukaryote-to-eukaryote by a process called secondary endosymbiosis (Archibald and Keeling, 2002). In this case (Figure 2), primary endosymbiosis refers to the engulfment and retention of the cyanobacterium by a eukaryote, which gave rise to so-called 'primary plastids' found in glaucocystophytes, red algae, green algae, and plants (which evolved from green algae). While this is a large and diverse collection of algae, it is only a fraction of the total diversity of photosynthetic eukaryotes. Most algal groups acquired photosynthesis by engulfing one of these primary algae (either a red or a green, never a glaucocystophyte), and converting this primary alga into an organelle by a process of reduction and gene transfer similar to what is seen in primary endosymbiosis. In most cases, this process resulted in a plastid distinguished only by extra membranes corresponding to the phagosomal membrane of the secondary host and the plasma membrane of the primary alga (dinoflagellates and euglenids have only three membranes and it is thought that the primary algal plasma membrane has been lost in these plastids). In two cases, however, a relict algal nucleus with a miniaturised genome also persisted between the second and third membranes (McFadden *et al.*, 1997).

An understanding of secondary endosymbiosis helps explain plastid diversity and shows quite clearly how plastids can be distributed across the tree of eukaryotes, but it leaves a few questions unsolved, and raises a few new ones. There has long been a debate over whether plastids ultimately trace back to a single primary endosymbiosis or whether they arose by several independent endosymbiotic events involving different cyanobacteria. Currently, two lines of evidence very strongly favour the former hypothesis. First, in practically all phylogenies that have been constructed, the plastids form a single clade within the cyanobacteria. Second, certain characteristics of the plastid proteomes and genomes are unique to plastids. Such features include the rRNA inverted repeat, the so-called plastid superoperon, and three-helix light-harvesting antennae proteins, none of which are found in cyanobacteria (Palmer, 2003). Between these characteristics and the phylogenetic trees, there is no good evidence for multiple origins of plastids

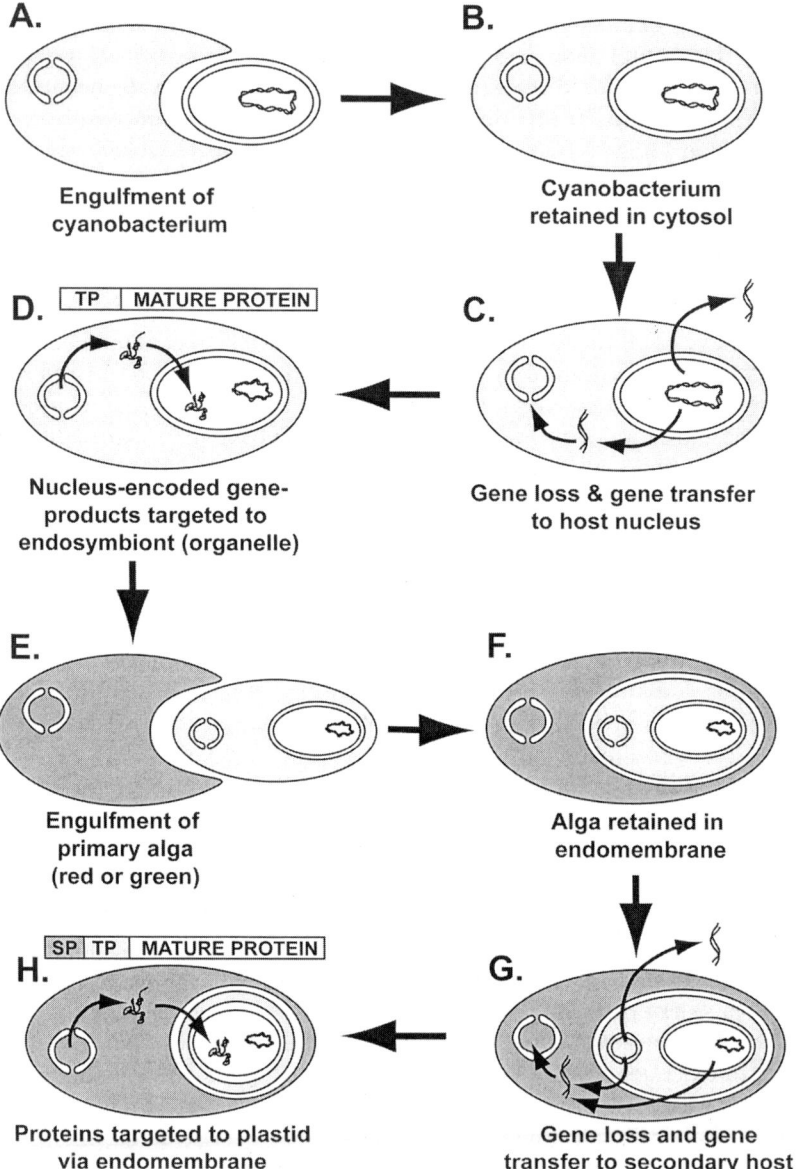

Figure 2 Schematic of the endosymbiotic origins and spread of plastids. (**A–D**) In events similar to the origin of mitochondria (Figure 1), a heterotrophic eukaryote engulfed and retained a cyanobacterium. The endosymbiont and host were integrated in a process involving many gene losses and gene transfers, and the

(see Stiller *et al*., 2003), although one interesting potential exception is a little known amoeba called *Paulinella chromatophora*, which may have independently acquired its plastid (Kies, 1974). In addition, the discovery that plastids could be transferred from one lineage to another has led to considerable debate as to how many times this has happened and what kinds of cells were involved. Current evidence favours three secondary endosymbiotic events: two involving green algae and one involving a red alga (see Figure 3), the latter being discussed in more detail below.

Secondary endosymbiosis also has profound implications for protein trafficking. Primary plastids, like mitochondria, have transferred the bulk of their genetic material to the host nucleus, and the protein products of these genes are post-translationally targeted to the plastid. There is some variation in the method of translocation, but typically a transit peptide is recognised by membrane proteins called TICs and TOCs (*T*ranslocation *I*nner *C*hloroplast membrane and *T*ranslocation *O*uter *C*hloroplast membrane, respectively), and the protein is moved across both membranes using these protein complexes and several molecular chaperones, much like protein translocation in mitochondria (Cline and Henry, 1996; Hiltbrunner *et al*., 2001). Secondary plastids, however, are not actually in the cytoplasm of the secondary host, but are in the endomembrane system (the outermost membrane is derived from the phagosome of the secondary host). Proteins expressed in the cytosol of these cells are not exposed to the TOCs, and could not be targeted with a transit peptide alone. All algae with secondary plastids have overcome this problem in the same general fashion: plastid-targeted proteins encode complex, bipartite leaders with a signal peptide followed by a transit peptide (some have additional information to move the

---

establishment of a protein-targeting apparatus similar to that of mitochondria. These plastids (**D**) are called primary plastids and are found in glaucocystophytes, red algae, green algae, and land plants. Proteins are targeted to primary plastids using a system much like that of mitochondria, and these proteins have N-terminal transit peptides as shown in (**D**). In secondary endosymbiosis (**E–F**), a second heterotrophic eukaryote then engulfs and retains one of these primary plastid-containing algae, and the eukaryotic algae is itself reduced to an organelle. This involves the transfer of many genes from the algal nucleus to the new host nucleus (**G**), and the establishment of a more complex protein-targeting system (**H**) to direct the protein products of these genes to the plastid via the endomembrane system using a signal peptide (SP), as well as the original transit peptide (TP). A schematic of a protein with such a complex leader is shown in (**H**). Secondary plastids are found in euglenids, chlorarachniophytes, cryptomonads, heterokonts, haptophytes, dinoflagellates, and apicomplexans. All plastids contain a small genome, and in two cases, the secondary endosymbiotic algal nucleus has been retained in a much degenerated state.

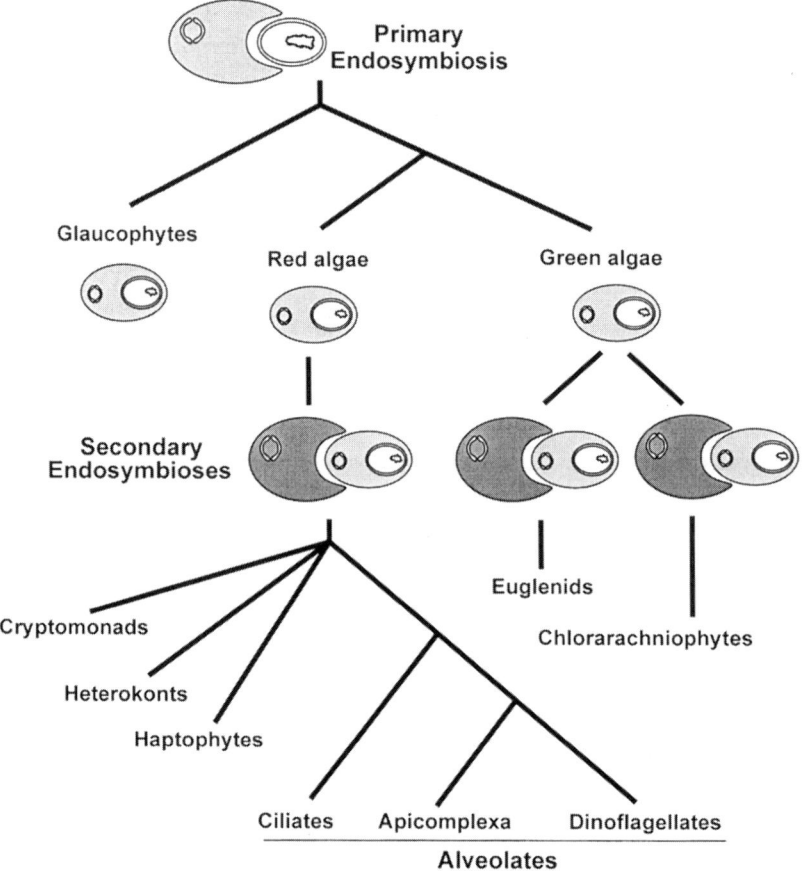

Figure 3 Schematic of the evolutionary history of plastids in eukaryotes. Primary endosymbiosis involving a cyanobacterium (Top) led to glaucocystophytes, red algae, and green algae (plants are derived from green algae). Three secondary endosymbiosis (Bottom) involving two independent green algae (Right) and one red alga (Left) resulted in the remaining plastid-containing eukaryotes. The single origin of all red algal secondary plastids has been demonstrated phylogenetically, and has important implications for the plastids of several parasitic protists.

protein to the thylakoid within the plastid). The signal peptide directs the partially translated protein and ribosome to the rough endoplasmic reticulum, where the protein is inserted into the endomembrane system co-translationally, thus crossing the outermost membrane (McFadden, 1999). The transit peptide ensures that the protein can pass through the two inner, cyanobacterium-derived membranes, but how the protein crosses the membrane derived from the primary algal plasma membrane is unknown. This added complexity in targeting is a signature for nuclear genes encoding proteins targeted to secondary plastids.

3. CRYPTIC ORGANELLES AND HOW TO FIND THEM

In most eukaryotes, mitochondria appear normal and function in the expected way. Similarly, virtually all algae are photosynthetic and their plastids are easily distinguishable and function normally. Even in cases where an alga has lost photosynthesis secondarily, the plastid is generally easily recognisable (Sepsenwol, 1973; Siu et al., 1976; Sekiguchi et al., 2002). In certain cases, however, the evolutionary specialisation of a group has been more extreme, and one reflection of this can be a reduction or transfiguration of the organelle beyond recognition. Very often these cytologically unrecognisable, or cryptic organelles were overlooked, and the organism was considered to be amitochondriate or aplastidial until closer inspection revealed an organelle drastically different in morphology and biochemistry than its antecedents.

One mode of life that often precipitates extreme evolutionary adaptation is parasitism. Many parasites are otherwise quite normal, but certain groups have evolved extremely sophisticated modes of parasitism, increasingly adapting to infect other organisms or other cells, typically at the expense of autonomy. Often, but not always, the most highly adapted parasites are extremely dependent upon their host. As a result they have discarded a number of biochemical capabilities, significantly altered their cellular morphology, and re-tooled their metabolism to take advantage of their new niche. In some cases it is difficult to know whether the adaptation to parasitism drove these changes, or if these cells were strange free-living organisms that later adapted to parasitising other cells. In either case, parasites offer an abundance of variation in many of the core components of the eukaryotic cell, not the least of which are organelles.

In the case of the mitochondrion, the otherwise ubiquitous nature of the organelle has always drawn immediate attention to any eukaryote that lacks it (Müller, 1992; Embley and Hirt, 1998; Sogin and Silberman, 1998; Lang et al., 1999; Keeling and Fast, 2002). Moreover, many amitochondriates are important parasites, in part because one of the abundant anoxic niches available to amitochondriates is inside other cells or organisms. Accordingly, there has been considerable effort put into actively looking for mitochondria in many putative amitochondriate parasites. Electron microscopy and biochemical analysis have both been applied to many of the amitochondriate parasites, but these organisms are highly adapted to living in anoxic or microaerophilic environments, so it is perhaps (in retrospect) not surprising that mitochondria were often not readily apparent by either means. Mitochondrial DNA was also sought, but in the vast majority of cases was not found (see below). In a special relative of mitochondria, hydrogenosomes, the actual organelle was discovered and well characterised cytologically and biochemically (Lindmark and Müller, 1973), but these are so unlike mitochondria in form and function that their relationship to mitochondria was not obvious at the time (Müller, 1973, 1993, 1997).

Ultimately, the secrets of cryptic mitochondria were given up by the nuclear genome of the host. Recall that the genomes of normal mitochondria are highly reduced: those genes that have been retained in most mitochondrial genomes encode proteins involved in gene expression or respiration, and these genes (and therefore the genome itself) would probably be lost relatively quickly if the organism adapted to an anoxic environment. However, most mitochondrial genes were transferred to the host nucleus and the proteins are post-translationally targeted to the organelle (Figure 1). It was reasoned that even if the organelle was significantly altered (and perhaps even if completely lost) some of these mitochondrion-derived proteins would retain an indispensable activity and the genes would remain in the nucleus (Clark and Roger, 1995). The genes that were most widely chosen to search for clues of a cryptic mitochondrion were molecular chaperones (heat shock proteins) that are part of the protein import system of all mitochondria. In most cases, genes encoding the 60 kDa chaperonin (Cpn60) or the 70 kDa class of heat shock protein (HSP70) were the critical markers for missing mitochondria, although HSP10, pyridine nucleotide transhydrogenase, and pyridoxal-5'-phosphate-dependent cysteine desulfurase also played important roles in the search (Clark and Roger, 1995; Bui *et al.*, 1996; Germot *et al.*, 1996, 1997; Horner *et al.*, 1996; Hirt *et al.*, 1997; Peyretaillade *et al.*, 1998; Roger *et al.*, 1998; Morrison *et al.*, 2001; Tachezy *et al.*, 2001; Williams *et al.*, 2002). Chaperones involved in protein import were logical choices because their activity is critical, no matter what the function of the organelle may be, and because they are typically highly conserved and well-sampled genes that are amenable to molecular phylogenies. Finding these genes and demonstrating that they are phylogenetically related to other mitochondrial homologues, provides compelling evidence that the organisms in which they are found are descended from mitochondriate ancestors, but it does not necessarily prove that they still harbour the organelle, despite the fact that in many cases the proteins were found to encode N-terminal leaders that resembled transit peptides

(Clark and Roger, 1995; Bui *et al*., 1996). The most important evidence for the presence of the organelle itself comes from using the proteins as markers for the organelle by immunocytochemistry. By raising antibodies against the newly characterised mitochondrial chaperones it is possible to determine where in the cell these proteins accumulate, and to find the organelle (e.g., Mai *et al*., 1999; Tovar *et al*., 1999; Williams *et al*., 2002).

The situation with plastids is much different than that of mitochondria, since plastids are restricted in distribution by virtue of their relatively late introduction into eukaryotes (Archibald and Keeling, 2002). Accordingly, a plastid-lacking eukaryote does not attract much attention based on that character alone unless it is demonstrated to have evolved from photosynthetic ancestors. Therefore, cryptic plastids have been discovered by two different routes: an unusual organism is found to have evolved from an alga and the plastid is then sought directly, or a cryptic plastid is discovered quite by chance in an organism where it is not expected to exist.

Also, in contrast to the situation with mitochondria, outright loss of the plastid genome seems to be extremely rare, even in the most highly derived plastids (although see Nickrent *et al*., 1997b). Why this is the case is not clear, but in general, plastid genomes are less reduced than those of mitochondria and encode genes relating to a greater diversity of metabolic processes than mitochondrial genomes (Wolfe *et al*., 1992; Nickrent *et al*., 1998; Palmer and Delwiche, 1998; Gray, 1999). Most organelle genes encode components of the gene expression machinery, so they are not intrinsically essential (i.e., if the genome is lost, they are no longer essential). In the mitochondria, most other genes are for proteins involved in respiration, so if respiration were lost, the genome would also presumably be lost. In contrast, plastid genomes generally encode proteins involved in a number of processes other than transcription, translation, and photosynthesis, which may make it more difficult for a plastid genome to be eliminated. The retention of a genome also has important practical implications. It means that the cell retains many more direct and obvious molecular clues with which to investigate the presence of a plastid, and the localisation of the organelle can be much more straightforward technically, since the genome or its rRNA gene products can be the target for localisation (which requires DNA probes) rather than proteins encoded by nuclear genes (which requires antibodies).

Even when a cryptic organelle is detected and identified, however, the mystery is only partly solved, since the major question is, "what is it doing in the cell?" Below, we will outline the history of the discovery of several cryptic mitochondria and plastids in a variety of important parasitic protists and fungi, and we will bring together the (sometimes scant) evidence for what role these unusual organelles play in their hosts.

4. CASE HISTORIES – MITOCHONDRIA

4.1. Hydrogenosomes

Hydrogenosomes were first described biochemically in 1973 in the parabasalian parasite, *Tritrichomonas foetus* (Lindmark and Müller, 1973). Although the structure corresponding to this organelle had been previously recognised in other trichomonad Parabasalia, it was assumed to be an aberrant peroxisome or mitochondrion. However, Lindmark and Müller recognised that the organelle housed a unique biochemistry, in which molecular hydrogen was produced through the oxidation of pyruvate, distinguishing it from both peroxisomes and mitochondria. Hydrogenosomes have subsequently been discovered to be common to all parabasalids, and have also been found in a variety of other phylogenetically distantly related organisms, including certain anaerobic ciliates and chytrid fungi. Ultrastructural data also suggest that hydrogenosomes may be present in the amoeboflagellate heterolobosean *Psalteriomonas lanterna* (Broers *et al.*, 1990), the euglenozoan *Postgaardi mariagerensis* (Simpson *et al.*, 1997), *Trimastix pyriformis* (O'Kelly *et al.*, 1999), and the enigmatic flagellate *Carpediemonas membranifera* (Simpson and Patterson, 1999), which is closely related to diplomonads and retortamonads (Simpson *et al.*, 2002). However, there is currently little biochemical data about the activities of these organelles.

All hydrogenosome-bearing organisms are linked by the ecological trait of living in anaerobic or microaerobic environments and, where it has been examined, the organelle produces ATP using hydrogen rather than oxygen as the terminal electron acceptor (Müller, 1993). The hydrogenosomes from each of these organisms contain comparable, though not identical biochemistry (Hackstein *et al.*, 1999). The long history of hydrogenosome research indicates that these are not, strictly speaking, "cryptic" organelles, however, their evolutionary origins and the relationships between hydrogenosomes in various groups have been a source of much debate.

4.1.1. Ciliate hydrogenosomes

While most ciliates are aerobic and contain typical mitochondria, a small number of ciliates have been found to contain hydrogenosomes (Fenchel and Finlay, 1995). These anaerobes are typically found in anoxic freshwater and marine sediments or in the rumen or caecum, where they exist in complex associations with their animal hosts and other anaerobic protists and fungi. The hydrogenosomes of these ciliates have not been studied in detail, but their metabolism has been worked out to some extent (Müller, 1993), and their hydrogenosomes have received ever greater attention in recent years.

It is evident that the hydrogenosome-containing ciliates and chytrid fungi both evolved from mitochondriate ancestors, as most ciliates and fungi are mitochondriate and the hydrogenosome-containing members are phylogenetically positioned well within these otherwise mitochondriate groups. Given that the hydrogenosomal ciliates and fungi all lack mitochondria, it has generally been considered most parsimonious that the hydrogenosomes in these groups evolved from pre-existing mitochondria (Embley and Hirt, 1998). This is especially clear in ciliates, since their hydrogenosomes are bound by a double membrane and ultrastructurally similar to the mitochondrion (Embley *et al.*, 1997). Molecular phylogenies based on SSU rRNA have shown that hydrogenosome-bearing ciliates do not form a monophyletic group, but instead have originated independently several times in ciliate evolution. This is further evidence favouring the hypothesis that ciliate hydrogenosomes evolved from mitochondria, as several convergent endosymbioses are conceivably unparsimonious (Embley *et al.*, 1995).

More recently, definitive evidence for the mitochondrial origin of a ciliate hydrogenosome has been produced, as a genome has been discovered in the hydrogenosome of *Nyctotherus ovalis*. This was demonstrated by showing co-localisation of the staining pattern of an anti-hydrogenase antibody with an anti-DNA antibody within the hydrogenosome (Akhmanova *et al.*, 1998). This evidence was corroborated by the amplification of a eubacterial-like SSU rRNA gene from several strains of *Nytotherus ovalis*, which branched with mitochondrial SSU genes from other ciliates in phylogenetic analyses (Hackstein *et al.*, 1999). One would expect that this SSU rRNA would also localise to the hydrogenosome of *Nyctotherus* and that it would be encoded on the hydrogenosomal genome. However, the location of the SSU transcript is still unknown and it is not known which genes are encoded in the *Nyctotherus* hydrogenosomal genome. Characterisation of this genome promises to be fascinating, as other hydrogenosomes apparently do not possess genomes (Clemens and Johnson, 2000). It will be interesting to see what genes were retained and why the genome was maintained as the *Nyctotherus* mitochondrion was metabolically transformed (Palmer, 1997).

4.1.2. Chytrid hydrogenosomes

As with ciliates, most chytrid fungi are aerobic and contain mitochondria, but a small group of anaerobic chytrids is found in the alimentary tract of large herbivorous mammals, the best-studied being *Neocallimastix*. In the case of the chytrids, the debate has largely focused on whether the hydrogenosome evolved from a mitochondrion or a peroxisome. The chytrid hydrogenosome was initially thought to be bound by a single membrane until it was shown to be double membrane-bounded using freeze fracture and standard electron microscopy using membrane preserving methods (Benchimol et al., 1997; van der Giezen et al., 1997). Thus, in this respect, it is morphologically similar to the mitochondrion and not the peroxisome. Peroxisomal targeting signals were thought to be present in Neocallimastix hydrogenase proteins based on cross-reactivity between this hydrogenosomal protein and an antibody raised to the peroxisomal targeting signal 1 (PTS1), a motif consisting of a C-terminal SKL motif (Marvin-Sikkema et al., 1993). This was interpreted as indicating an evolutionary relationship between the fungal hydrogenosome and peroxisome. However, sequencing of the Neocallimastix hydrogenase gene revealed that it did not encode an SKL motif, but rather an N-terminal extension resembling a targeting peptide (Davidson et al., 2002). Furthermore, Neocallimastix hydrogenosomal proteins have been shown to contain targeting peptides recognised by the yeast mitochondrial import system. When the yeast Hansenula polymorpha was transfected with the Neocallimastix hydrogenosomal malic enzyme, it was imported into the mitochondrion rather than the peroxisome (van der Giezen et al., 1998). Now, phylogenetic analyses have also shown that the Neocallimastix AAC transporter, a hydrogenosomal membrane protein responsible for the import of ADP and export of ATP, is closely related to the mitochondrial AAC carrier (van der Giezen et al., 2002). Taken together, the evidence strongly favours an hypothesis where the fungal hydrogenosome has in fact evolved from a mitochondrion.

4.1.3. Parabasalian hydrogenosomes

Parabasalia are a very large and extremely diverse group of protists, virtually all of which are found in some sort of association with animals. Most Parabasalia are innocuous or even essential to their hosts' survival (e.g., in the termite hindgut environment: Yamin, 1979), but others are important pathogens. *Trichomonas vaginalis* is a sexually transmitted human parasite infecting 170 million people each year and, as such, is the most common non-viral sexually transmitted disease. It causes genitourinary tract infections and increases susceptibility to HIV infection and cervical cancer, as well as increasing the risk of perinatal complications (Petrin et al., 1998). The current drug of choice for treatment is metronidazole. This drug's specificity relies on activation by the hydrogenosomal enzyme pyruvate:ferredoxin oxidoreductase (PFOR) (Müller and Lindmark, 1976).

The parabasalian hydrogenosome was the first to be characterised, and is currently the best studied in terms of biochemistry and cell biology. Initial reports suggested that parabasalid hydrogenosomes were bound by a single membrane (Lindmark and Müller, 1973), but it is now known that they are bound by a double membrane, as is the case with mitochondria (Benchimol and DeSouza, 1983). Furthermore, although one uncorroborated study did report circular DNA in trichomonad hydrogenosomes (Cerkasovova et al., 1976), it is now clear that those parabasalian hydrogenosomes that have been most closely examined lack DNA (Clemens and Johnson, 2000). The metabolism of parabasalian hydrogenosomes has been worked out in detail, and all the key metabolic enzymes from *T. vaginalis* have been characterised and their genes sequenced. These genes possess amino-terminal extensions that look something like mitochondrial targeting transit peptides, although they are by no means canonical (Bradley et al., 1997). Even with these data, however, the origin of the organelle was not entirely clear. Unlike the chytrids and ciliates, there was no certain indication that Parabasalia evolved from mitochondrial ancestors, so our understanding of the origin of the parabasalian hydrogenosome has been strongly influenced by our understanding of the evolutionary history of the host cell.

In contrast to ciliates and chytrids, parabasalids were once widely considered to be primitively amitochondriate organisms, members of the Archezoa, supposedly diverging from other eukaryotes prior to the endosymbiotic origin of mitochondria (Cavalier-Smith, 1983). Based on this perceived phylogenetic position, it was thought that the parabasalian hydrogenosome must have originated independently of the mitochondrion, either by an endogenous or alternative endosymbiotic origin. Indeed, even before the Archezoa hypothesis was formulated, it was suggested that hydrogenosomes might have originated from an endosymbiotic event with a *Clostridium*-like bacterium, as the enzymes PFOR and hydrogenase were not thought to occur in any other eukaryotes (Whatley et al., 1979).

In order to challenge the hypothesis that parabasalids originated before the mitochondrial endosymbiosis, and to see if an evolutionary link existed between parabasalian hydrogenosomes and typical mitochondria, molecular evidence for a cryptic mitochondrion was sought in the *Trichomonas* nuclear genome. Three chaperone genes were the targets of this search: *cpn*60, *hsp*70, and *hsp*10 (Bui et al., 1996; Germot et al., 1996; Horner et al., 1996; Roger et al., 1998). Recall that, in most eukaryotes, these genes are encoded in the nucleus, but their protein products are targeted to the mitochondrion. Phylogenetic analyses of these genes also show that they are closely related to alpha-proteobacterial homologues, as expected of mitochondrial genes. Accordingly, although none of these genes has been characterised in any mitochondrial genome, these genes are reasonably inferred to be derived from the mitochondrial endosymbiont. All three genes were characterised in *Trichomonas*, and in phylogenetic analyses all three were found to cluster with their mitochondrial counterparts (e.g., Figure 4). This provided strong evidence that *Trichomonas* evolved from a lineage bearing a mitochondrion

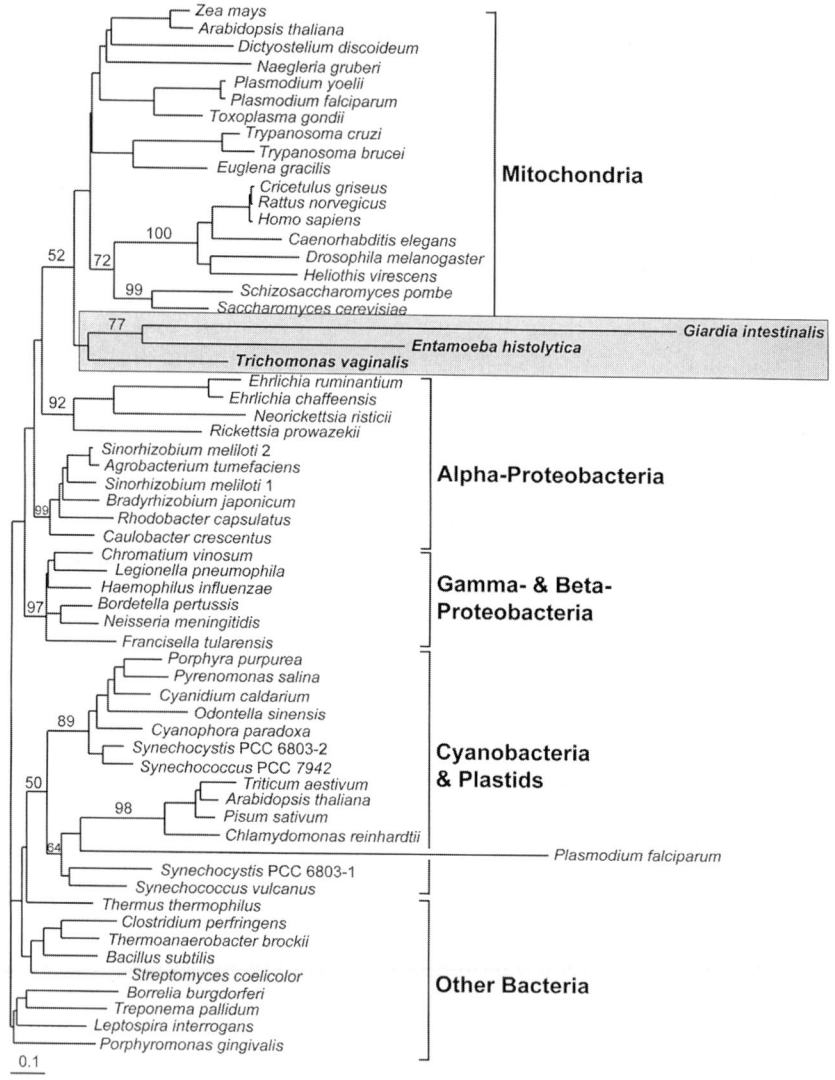

Figure 4 Phylogenetic tree of eubacterial, mitochondrial, and plastid chaperonin 60 (Cpn60). The tree is a weighted neighbour-joining tree (Bruno *et al.*, 2000) based

or mitochondrial ancestor, as these genes were interpreted as having been inherited from the mitochondrial endosymbiont and subsequently transferred to the nuclear genome, as in other eukaryotes. Ensuing research reinforced this conclusion by the characterisation of a mitochondrial-like pyridoxal-5'-phosphate-dependent cysteine desulfurase (IscS) from Parabasalia (Tachezy et al., 2001).

While the phylogenetic data argue very strongly for a mitochondriate ancestry of Parabasalia, they do not strictly address the origin of hydrogenosomes. This link was first established by demonstrating that the mitochondrial-like Cpn60 protein was co-extracted with proteins from hydrogenosomal fractions of *Trichomonas* cells, showing that the product of a gene of mitochondrial endosymbiont origin was localised to the hydrogenosome (Bui et al., 1996). Hydrogenosomal homologues of ADP-ATP carrier proteins were subsequently discovered and shown to be related to mitochondrial homologues. In addition, these proteins were shown to target yeast mitochondria in a manner similar in nature to the targeting of yeast homologues (Dyall et al., 2000). However, the most decisive evidence for a mitochondrial origin for the parabasalian hydrogenosome promises to come from the characterisation and analysis of certain elements of the protein-targeting machinery. Most proteins that have been analysed to date are derived from the genome of the alpha-proteobacterial endosymbiont, but some of the mitochondrial protein translocation factors were established only after the endosymbiosis. These proteins are not found in the bacterial antecedent of the organelle and are of host origin. The translocation system of *Trichomonas* is now being analysed biochemically, and distinctive features have been found in common with mitochondrial protein translocation (Bradley et al., 1997), suggesting that mitochondrion-specific proteins, such as the processing peptidase, may be present in *Trichomonas*. Characterising these proteins, and showing they are phylogenetically related to mitochondrial homologues, will provide even stronger evidence that the parabasalian hydrogenosome evolved from a fully integrated mitochondrion.

---

on gamma-corrected distances (Strimmer and von Haeseler, 1996). Numbers at nodes refer to bootstrap support for that grouping, only high or important bootstrap values are given. Major lineages are bracketed and named to the right. The tree shows the mitochondrial homologues related to alpha-proteobacteria and plastid homologues related to cyanobacteria, as expected. The relatively divergent *cpn*60 genes from the erstwhile "amitochondriates" *Trichomonas*, *Entamoeba*, and *Giardia* (boxed and shaded) all branch with the mitochondrial clade, providing evidence for a mitochondrial ancestry of each of these protist parasites. Note also the plastid Cpn60 from the apicomplexan *Plasmodium*.

4.2. Entamoebid Mitosomes

The Archamoebae is a second group of purportedly amitochondriate protists proposed to belong to the Archezoa. Archamoebae encompasses the amitochondriate amoebae *Entamoeba*, *Pelomyxa* and mastigamoebids, of which the human parasite *Entamoeba histolytica* is the best studied. These groups were assumed to represent another ancient eukaryotic lineage on the basis of their "primitive" features: the absence of an identifiable mitochondrion, peroxisomes or Golgi dictyosomes (Cavalier-Smith, 1991a), and their "bacterial-like" fermentative metabolism using the enzymes PFOR, ferredoxin, and alcohol dehydrogenase (Reeves, 1984; Müller, 1992). This primitive cell structure led to their inclusion within the Archezoa, but they were soon removed from this group as SSU rRNA phylogenies showed that *Entamoeba* diverged from other eukaryotes after several mitochondrial groups. This position was debated extensively by comparing phylogenies based on several different molecular markers (Hasegawa *et al.*, 1993; Leipe *et al.*, 1993), and there was even considerable evidence that the various groups of Archamoebae were not related to one another (for review see Keeling, 1998). Now there is a growing consensus that Archamoebae branch with certain other amoebae and slime moulds at some position near the divergence of animals and fungi (Bapteste *et al.*, 2002). Interestingly, recent data on the root of the eukaryotic tree have brought this debate full circle in one sense, by leading to the revival of the notion that Archamoebae are near the base of eukaryotes (Stechmann and Cavalier-Smith, 2002). Even while the position of *Entamoeba* was under scrutiny, it was assumed to lack mitochondria, and that a mitochondriate origin for the Archamoebae only meant that the organelle had been lost secondarily. However, suggestions of the presence of a mitochondrion can be found in early *Entamoeba* literature: mitochondria-like organelles were found through the use of vital stains, and the activity of the mitochondrial enzyme pyridine nucleotide transhydrogenase (PNT) had been demonstrated (Causey, 1925; Harlow *et al.*, 1976). Subsequently, mitochondrial gene sequences were sought and found in the nuclear genome of *Entamoeba*, proving its mitochondrial ancestry decisively. In this case, gene sequence from PNT and mitochondrial *cpn*60 were initially characterised (Clark and Roger, 1995), and both were shown to be closely related to mitochondrial homologues in phylogenetic analysis (e.g., Figure 4). This was recently supported by the characterisation of a mitochondrial HSP70 in *Entameoba* (Bakatselou *et al.*, 2000). Interestingly, all three genes encode N-terminal extensions with some similarity to transit peptides (Clark and Roger, 1995; Bakatselou and Clark, 2000), indicating the possible presence of mitochondrial-targeting peptides, fueling the idea that a remnant mitochondrion may still be present in *Entamoeba*.

This possibility was tested by localisation studies using the Cpn60 protein – the protein with the most compelling evidence for mitochondrial ancestry from phylogenetics. This protein was independently shown by two groups to localise to a relatively nondescript, double membrane-bounded structure (Mai *et al.*, 1999; Tovar *et al.*, 1999). In one study, cells were transfected with a mtCpn60-c-*myc* fusion protein, and the protein was shown to be localised to a single 1 μm diameter structure using anti-c-myc antibodies (Tovar *et al.*, 1999). When the first 15 amino acids of the sequence were removed, it was shown that localisation was disrupted, but then restored by the addition of a trypanosome mitochondrial transit peptide. A second study showed that antibodies raised directly against the *Entamoeba* Cpn60 localised to a small, low copy number organelle in a similar pattern (Mai *et al.*, 1999). This cryptic mitochondrion was dubbed either the mitosome or crypton (Mai *et al.*, 1999; Tovar *et al.*, 1999). Interestingly, the mitosome has also been shown to stain with the DNA-binding fluorochromes, propidium iodide, sytox green, and acridine orange, which has been taken to mean that it has retained a genome (Ghosh *et al.*, 2000). Whether this does represent the presence of DNA and, if so, exactly what this genome would encode, are both intriguing questions that await molecular characterisation. It is also difficult to predict the possible function of the organelle; presently Cpn60 is the only protein that has been localised to the structure, although it seems very likely that PNT and HSP70 are also mitosomal based on their phylogenetic positions. In addition, PFOR has been reported to be associated with discrete cytoplasmic structures, possibly mitosomes (Rodriguez *et al.*, 1998). In contrast, two other key enzymes in the metabolism of *Entamoeba*, ADH and ferredoxin, have been demonstrated to reside in the cytosol (Mai *et al.*, 1999). This story is obviously very complex, and only a small part of the picture is available at present. Nevertheless, it seems that the core carbon metabolism of *Entamoeba* is not compartmentalised to the extent that is seen in parabasalian hydrogenosomes, and that the mitosome is probably retained for a limited number of functions, whatever they may be.

4.3. Microsporidian Mitosomes

Microsporidia are intracellular parasites that infect a broad range of animals, including immunocompromised humans, as well as a few species of ciliate and gregarine Apicomplexa. In humans, they cause a variety of ailments, most commonly diarrhoea, but also keratitis, sinusitis, encephalitis, and myositis (Weber et al., 1994). The extracellular spore stage of microsporidia is dominated by a highly specialised and unique infection apparatus (Vávra and Larsson, 1999). Most obvious is the polar tube, which is used to pierce host cells and transmit the parasitic sporoplasm to the cytosol of the host. In contrast to the highly ordered and easily recognisable spore, electron microscopy of meront stages of microsporidia, an intracellular stage representing most of the actively dividing microsporidia, shows these cells to be amorphous with few obvious ultrastructural features, lacking mitochondria, centrioles, peroxisomes, and microtubules (Vávra and Larsson, 1999).

The apparent simplicity of microsporidia was interpreted as primitive, and thus microsporidia were also included in the Archezoa and considered to be primitively amitochondriate (Cavalier-Smith, 1983). The first molecular phylogenies based on SSU rRNA, together with a prokaryotic-like fused 5.8S-LSU rRNA structure, supported the hypothesis that microsporidia were one of the first branches of the eukaryotic tree (Vossbrinck and Woese, 1986; Vossbrinck et al., 1987). Ensuing phylogenies of microsporidian EF-1α and EF-2 genes further substantiated this position (Kamaishi et al., 1996a, b), and established microsporidia as a "textbook" example of early branching eukaryotic cells (Tudge, 2000; Madigan et al., 2002). However, rigorous phylogenetic analyses based on several protein-coding genes have now shown microsporidia to be related to fungi, and not at all ancient (Edlind et al., 1996; Keeling and Doolittle, 1996; Fast et al., 1999; Hirt et al., 1999; Keeling et al., 2000; Van de Peer et al., 2000; Keeling, 2003). As fungi are clearly mitochondriate, microsporidia must have evolved from a mitochondriate ancestor, and either lost the organelle or retained a cryptic mitochondrion. The first direct evidence for a mitochondriate ancestry came, once again, from genes encoding mitochondrial HSP70 (e.g., Figure 5). This mitochondrion-derived gene was characterised from the nuclear genomes of several species of microsporidia (Germot et al., 1997; Hirt et al., 1997; Peyretaillade et al., 1998), and was soon supported by the characterisation of α and β subunits of mitochondrial pyruvate dehydrogenase (PDH) (Fast and Keeling, 2001), putting to rest the idea that microsporidia were primitively amitochondriate (Figure 6). The complete sequence of the E. cuniculi genome added several more putative mitochondrial genes to this list, some of which have interesting implications. In addition to HSP70 and the two subunits of PDH, the Encephalitozoon genome was found to encode homologues of TIM22, TOM70, Erv1p, Atm1p (a homologue of the ABC transporter), ferredoxin (Yah1p), manganese superoxide dismutase (MnSOD), Nifu1p-like protein, frataxin (Yfh1p), and mitochondrial glycerol-3-phosphate dehydrogenase (Katinka et al., 2001). This extensive list allowed Katinka et al. (2001) to hypothesise

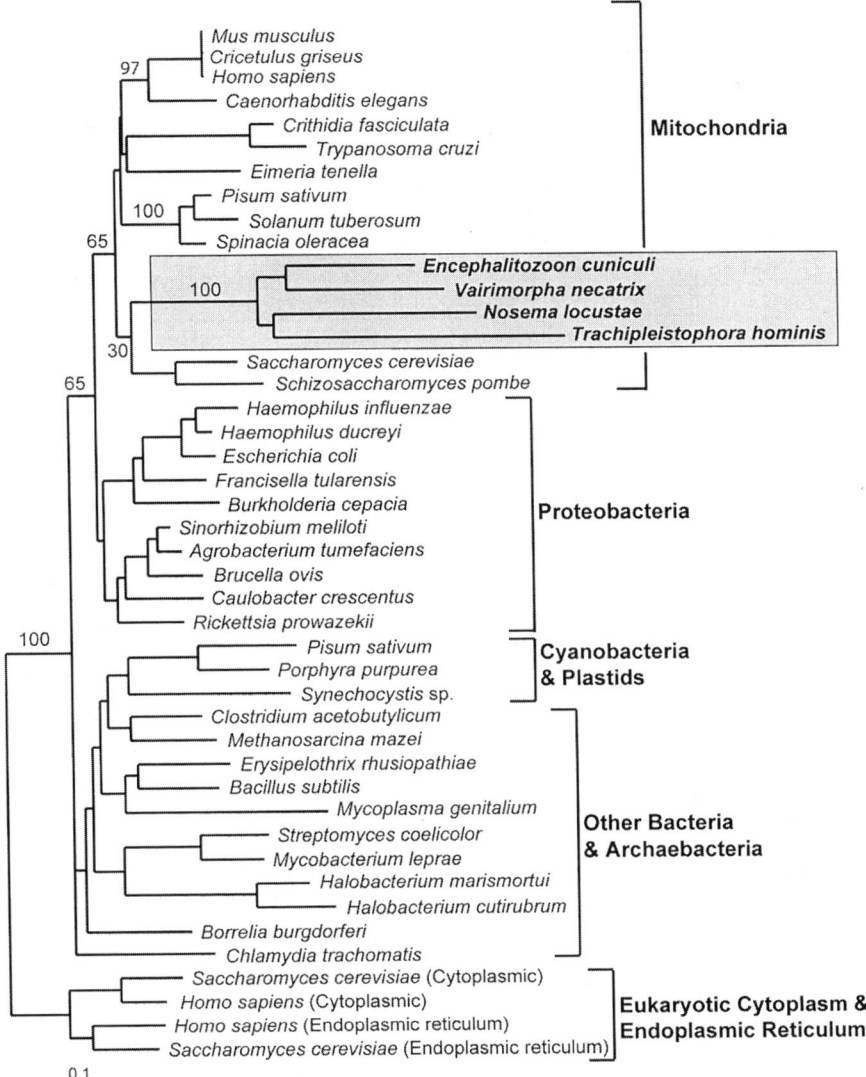

Figure 5 Phylogenetic tree of eubacterial, mitochondrial, and plastid HSP70 proteins. The tree is constructed in the same way as that shown in Figure 4. Note that the sequences from four microsporidian sequences (boxed and shaded) all branch within the mitochondrial clade, and branch weakly with fungi.

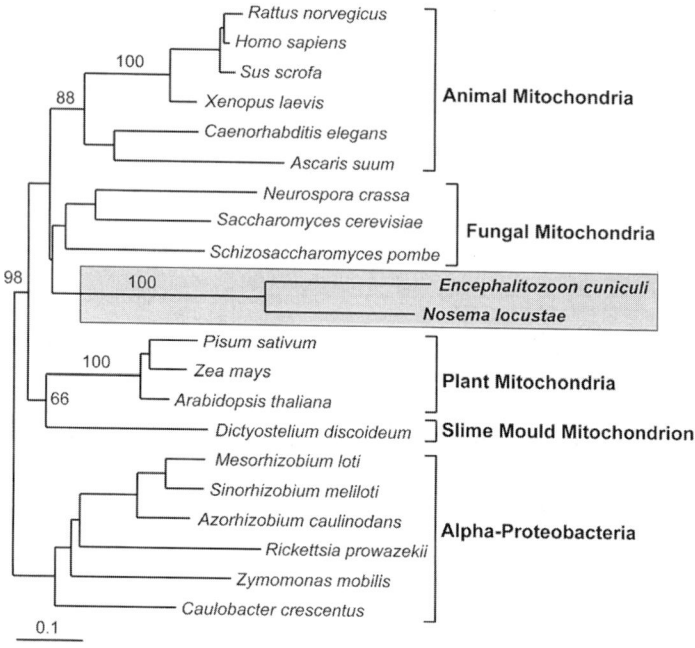

Figure 6 Phylogenetic tree of the beta subunit of pyruvate dehydrogenase (PDH) E1 from alpha-proteobacteria and mitochondria. The tree is constructed in the same way as that shown in Figure 4. Note that the microsporidian sequences (boxed and shaded) are related to mitochondrial homologues, and branch weakly with the fungi. This tree is an example of how phylogeny can be used to show the mitochondrial origin of a protein in microsporidia.

the existence of a remnant mitochondrion-derived organelle in microsporidia, which they called a mitosome. However, there was still no physical evidence for the presence of an organelle, and analysis of the N-termini of putatively mitochondrial proteins from microsporidia led to contradictory conclusions (Fast and Keeling, 2001; Katinka *et al.*, 2001), so it remained uncertain whether microsporidia had actually retained a mitochondrial organelle. The production of a specific antibody to the mitochondrial HSP70 from the microsporidian *Trachipleistophora hominis* allowed the localisation of the first mitochondrion-derived protein in microsporidia (Williams *et al.*, 2002). The HSP70 was found to localise to double membrane-bounded structures, which are abundant but extremely small, at just 90 nm at their broadest. These structures are relict mitochondria-derived organelles, and probably represent the smallest mitochondria presently known (Williams *et al.*, 2002). Even in the smallest eukaryote, *Ostreococcus tauri*, a picoplanktonic alga, which measures a mere 1.5 µm in diameter, mitochondria are twice the size of the structures seen in the microsporidian cells (Chretiennot-Dinet *et al.*, 1995).

Like other putatively mitochondrial proteins in microsporidia, mitochondrial-targeting signals could not be detected in the *T. hominis* HSP70 sequence. This is unlike the situation in trichomonad hydrogenosomes or *Entamoeba* mitosomes, where recognisable mitochondrial-targeting signals have been retained. Despite this, the *T. hominis* HSP70 protein is apparently successfully translocated across the organelle's double membrane into the matrix. This raises interesting questions regarding the nature of the import system present in the microsporidian mitochondrial-derived organelle. It is clearly homologous to other mitochondrial-targeting systems, since the *Encephalitozoon* genome encodes homologues of mitochondrial translocation complex proteins (e.g., TIM22, TOM70), but the system is also clearly different. Most of the translocation complex homologues have not been detected in the *Encephalitozoon* genome (including a stromal peptidase), and the transit peptides are typically very reduced or even absent, as discussed above (although some genes have very long leaders that do resemble transit peptides, e.g., glycerol-3-phosphate dehydrogenase). Microsporidia are well known to have streamlined much of their biology in adapting to their parasitic lifestyle (Keeling, 2001; Keeling and Fast, 2002) and it is possible that this process has extended to the mitochondrial protein import machinery, resulting in a unique system that could potentially no longer rely on cleaved N-terminal transit peptides.

The possible functions of this organelle also remain somewhat unclear, although the complete genome of *Encephalitozoon* certainly allows for informed speculation. As microsporidia are anaerobes and have previously been shown to lack oxidative phosphorylation and TCA cycle (Weidner *et al.*, 1999), none of the proteins related to these pathways are found in the genome. A high proportion (five) of the identified mitochondrial proteins retained by *E. cuniculi* is involved in iron–suphur cluster assembly, making this a leading candidate function for the highly reduced organelle (Katinka *et al.*, 2001). In addition, the presence of PDH subunits is interesting. In typical mitochondria, PDH complex (composed of PDH E1, E2, and E3) catalyses the first step in the breakdown of pyruvate, resulting in acetyl-CoA which is passed into the TCA cycle (Pronk *et al.*, 1996). In the "amitochondriates" like *Trichomonas*, *Entamoeba*, and *Giardia*, PDH is absent and pyruvate is activated by PFOR (Müller, 1992), the target for the most widely used drugs against these parasites. Microsporidia appear to lack PFOR, contain PDH E1, but lack E2 and E3 (Fast and Keeling, 2001;

Katinka et al., 2001). They also contain ferredoxin, and ferredoxin:NADH oxidoreductase (Katinka et al., 2001). Taken together, this suggests that there is some energy-generating metabolism of pyruvate taking place in the mitosome, but of a unique nature; resembling neither typical mitochondrial metabolism or the metabolism that has been characterised in other "amitochondriates".

4.4. Mitochondria in Diplomonads?

The diplomonads, of which *Giardia intestinalis* (syn. *lamblia*) is the best studied member, are a fairly small group of anaerobes and microaerophiles that are predominantly parasites of a variety of animals, but also include some free-living genera (Brugerolle and Lee, 2002). Diplomonads are another textbook example of "primitive" amitochondriate eukaryotes, often lauded as the first lineage of eukaryotes (Sogin et al., 1989; Sogin and Silberman, 1998). This evolutionary position has been substantiated by many molecular phylogenies (Sogin et al., 1989; Hashimoto et al., 1994, 1995), but has also come under scrutiny in recent years because of evidence that other purportedly "deep" lineages are really misplaced long-branches in the eukaryotic tree, and the observation that diplomonad sequences are also generally very divergent (Embley and Hirt, 1998). Like *Trichomonas* and *Entamoeba*, *Giardia* relies on anaerobic rather than aerobic energy metabolism and uses the key enzyme PFOR. Indeed, the metabolism characterised in *Giardia* is very similar to that of *Entamoeba* (Müller, 1992), despite clear evidence that the two are very distantly related. The important metabolic enzymes that have been studied in *Giardia* have not been shown to act within a specialised compartment. Diplomonads were also considered to be Archezoa, and accordingly thought to have never had a mitochondrion (Cavalier-Smith, 1983). However, as for *Trichomonas*, microsporidia, and *Entamoeba*, molecular phylogenetic evidence for a mitochondrial ancestry has been found in diplomonads.

The evidence for a mitochondriate origin of diplomonads actually goes back some time, but the early evidence was weak and, in retrospect, probably not accurately interpreted. The first such evidence came from phylogenies of the glycolytic enzyme glyceraldehyde-3-phosphate dehydrogenase (GAPDH). Although GAPDH does not act in the mitochondrion of modern eukaryotes (with one very odd exception: the diatom *Phaeodactylum tricornutum*: Liaud et al., 2000), phylogenies were interpreted as showing that the eukaryotic enzyme was derived from the mitochondrial endosymbiont (Martin et al., 1993). The *Giardia* GAPDH was like other eukaryotic homologues, and it was therefore interpreted as evidence for a mitochondrial ancestry of diplomonads. However, the mitochondrial origin of eukaryotic GAPDH genes has not withstood further phylogenetic analysis (Kurland and Andersson, 2000), so the *Giardia* GAPDH does not likely reflect a mitochondriate origin of diplomonads after all. Similar arguments have been made for another glycolytic enzyme, triosephosphate isomerase (TPI) (Keeling and Doolittle, 1997). Once again, however, the mitochondrial origin of TPI has not been substantiated by further analysis (Kurland and Andersson, 2000).

Much stronger evidence for a mitochondriate ancestry of diplomonads came from the dual-function valyl-tRNA synthetase of eukaryotes (which functions in both the cytosol and mitochondrion). Phylogenetic analysis suggested that this enzyme was also derived from the mitochondrial endosymbiont (Brown and Doolittle, 1995), including the *Giardia* homologue (Hashimoto *et al.*, 1998). This case is stronger than either TPI or GAPDH, but still suffers from a somewhat indirect connection to the mitochondrion. More direct evidence for mitochondria was sought, as in other Archezoa, using the mitochondrial chaperonin, Cpn60. In contrast to the chain of discoveries in other Archezoa, however, the first data for putative mitochondrial chaperonins in *Giardia* came from localisation experiments using heterologous antibodies to Cpn60 raised against human or rodent mitochondrial Cpn60, or prokaryotic Cpn60. Labelling of *Giardia* cells with such heterologous antibodies suggested that a prokaryotic Cpn60 was present in the cell and was localised in a punctate fashion, but not to discrete compartments (Soltys and Gupta, 1994). Subsequently, genes for *cpn*60, *hsp*70, and IscS were characterised in *Giardia*, and were shown to be phylogenetically related to mitochondrial homologues (Roger *et al.*, 1998; Morrison *et al.*, 2001; Tachezy *et al.*, 2001), confirming that the diplomonads did have a mitochondriate ancestry (e.g., Figure 4).

However, the presence or absence of a mitochondrion has not yet been rigorously tested in *Giardia*, and at present available evidence is conflicting. First, none of the proteins shown to be phylogenetically derived from the mitochondrial endosymbiont encodes leaders that could unambiguously be interpreted as transit peptides (Roger *et al.*, 1998; Horner and Embley, 2001; Morrison *et al.*, 2001; Tachezy *et al.*, 2001). Moreover, localisation data from heterologous antibodies to Cpn60 give slightly different results depending on the antibody used, although none have produced convincing evidence for an organelle (Soltys and Gupta, 1994; Roger *et al.*, 1998). Similarly, thorough inspection of electron micrographs of *Giardia* has not shown the presence of a typical mitochondrion, although early literature on *Giardia* did document sightings of structures which may resemble reduced mitochondria, and hypothesised that their unusual structure may be due to the low oxygen environment to which *Giardia* is adapted (Cheissin, 1965).

Nevertheless, areas of the *Giardia* cell have also been shown to react with CTC, a marker of respiratory chain activity. Similarly, the membrane potential detecting dye, Rhodamine 123, also labelled isolated zones, indicating discrete, membrane-bounded areas within the cell with an electron potential (Lloyd *et al.*, 2002).

Interestingly, a little-studied flagellate, *Carpediemonas membranifera*, has recently been described and shown to have a membrane-bounded organelle reminiscent of a hydrogenosome, although there are no biochemical data on the function of this organelle (Simpson and Patterson, 1999). The phylogenetic position of *Carpediemonas* has now been analysed using alpha-tubulin, beta-tubulin, and SSU rRNA, and intriguingly it was found to be strongly supported as the sister group to retortamonads and diplomonads (Simpson *et al.*, 2002). This free-living flagellate has the potential to hold many of the answers to the evolutionary fate of the mitochondrion in diplomonads like *Giardia*, but it will be difficult to draw any conclusions until the *Carpediemonas* organelle is better characterised. Until then, the data for or against an organelle in *Giardia* remain inconclusive, and if such an organelle does exist, its function is entirely unclear. The characterisation of iron–sulphur cluster assembly proteins (Tachezy *et al.*, 2001) may indicate that, as in other eukaryotes, iron–sulphur centres are created within the mitochondrion and exported for use in cytosolic iron–sulphur proteins. However, this is very speculative as definitive evidence for an organelle is still lacking and, by extension, no protein has been localised to an organelle in *Giardia*, both of which are necessary to understand the function of a cryptic organelle.

5. CASE HISTORIES – PLASTIDS

5.1. Apicomplexa

The Apicomplexa is a phylum entirely composed of obligate intracellular parasites, some of which cause extremely important diseases of both medical and commercial concern. The most serious of these is malaria, caused by members of the genus *Plasmodium*, which remains one of the most deadly infectious diseases throughout much of the world today.

In the mid-1970s, electron microscopic observations of the extrachromosomal DNA of an avian malaria parasite revealed the presence of a circular element that contained a cruciform structure indicative of an inverted repeat (Kilejian, 1975). This genome was reasonably interpreted as the mitochondrial genome (Kilejian, 1975; Gardner *et al.*, 1988), but subsequent sequencing of genes from this element revealed that, although they were eubacterial in nature, they did not bear any particular resemblance to homologues from other mitochondrial genomes. Instead, and surprisingly, these genes were found to be most similar to homologues from plastid genomes (Gardner et al., 1991, 1993), a finding corroborated by the discovery of certain genes that are never found in other mitochondrial genomes, but are common to plastid genomes (Gardner et al., 1994; Williamson et al., 1994).

Overall, the characteristics of this genome led Iain Wilson and colleagues to propose that the 35 kb circle, as it was known, was a plastid genome (Gardner et al., 1991). This rather unusual idea was significantly supported by the discovery and characterisation of a second extrachromosomal element in Apicomplexa. This was a 6 kb linear element that encoded genes for coxI, coxIII, and cob, all mitochondrial proteins, as well as fragments of mitochondrial rRNA genes (Vaidya et al., 1989; Feagin et al., 1991). This small and unusual element was shown to co-fractionate with the mitochondrion (Wilson et al., 1992), revealing it to be the most reduced and odd of all mitochondrial genomes known (Gray et al., 1999).

Ultimately, the plastid nature of the 35 kb circle was made crystal clear by the complete sequencing of the *Plasmodium falciparum* element (Wilson et al., 1996), and later that of *Toxoplasma gondii* (Köhler et al., 1997). Although the genome was reduced and obviously lacked any genes directly related to photosynthesis, the complete sequence revealed a number of characteristics known to be unique to plastid genomes. These include the rRNA operon inverted repeat (which was originally observed by electron microscopy), and the plastid "superoperon (Wilson et al., 1996). At the same time, plastid genomes were also shown to be present in a diversity of Apicomplexa (Wilson et al., 1996; Lang-Unnasch et al., 1998), showing that the genome is a widespread characteristic of the group.

The actual organelle was identified by localising transcripts from plastid-encoded SSU rRNA (McFadden et al., 1996), and by localising the genome itself (Köhler et al., 1997). Not surprisingly, the plastid bears little resemblance to the canonical chloroplast; instead, it is a relatively non-descript multimembrane structure typically found once per cell. The first description of the *Toxoplasma* plastid suggested it was surrounded by two or perhaps three membranes (McFadden et al., 1996), however it was later shown to be a four-membrane structure (Köhler et al., 1997), suggesting a secondary origin (see Figure 2). This has subsequently come under dispute once again with the suggestion that there are actually three plastid membranes (Hopkins et al., 1999), but this has not been corroborated. Interestingly, the structure finally identified as the plastid had been documented on several occasions and given several different names by earlier electron microscopists (McFadden and Waller, 1997). Understandably, none had guessed its origin or had any idea of its function.

Indeed, the function of the organelle was still unclear even after both the *Plasmodium* and *Toxoplasma* plastid genomes had been completely sequenced: both genomes contained a large number of genes encoding proteins involved in gene expression, but none that stood out as obviously functionally critical genes that could explain the retention of the organelle (Wilson *et al*., 1996; Köhler *et al*., 1997). These obligate intracellular parasites are obviously not photosynthetic, so what does a plastid do in Apicomplexa? It was shown very early on that the plastid was essential to the parasites, although it is interesting to note that knocking out plastid translation did not kill the parasites immediately, but disrupted subsequent infection cycles (Fichera and Roos, 1997; He *et al*., 2001). The answer to this puzzle lay in the biochemical flexibility of plastids, and the nuclear genome of the parasites. In plants and algae, plastids are nearly always thought of in association with photosynthesis, as this is the major biochemical activity of the organelle and the characteristic that most obviously sets these organisms apart from other eukaryotes. However, plastids are more than simply photosynthetic sugar factories. In addition to photosynthesis, and the biochemical pathways responsible for synthesising the chemical components necessary for photosynthesis, plant and algal plastids have a number of other important jobs. The best understood are the synthesis of fatty acids, isopentenyl diphosphate (the core unit of isoprenoid compounds), heme, and some amino acids, although a number of other important roles are also known (Herrmann, 1995; Harwood, 1996; Rohdich *et al*., 2001). Recall that, during the early stages of endosymbiosis, most of the plastid genes were moved to the host nucleus, with their products being post-translationally targeted back to the organelle. Accordingly, the majority of the functionally critical proteins are encoded by plastid-derived genes in all plants and algae, but many of these genes are nucleus-encoded. Searching the nuclear genomes of *Plasmodium* and *Toxoplasma* soon revealed the presence of a number of genes for plastid-targeted proteins, which were identified based on their phylogenetic relationship with plastid and cyanobacterial homologues (e.g., Figure 7). The first such proteins to be identified were involved in fatty acid biosynthesis (Waller *et al*., 1998). These proteins were shown to be localised to the plastid by immunocytochemistry, and to encode N-terminal leader peptides with signal and transit-peptide domains (as expected for a secondary plastid) that were sufficient to target Green Fluorescent Protein (GFP) to the plastid in transfected cells (Waller *et al*., 1998). Subsequently, the nuclear genome has also been found to harbour genes for plastid-targeted proteins making up complete pathways for fatty acid, isopentenyl diphosphate, and heme biosynthesis (Waller *et al*., 1998;

CRYPTIC ORGANELLES IN PARASITIC PROTISTS AND FUNGI

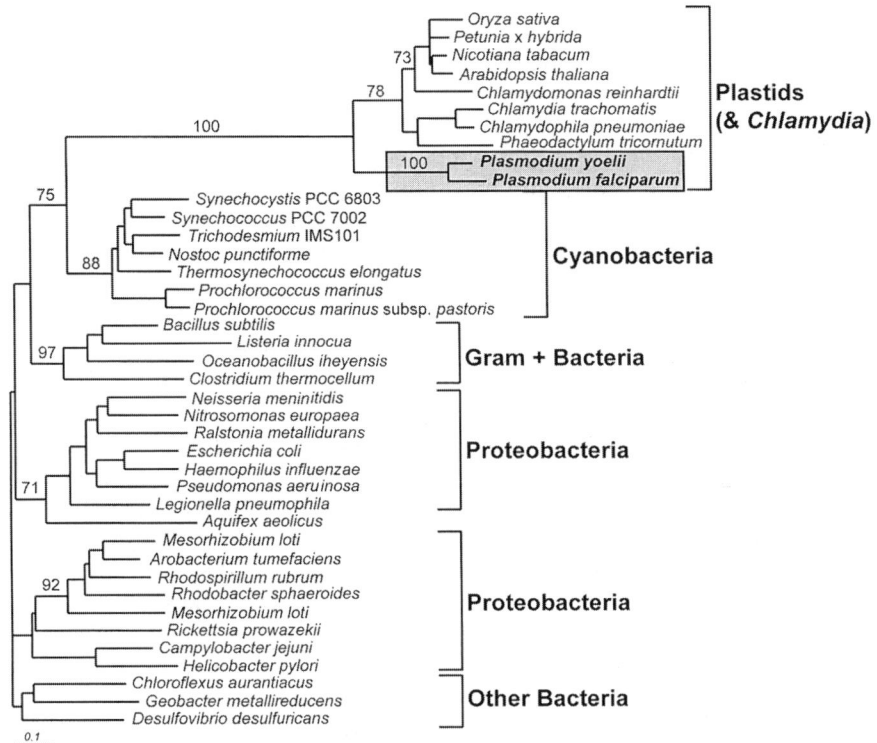

Figure 7 Phylogenetic tree of eubacterial and plastid FabI (enoyl-ACP reductase). The tree is constructed in the same way as that shown in Figure 4. The (divergent) plastid homologues branch with the cyanobacteria, as expected, and the *Plasmodium* genes (boxed and shaded) branches within the plastid with very high statistical support. Curiously, the chlamydial FabI appears to be derived from a plastid gene, which has been noted previously. This tree is an example of how phylogeny can be used to show the plastid origin of a protein in *Plasmodium*, and shows that Type I fatty acid biosynthetic enzymes are plastid-derived in Apicomplexa.

Jomaa *et al.*, 1999; Gardner *et al.*, 2002; Sato and Wilson, 2002). The Apicomplexa also have genes for aromatic amino acid biosynthesis (Roberts *et al.*, 1998), but these are not plastid derived (Keeling *et al.*, 1999) and are not localised to the plastid (Fitzpatrick *et al.*, 2001). The biosynthesis of fatty acids, isopentenyl diphosphate, and heme can explain why the Apicomplexa retains a plastid. When the plastid originated in this lineage, it assumed essential functions other than photosynthesis (either these pathways did not exist in the pre-plastid host, or they were made redundant by the plastid pathway and were lost in the host). As the ancestors of Apicomplexa progressively adapted to intracellular parasitism and the plastid lost its role in photosynthesis, it could not simply be lost unless another source of these essential compounds could be secured. As parasites, the Apicomplexa absorbs many nutrients from their hosts, but apparently the compounds synthesised by the plastid are either not acquired from the host at some stage of the life cycle, or some particular variants need to be synthesised for specialised purposes.

At about the time that the first sequences from the apicomplexan plastid genome became available, the evolutionary position of the Apicomplexa was also being examined using nuclear gene sequences. Apicomplexa had been classified for convenience (Kudo, 1947) with other "spore-forming parasites" such as microsporidia and actinosporidia (apicomplexans do not form spores). The phylogenies of nuclear SSU and 5S rRNA, however, showed a very strongly supported relationship between Apicomplexa, ciliates, and dinoflagellate algae (Gajadhar et al., 1991; Wolters, 1991). Subsequent phylogenies based on a variety of protein-coding genes have lent further strong support for this group (Fast et al., 2002; Saldarriaga et al., 2003), and ultrastructural studies have revealed that these three lineages share a number of structural features. This group is now known as Alveolata, and is one of the best-supported major protist groupings presently defined. Within the alveolates, the Apicomplexa appear to be more closely related to the dinoflagellates than either are to ciliates (Fast et al., 2002). Dinoflagellates are a very diverse group of predators, parasites, and most interestingly, phototrophs. About one half of the described dinoflagellates are photosynthetic (Taylor, 1987), and their plastids were acquired by secondary endosymbiosis with a red alga (Zhang et al., 2000).

The relationship between Apicomplexa and dinoflagellates led to obvious questions as to whether their plastids arose from a single common endosymbiosis in their ancestor, or if they arose independently by two secondary endosymbiotic events. The former alternative requires fewer independent plastid origins, and indeed, it has been suggested that the plastid in both groups arose much earlier in a common ancestor of all alveolates and a number of other groups with secondary plastids, namely heterokonts, haptophytes, and cryptomonads (Cavalier-Smith, 1999). On the other hand, no plastids have been found in the apparent deep diverging lineages of both dinoflagellate and apicomplexan lineages (Saldarriaga et al., 2001; Kuvardina et al., 2002), so it was also reasonable to assume that each group acquired its plastid individually and recently (Taylor, 1999). In addition, there has been some debate over whether the apicomplexan plastid was derived from a red or a green alga. In favour of a red algal origin, the plastid genome bears some characters in common with red algal plastid genomes, most notably the structure of the plastid super-operon, and several genes have also shown weak support for a red algal affinity (Gardner et al., 1994; McFadden and Waller, 1997; Blanchard and Hicks, 1999). In contrast, the phylogeny based on the plastid tufA gene weakly showed a green algal affinity (Köhler et al., 1997). A green algal origin of the apicomplexan plastid is inconsistent with a single early origin for apicomplexan and dinoflagellate plastids since the ancestral dinoflagellate plastid is clearly derived from a red alga by secondary endosymbiosis (Zhang et al., 2000). The obvious way to solve each of these problems is to infer phylogenetic trees using plastid gene sequences to see if the dinoflagellates and Apicomplexa form a unique clade nested within the red algae. Plastid gene sequences from dinoflagellates were only recently characterised, and were found to reside on single gene mini-circles rather than the expected single circular chromosome as found in other plastids (Zhang et al., 1999). Unfortunately, nearly all genes found in dinoflagellate plastids to date have encoded proteins specific for photosynthetic functions (Zhang et al., 1999; Barbrook and Howe, 2000; Barbrook et al., 2001; Hiller, 2001), so there is little plastid information to compare between Apicomplexa and dinoflagellates. To make matters worse, both apicomplexan and dinoflagellate plastid genes are very divergent, which can potentially lead to difficulties in phylogenetic inferences (Zhang et al., 2000). Nonetheless, in gene trees based on plastid SSU and LSU rRNA (the only comparable genes known today), the apicomplexan and dinoflagellates form a clade (Zhang et al., 2000). In contrast to the plastid-encoded genes, genes for plastid-targeted proteins in both groups appear to be relatively conserved, leading to a large source of potential genes for comparison. The first (and presently the only) plastid-targeted gene to be sampled from both groups is GAPDH. GAPDH is found in both the cytosol, where it is largely NADH-dependent and catabolic, and the plastid, where it is either NADH or NADPH-dependent, and anabolic. In plants, green algae, red algae, and *Euglena*, the cytosolic gene is related to other eukaryotic GAPDH genes and the plastid gene is related to cyanobacterial homologues, as expected. However, in Apicomplexa and dinoflagellates, both GAPDH homologues are related to eukaryotic cytosolic genes, and most importantly, both the cytosolic and plastid genes from both groups form clades and are each most closely related to one another than to any other GAPDH (Fast et al., 2001). The situation in dinoflagellates is perhaps more complex, since putative plastid-targeted genes resembling canonical plastid-targeted genes have also been described (Fagan and Hastings, 2002). Where these genes came from is not certain, but it appears most likely that the cytosolic GAPDH in some ancestor of Apicomplexa and dinoflagellates duplicated and the protein product of one copy was targeted to the plastid where it took over the function of the existing cyanobacterium-derived plastid gene. This kind of event, called "endosymbiotic gene replacement" is known to happen (e.g., Brinkmann and Martin, 1996), but is very unusual and distinctive. Such a rare event, together with the high degree of similarity between the plastid homologues from both groups, provides very strong evidence that the plastids found in these two lineages originated by a single secondary endosymbiosis in a common ancestor of Apicomplexa and dinoflagellates. More interesting still, it has been shown that other photosynthetic groups with red algal secondary endosymbionts, namely cryptomonads, heterokonts (Fast et al., 2001) and haptophytes (J. T. Harper and P. J. Keeling, unpublished data), also acquired their plastid GAPDH from this gene duplication. Thus, this plastid was transferred in the common ancestor of all these groups, long before the origin of Apicomplexa (Figure 3). This finding has significant impact on our understanding of a number of other parasites. The human parasite *Blastocystis* and oomycete parasites such as *Pythium* and *Phytopthora*, belong to the heterokonts (Nakamura et al., 1996; Arisue et al., 2002). These have not traditionally been considered to possess a plastid, but the ancient origin of the plastid in their photosynthetic relatives suggested by the GAPDH data implies that these organisms might indeed possess cryptic plastids (see Andersson and Roger, 2002). To date no direct evidence for such an organelle exist, but this is not surprising since little molecular data are known for most of these organisms.

The single, early, red algal origin of the plastids in Apicomplexa and other chromalveolates has recently been challenged once again by the finding of a curious split CoxII in Apicomplexa. CoxII is a mitochondrial protein that is encoded in the mitochondrial genome of most eukaryotes, but in certain green algae the gene has been transferred to the nucleus and is split into two subunits, *cox*2a and *cox*2b (Perez-Martinez et al., 2001). Interestingly, the genomes of *Plasmodium* and *Toxoplasma* both encode a similarly split form of *cox*2, and it has been argued that this demonstrates that the apicomplexan plastid is derived from a green alga (Funes et al., 2002). While this certainly is an intriguing discovery, whether it relates to the origin of the plastid is far from obvious. For these two *cox*2 subunits to have originated from a green algal endosymbiont, they would have had to have been transferred from the nuclear genome of the endosymbiont to the nuclear gene of the secondary host, presumably replacing an ancestral gene encoded in the mitochondrial genome. This is not impossible to imagine, but the complete genome of *Plasmodium* has not been reported to encode any green algal genes other than the two *cox*2 subunits (Gardner et al., 2002), begging the important question, "why were both subunits of *cox*2 transferred, but no other green algal genes?" Moreover, the relationship between Apicomplexa and other alveolates is now irrefutable. The evidence for the close relationship between heterokonts, haptophytes and cryptomonads (Yoon et al., 2002), and their relationship to alveolates is growing stronger as molecular data continue to accumulate (Baldauf et al., 2000; Fast et al., 2001). This nests the Apicomplexa phylogenetically within a group of protists with secondary endosymbionts of clear red algal ancestry (Figure 3). Accordingly, the conclusion that apicomplexan plastids came from the same endosymbiosis as other chromalveolate plastids, strongly supported by GAPDH (Fast et al., 2001), requires no special or spectacular events. In contrast, a green algal origin of the apicomplexan plastid requires more explanation. One suggestion is that an ancestral red algal plastid was replaced with a new green one (Palmer, 2003). This kind of event has happened several times in dinoflagellates (Tengs et al., 2000; Saldarriaga et al., 2001), and is therefore a plausible explanation where both gene trees are correct. However, given that the plastid genome itself retains a few characteristics suggesting a red algal origin, the possibility that the *cox*2 data are misleading should be considered seriously. Indeed, a recent reanalysis of *cox*2 including the phylogenetically critical genes from ciliates showed that the apicomplexan and ciliate *cox*2 genes were closely related, as expected if no lateral transfer took place (R. F. Waller, P. J. K., G. van Dooren, and G. I. McFadden, unpublished results). Moreover, the ciliate *cox*2 gene contains a 300 amino acid insertion and exactly the position where the split took place, and hydrophobicity analyses suggest that the splitting of the *cox*2 protein might have occurred twice independently to allow the protein to be translocated across the mitochondrial membrane. Altogether, we consider it unlikely that apicomplexan and green algal *cox*2 genes share a close common ancestor, despite their shared structural characteristics.

The discovery of a relict plastid in Apicomplexa is of tremendous interest for reasons relating to basic questions of evolutionary biology (work on the apicomplexan plastid led to important discoveries about the history of secondary endosymbiosis) and cell biology. Ironically, the apicomplexan plastid is by far the most recently discovered plastid, and yet it is already the best characterised secondary plastid and is becoming a model system in which to study plastid targeting at the molecular level (Waller et al., 2000; Foth et al., 2003). However, the plastid is also interesting for practical reasons. Plastids, being derived from prokaryotic cyanobacteria, are full of biochemical processes and proteins not found in animals, and are accordingly potentially useful targets for therapeutics. A number of known antibiotics target differences in basic housekeeping activities such as DNA replication, transcription and translation, and several of these have been successfully tested on Apicomplexa (Fichera and Roos, 1997; McFadden and Roos, 1999; Ralph et al., 2001). Indeed, some drugs were known to kill the malaria parasite, but their mode of activity was not understood until the plastid was discovered and the target of the drug found within the plastid. More interesting still, the plastid uses a Type II, or dissociable fatty acid synthase (FAS) while animals use a very dissimilar Type I FAS (Waller et al., 1998, 2003). Similarly, the plastid uses the so-called non-mevalonate isopentenyl diphosphate biosynthesis pathway while animals uses the non-homologous mevalonate pathway (Jomaa et al., 1999). Both these pathways include a number of enzymes that are the target of well-known drugs that have been successfully tested in Apicomplexa. The plastid can thus be seen as one of the major fundamental differences between the cell biology and biochemistry of these protists and their animal hosts. These differences are the chinks in the armour of any parasite, and should be exploited fully.

5.2. Green Algal Parasites *Helicosporidium* and *Prototheca*

The Helicosporidia are another group of algal parasites, but the story of their cryptic plastid is very different and is just beginning to be told. *Helicosporidium parasiticum* was first described in 1921, but was so unlike any other group of eukaryotes that its evolutionary position was uncertain (Keilin, 1921). The infectious stage of the organism is a cyst containing four cells, three ovoid, internal cells, and a single peripheral cell wound about the others in a helix. *Helicosporidium* is an insect gut parasite and infects when the spore dehisces, or ruptures inside the gut of its host, releasing the helical cell. This cell penetrates a host epithelial cell, and eventually moves through the epithelium to invade the hemolymph, where it develops into a replicating vegetative form (Kellen and Lindegren, 1974; Boucias et al., 2001).

Originally, these parasites were classified separately from any other eukaryotic group (Keilin, 1921), but they have also been considered to be fungi, or related to other "spore-forming" protists, as were Apicomplexa (Kudo, 1947). Recently, however, the first molecular data from Helicosporidia have shown that none of these affinities is true. Instead, these unique parasites are actually green algae. Phylogenies based on SSU and LSU rRNA, as well as actin and beta-tubulin all show this with strong support (Tartar et al., 2002), and the phylogeny of SSU rRNA demonstrates that *Helicosporidium* is a close relative of another algal parasite, *Prototheca*. Unlike *Helicosporidium*, *Prototheca* is predominantly a vertebrate pathogen, and is generally associated with cutaneous infections, although it has been found to infect numerous other tissues (Cho et al., 2002; Piyophirapong *et al.*, 2002; Thiele and Bergmann, 2002). Interestingly, the green algal nature of *Prototheca* was recognised immediately at the time of its original description in 1894, although it was later regarded as a fungus (cited in Butler, 1954; El-Ani, 1967). Although others have argued it to be a fungus since then, there is now an abundance of data, molecular and otherwise, showing *Prototheca* is in fact a close relative of the green algal genus *Chlorella* (Huss and Sogin, 1990). One important exception to this is a misnamed organism, *Prototheca richardsi*, which has been shown to be an icthyosporian, a strange group of parasites closely related to animals and fungi (Baker *et al.*, 1999).

Despite its non-photosynthetic, obligately heterotrophic nature, *Prototheca* is known to have retained its plastid, although little is known about its function. The plastid genome of *P. wickerhamii* has been shown to be severely reduced in size (at 54 100 bp), and recently nearly half of this genome was sequenced (Knauf and Hachtel, 2002). As expected, the genome is dominated by genes related to expression, but interestingly a suite of ATPase genes were also found. What the ATPase complex is doing in this organelle is not clear (Knauf and Hachtel, 2002), but it suggests that the role adopted by the plastid in these parasites should be very interesting indeed.

While little is known about the plastid of *Prototheca*, virtually nothing is known about the plastid of the much more highly derived helicosporidia. No organelle matching the description of a plastid has been observed in ultrastructural investigations, but there are now molecular data for a plastid genome. A plastid SSU rRNA has been characterised from *Helicosporidium* (Tartar *et al.*, 2003) and has been shown to be related to the plastid SSU from *Prototheca*, as expected. There are no data on the size of the genome, or on what role it might play in the biochemistry of the parasite. In addition, no data are available on nuclear-encoded plastid-targeted proteins in either *Prototheca* or *Helicosporidium*. Such data are potentially very interesting, as they could provide a point of comparison with the apicomplexan plastid and show how two independently derived parasites have re-tooled their plastid as they adapted to parasitism.

5.3. Plastids in Parasitic Plants

Although plants are clearly neither protists nor fungi, and we do not usually consider plants when thinking about pathogens and parasites, there are a number of parasitic angiosperms (Kuijt, 1969) whose plastids provide further interesting points of comparison. Parasitic plants are actually relatively diverse and abundant: at least eleven groups of angiosperms have independently evolved some mode of parasitism, which range considerably in their degree of host dependence (Nickrent *et al.*, 1998). Some of these are still photosynthetic during certain stages of their life cycle, leading one to expect their plastids to more or less resemble those of other plants. In contrast, the most extreme forms of parasitic plants are completely non-photosynthetic, deriving their nutrients and water solely from the invasion of host-plant tissue via a specialised root called a haustorium (Kuijt, 1969). In even these extreme cases, the parasites are clearly derived from photosynthetic plants, and structural evidence for a plastid has been known for some time (Mangenot and Mangenot, 1968; Dodge and Lawes, 1974; Kuijt *et al.*, 1985). The angiosperms, therefore, offer a wealth of potential comparative information on the fate of plastids in parasites, but unfortunately very little is known about these organelles.

Most of what we do know about the plastids of these plants comes from the plastid genome. The complete plastid genome has been sequenced for one parasitic angiosperm, *Epifagus virginiana*, or beechdrop (Wolfe *et al.*, 1992). The genome is virtually co-linear with that of tobacco, but is severely reduced, almost entirely due to the loss of all genes relating to photosynthesis and photorespiration. Nearly all the remaining genes in the *Epifagus* plastid genome encode proteins involved in transcription and translation of the genome itself. Only four identified genes are not involved in gene expression, and one of these genes encodes the beta subunit of the carboxytransferase component of acetyl-CoA carboxylase, which is involved in fatty acid biosynthesis (Wolfe *et al.*, 1992). This gene is at least part of the reason why the plastid genome is retained at all, and indicates that fatty acid biosynthesis is one of the residual functions of the relict plastid, as it is in apicomplexa. Interestingly, the plastid genome of an unrelated parasitic plant, *Cuscuta europeae*, has retained the gene for the large subunit of ribulose-1,5-bisphosphate carboxylase-oxygenase (Rubisco), although it has apparently lost CO_2 fixation (Machado and Zetsche, 1990).

It is noteworthy that all four genes in the *Epifagus* plastid genome that are potentially involved in processes other than transcription or translation are known to have been lost in at least one other plastid genome (Downie and Palmer, 1992; Wolfe *et al.*, 1992). If a plastid genome encoded only genes involved in its own expression, the genome could be lost. Therefore, it is possible that all functionally important genes have been transferred or lost in some parasitic plant plastids, allowing its genome to be dispensed with altogether. As seen above, this has happened in several instances with mitochondria, but mitochondrial genomes are far more specialised than plastid genomes: the genes they encode are typically restricted to those involved in transcription, translation, and respiration (Gray, 1999). If respiration is lost, the mitochondrial genome could easily be discarded.

In the case of plastids, the genes encoded in plastid genomes of red algae and their derivatives generally represent a far greater diversity of metabolic functions than those found in green algal and plant plastid genomes, which are typically dominated by genes for proteins and RNAs involved in transcription, translation, and photosynthesis (Palmer and Delwiche, 1998). Accordingly, one would predict that green algal or plant plastids would be more likely to dispense with their genomes if photosynthesis were lost (Palmer and Delwiche, 1998). Indeed, there is evidence that parasitic plants have the smallest plastid genomes by far, the smallest known being *Cytinus*, at a mere 20 kb (Nickrent *et al.*, 1997b). There are also suggestions based on hybridisation and PCR experiments that some parasitic plants like *Rafflesia* might have lost their plastid genome altogether (Nickrent *et al.*, 1997b, 1998). However, this is difficult to distinguish from a high level of divergence, which is seen in the plastid genes of many parasitic plants (Nickrent *et al.*, 1997a). If the *Rafflesia* plastid does lack a genome, it is the first plastid known to have taken the process of genome reduction to its conclusion.

6. FUTURE DIRECTIONS

The last decade has seen a complete reversal in the way we regard the fundamental nature of several protist and fungal parasites. In large part, this has come about through phylogenetic analyses, which have given us a better appreciation for the evolutionary history of these parasites and their organelles. Phylogenetics have re-written the history of entire parasitic groups (e.g., the microsporidia are now considered fungi rather than protozoa), their organelles (e.g., hydrogenosomes are now considered to be modified mitochondria: Embley *et al.*, 2003), and revealed the presence of organelles that were never expected to exist (e.g., the apicomplexan plastid). As always, however, some of the most exciting scientific discoveries create even more questions than they can answer.

In the case of mitochondria, we are now faced with a growing list of organelles that have been transformed beyond recognition, morphologically and biochemically. The evolutionary histories of many of these are now reasonably well characterised, but their function is, with the exception of hydrogenosomes, poorly understood. The kinds of proteins that have been localised to the various cryptic mitochondria are not altogether informative. In the case of *Entamoeba*, little can be said of function, while in the microsporidia the complete genome of *Encephalitozoon* has allowed some extremely valuable predictions (Katinka *et al.*, 2001), but these need to be verified by localising putative mitochondrial proteins. The same holds for other highly derived mitochondria that have not been discussed, in particular that of the apicomplexan parasite *Cryptosporidium*, where work is just beginning to uncover the unusual biochemistry of this mitochondrion (Riordan *et al.*, 1999; Rotte *et al.*, 2001). In general, until the complete organellar proteome is determined for each of these cases, it will be difficult to imagine what the cellular role of these organelles may be. In the extreme case of *Giardia*, it has yet to be established if there even is an organelle, or just a few organelle-derived proteins that have been recruited to the cytosol. In this instance, the first priority is obvious: localise the chaperones with homologous antibodies to see whether they can be attributed to a specific cellular compartment. If not, then the mitochondriate nature of *Giardia* will have to be evaluated from the soon-to-be-completed genomic data alone (McArthur *et al.*, 2000).

Another aspect of cryptic mitochondria that is emerging as an interesting field is the mechanism by which nuclear-encoded proteins are targeted to the organelle. This system has been worked out for mitochondria, but there are hints that the system used to target proteins to some of these highly derived organelles may be quite different. These hints are vague at present, for example, the possible absence of several translocation proteins in the *Encephalitozoon* genome (Katinka *et al.*, 2001) and the lack of obvious leaders on many microsporidian mitochondrial proteins (Fast and Keeling, 2001). In each of these cryptic organelles, this system would have evolved independently starting from a canonical mitochondrial import system, so each case should be looked at individually to see how it might have evolved.

Another important, but frequently overlooked, point to consider is the potential diversity of these organelles within each protist or fungal group. By necessity, we generalise about the function or even presence of an organelle in a group based on evidence from a single genus, for instance *Giardia* representing all diplomonads. In reality, these are large and diverse groups, and we often (as in the case of *Giardia*) focus on some of the strangest members. In most cases our generalisations are probably fairly accurate, but it is possible that we are missing an unexpected diversity of function with some of these unusual organelles, especially given that they have undergone a radical evolutionary change in function at least once in their past. It would be profitable to take a comparative approach to our examination of the cryptic mitochondria by surveying for the presence and function in diverse members of each group where they have been characterised.

Lastly, despite the great successes in identifying cryptic mitochondria in some of the most well-studied amitochondriates, there are still a number of eukaryotes where there remains no evidence for a mitochondrion. One example of a large hole in our knowledge is the oxymonads, a very poorly studied group of amitochondriates where practically no molecular data are known (Keeling and Leander, 2003). Given the past 10 years of organelle discovery, the safest bet would be that these remaining "amitochondriates" also contain cryptic mitochondria. Nevertheless, it is possible that one or more of these groups may lack the organelle, perhaps even ancestrally (although this is highly doubtful), and it is virtually certain that characterising cryptic mitochondria in each of these lineages will greatly enhance our understanding of how far, and in what directions, the metabolic adaptations of these organelles can extend.

Not surprisingly, the state of our knowledge of cryptic plastids is much different from that of mitochondria, and consequently the major questions are not the same. In the case of the apicomplexan plastid, a flurry of research on plastid function and protein targeting (Waller *et al.*, 1998, 2000; Jomaa *et al.*, 1999; Sato and Wilson, 2002; Foth *et al.*, 2003), helped along by a complete genome sequence (Gardner *et al.*, 2002), have quickly addressed many of the main questions in these areas. A clear picture of this plastid's evolutionary history has also made rapid advances, but remains a somewhat woollier issue. The single, ancient origin of a red algal plastid in all chromalveolates has gained considerable support with recent evidence, but needs to be confirmed by additional molecular data. This conclusion would suggest that many non-photosynthetic protists once did, and may still, contain a plastid (Fast *et al.*, 2001). These include many parasites such as *Cryptosporidium*, *Perkinsus*, *Blastocystis*, *Pythium*, and *Phytophthora*, as well as all ciliates. There is phylogenetic evidence for one potentially plastid-derived gene from *Phytophthora* (Andersson and Roger, 2002), but no clear-cut evidence for an organelle in any of these organisms, and there are arguments that *Cryptosporidium* lacks a plastid (Zhu *et al.*, 2000). On the other end of the spectrum of plastid diversity, the euglenids are known to be closely related to the trypanosomatid parasites (e.g., *Trypanosoma*), and there have been suggestions that the euglenid plastid might also be older than is currently recognised. One intriguing recent analysis of several *Trypanosoma* metabolic enzymes has suggested that they might be derived from an algal endosymbiont (Hannaert *et al.*, 2003). Indeed, Cavalier-Smith has proposed that a very large and diverse group of eukaryotes exists, dubbed the Cabozoa, which ancestrally contained a plastid of green algal origin (Cavalier-Smith, 1999, 2000). This group is proposed to encompass euglenids, trypanosomes, diplomonads, Parabasalia, chlorarachniophytes and a number of other lineages. There is presently no direct evidence for this, but the possibility should be confirmed or refuted in coming years as the generation of molecular data from diverse protist groups has accelerated significantly.

Aside from questions about the distribution and evolutionary origin of plastids, there are still many unanswered questions about the function of cryptic plastids in parasites other than Apicomplexa. The Helicosporidia are thought to contain a plastid genome (Tartar *et al.*, 2003), but nothing is known about it or the function of the organelle in which it resides. Similarly, little is known about the function of the *Prototheca* plastid, but some intriguing differences between it and that of Apicomplexa have already been identified (Knauf and Hatchel, 2002). An improved understanding of these systems will provide a remarkable opportunity to compare the evolution of a primary, green plastid with a secondary, red plastid in two unrelated groups of intracellular parasites that have inherited plastids by very different evolutionary routes. Even less is known about the function of plastids in most parasitic plants, and the genomes of these plastids are intriguing as they appear to be the most reduced of all plastid genomes, perhaps even having been eliminated altogether (Nickrent *et al.*, 1997b, 1998). In addition, there are a great number of algae that have lost photosynthesis for reasons other than adapting to parasitism. Most of these have simply become heterotrophs or osmotrophs, and it would be interesting to compare and contrast how their plastids have evolved with those of parasitic algae to see if any significant trends can be identified. The complete plastid genome of an osmotrophic euglenid, *Astasia longa*, has now been sequenced (Gockel and Hachtel, 2000). Like plastids in parasitic algae, its genome is highly reduced (about one-half the size of the plastid genome of the closely related *Euglena gracilis*) and virtually all of the remaining genes encode proteins involved in gene expression. Like parasites, the genome also contains a small number of genes related to other functions (in the case of *Astasia*, the gene for the Rubisco large subunit). Whether the plastid genomes of non-photosynthetic heterotrophs and osmotrophs like *Astasia* tend to retain different kinds of genes than those of parasites is presently unclear, but comparing these genomes could provide some interesting glimpses into the different ways that the plastid can adapt to the loss of photosynthesis.

Lastly, there are a number of other organelles of interesting or questionable origin in a diverse variety of protist parasites. For instance, the kinetoplastid glycosome is a membrane-bounded compartment for glycolysis and purine salvage (Parsons *et al.*, 2001) that is restricted to this group: in other eukaryotes these processes take place in the cytosol (although glycolytic enzymes have been found to be targeted to the mitochondria in some heterokonts: Liaud *et al.*, 2000). Glycosomes appear to have evolved from peroxisomes (Parsons *et al.*, 2001; Hannaert *et al.*, 2003), and exactly how they acquired their current metabolic pathways is an intriguing question. Several other organisms have cryptic organelles that look much like hydrogenosomes, but have not been characterised biochemically, for instance *Psalteriomonas, Postgaardi, Trimastix,* and *Carpediemonas* (Broers *et al.,* 1990; Simpson *et al.,* 1997, 2002; O'Kelly *et al.,* 1999; Simpson and Patterson, 1999). Characterising these organelles could show that the distribution of hydrogenosomes is even broader than initially imagined, or could reveal a new kind of cryptic organelle with metabolic functions different once again from those we now know. Considering only those facts that we already have, it is clear that the evolutionary trajectory of these unusual organelles can take many unpredictable paths. The metabolism that has been described consists of an odd mix of ancestral functions cobbled together with new enzymes from various sources, reflecting the action of evolution: building new machines with the pieces at hand, in an *ad hoc* fashion, and as the need arises.

ACKNOWLEDGEMENTS

This work was supported by a grant (MOP-42517) from the Canadian Institutes for Health Research (CIHR) and a New Investigator award from the Burroughs-Wellcome Fund (BWF). We would like to thank M. Embley and R. Hirt for helpful input, J. Archibald, N. Fast, J. Harper, M. van der Giezen, and M. W. Gray for critical reading of the manuscript, and two anonymous reviewers for insightful comments. PJK is a Scholar of the Canadian Institute for Advanced research, and a New Investigator of the BWF, the Michael Smith Foundation for Health Research, and the CIHR.

NOTE

A recent report (Tovar, J., Leon-Avila, G., Sánchez, L.B., Sutak, R., Tachezy, J., van der Giezen, M., Hernández, M., Müller, M. and Lucocq, J.M. 2003. Mitochondrial remnant organelles of *Giardia* function in iron-sulphur cluster metabolism. Nature, in press) has now provided unambiguous evidence for a relict mitochondrion in the diplomonad parasite, *Giardia*. This represents one of the last major groups of amitochondriate eukaryotes and further supports the conclusion that no known eukaryotes are ancestrally amitochondriate (Oxymonads are one group where no evidence is yet available, but they have not been thoroughly examined).

REFERENCES

Akhmanova, A., Voncken, F., van Alen, T., van Hoek, A., Boxma, B., Vogels, G., Veenhuis, M. and Hackstein, J.H. (1998). A hydrogenosome with a genome. *Nature* **396**, 527–528.
Altmann, R. (1890). *Die Elementarorganismen und ihre Beziehungen zu den Zellen.* Leipzig: Veit.
Andersson, J.O. and Roger, A.J. (2002). A cyanobacterial gene in nonphotosynthetic protists – an early chloroplast acquisition in eukaryotes? *Current Biology* **12**, 115–119.
Archibald, J.M. and Keeling, P.J. (2002). Recycled plastids: a 'green movement' in eukaryotic evolution. *Trends in Genetics* **18**, 577–584.
Arisue, N., Hashimoto, T., Yoshikawa, H., Nakamura, Y., Nakamura, G., Nakamura, F., Yano, T.A. and Hasegawa, M. (2002). Phylogenetic position of *Blastocystis hominis* and of stramenopiles inferred from multiple molecular sequence data. *Journal of Eukaryotic Microbiology* **49**, 42–53.
Bakatselou, C. and Clark, C.G. (2000). A mitochondrial-type hsp70 gene of *Entamoeba histolytica*. *Archives of Medical Research* **31**, S176–S177.
Bakatselou, C., Kidgell, C. and Clark, C.G. (2000). A mitochondrial-type hsp70 gene of *Entamoeba histolytica*. *Molecular and Biochemical Parasitology* **110**, 177–182.
Baker, G.C., Beebee, T.J. and Ragan, M.A. (1999). *Prototheca richardsi*, a pathogen of anuran larvae, is related to a clade of protistan parasites near the animal-fungal divergence. *Microbiology* **145**, 1777–1784.
Baldauf, S.L., Roger, A.J., Wenk-Siefert, I. and Doolittle, W.F. (2000). A kingdom-level phylogeny of eukaryotes based on combined protein data. *Science* **290**, 972–977.
Bapteste, E., Brinkmann, H., Lee, J., Moore, D., Sensen, C., Gordon, P., Durufle, L., Gaasterland, T., Lopez, P., Müller, M. and Philippe, H. (2002). The analysis of 100 genes supports the grouping of three highly divergent amoebae: *Dictyostellium, Entamoeba,* and *Mastigamoeba*. *Proceedings of the National Academy of Sciences, USA* **99**, 1414–1419.
Barbrook, A.C. and Howe, C.J. (2000). Minicircular plastid DNA in the dinoflagellate *Amphidinium operculatum*. *Molecular and General Genetics* **263**, 152–158.
Barbrook, A.C., Symington, H., Nisbet, R.E., Larkum, A. and Howe, C.J. (2001). Organisation and expression of the plastid genome of the dinoflagellate *Amphidinium operculatum*. *Molecular Genetics and Genomics* **266**, 632–638.
Benchimol, M. and De Souza, W. (1983). Fine structure and cytochemistry of the hydrogenosome of *Tritrichomonas foetus*. *Journal of Protozoology* **30**, 422–425.
Benchimol, M., Durand, R. and Almeida, J.C. (1997). A double membrane surrounds the hydrogenosomes of the anaerobic fungus *Neocallimastix frontalis*. *FEMS Microbiology Letters* **154**, 277–282.
Blanchard, J.L. and Hicks, J.S. (1999). The non-photosynthetic plastid in malarial parasites and other apicomplexans is derived from outside the green plastid lineage. *Journal of Eukaryotic Microbiology* **46**, 367–375.
Blankenship, R.E. (1994). Protein structure, electron transfer and evolution of prokaryotic photosynthetic reaction centers. *Antonie Van Leeuwenhoek* **65**, 311–329.

Boucias, D.G., Becnel, J.J., White, S.E. and Bott, M. (2001). In vivo and in vitro development of the protist *Helicosporidium* sp. *Journal of Eukaryotic Microbiology* **48**, 460–470.

Bradley, P.J., Lahti, C.J., Plumper, E. and Johnson, P.J. (1997). Targeting and translocation of proteins into the hydrogenosome of the protist *Trichomonas*: similarities with mitochondrial protein import. *EMBO Journal* **16**, 3484–3493.

Brinkmann, H. and Martin, W. (1996). Higher plant chloroplast and cytosolic 3-phosphoglycerate kinases: A case of endosymbiotic gene replacement. *Plant Molecular Biology* **30**, 65–75.

Broers, C.A.M., Stumm, C.K., Vogels, G.D. and Brugerolle, G. (1990). *Psalteriomonas-Lanterna* gen-nov, sp-nov, a free-living ameboflagellate isolated from fresh-water anaerobic sediments. *European Journal of Protistology* **25**, 369–380.

Brown, J.R. and Doolittle, W.F. (1995). Root of the universal tree of life based on ancient aminoacyl-tRNA synthetase gene duplications. *Proceedings of the National Academy of Sciences, USA* **92**, 2441–2445.

Brugerolle, G., and Lee, J.J. (2002). Order Diplomonadida. In: *An Illustrated Guide to the Protozoa* (J.J. Lee, G.F. Leedale and P. Bradbury, eds), Vol. 2, pp. 1125–1135. Lawrence, Kansas: Society of Protozoologists.

Bruno, W.J., Socci, N.D. and Halpern, A.L. (2000). Weighted neighbor joining: a likelihood-based approach to distance-based phylogeny reconstruction. *Molecular Biology and Evolution* **17**, 189–197.

Bui, E.T., Bradley, P.J. and Johnson, P.J. (1996). A common evolutionary origin for mitochondria and hydrogenosomes. *Proceedings of the National Academy of Sciences, USA* **93**, 9651–9656.

Butler, E.E. (1954). Radiation-induced chlorophyll-less mutants of *Chlorella*. *Science* **120**, 274–275.

Causey, D. (1925). Mitochondria and Golgi bodies in *Endamoeba gingivalis* (Gros) Brumpt. *University of California Publications in Zoology* **32**, 1–18.

Cavalier-Smith, T. (1983). A 6-kingdom classification and a unified phylogeny. In: *Endocytobiology Vol II.* pp. 1027–1034. Berlin – New York: Walter de Gruyter & Co.

Cavalier-Smith, T. (1987). The simultaneous symbiotic origin of mitochondria, chloroplasts, and microbodies. *Annals of the New York Academy of Sciences* **503**, 55–71.

Cavalier-Smith, T. (1991a). Archamoebae: the ancestral eukaryote? *BioSystems* **25**, 25–38.

Cavalier-Smith, T. (1991b). The evolution of cells. In: *Evolution of Life* (S. Osawa and T. Honjo, eds). pp. 271–304. Tokyo: Springer-Verlag.

Cavalier-Smith, T. (1999). Principles of protein and lipid targeting in secondary symbiogenesis: euglenoid, dinoflagellate, and sporozoan plastid origins and the eukaryote family tree. *Journal of Eukaryotic Microbiology* **46**, 347–366.

Cavalier-Smith, T. (2000). Membrane heredity and early chloroplast evolution. *Trends in Plant Science* **5**, 174–182.

Cavalier-Smith, T. (2002). The neomuran origin of archaebacteria, the negibacterial root of the universal tree and bacterial megaclassification. *International Journal of Systematic and Evolutionary Microbiology* **52**, 7–76.

Cerkasovova, A., Cerkasov, J., Kulda, J. and Reischig, J. (1976). Circular DNA and cardiolipin in hydrogenosomes, microbody-like organelles of trichomonads. *Folia Parasitologica (Praha)* **23**, 33–37.

Cheissin, E.M. (1965). Ultrastructure of *Lamblia duodenalis* 2. The locomotory apparatus, axial rod and other organelles. *Archiv für Protistenkunde* **108**, 8–18.

Cho, B.K., Ham, S.H., Lee, J.Y. and Choi, J.H. (2002). Cutaneous protothecosis. *International Journal of Dermatology* **41**, 304–306.

Chretiennot-Dinet, M., Courties, C., A., V., Neveux, J., Claustre, H., Lautier, J. and Machado, M. (1995). A new marine picoeukaryote: *Ostreococcus tauri* gen. et sp.nov. (Chlorophyta, Prasinophycae). *Phycologia* **34**, 285–292.

Clark, C.G. and Roger, A.J. (1995). Direct evidence for secondary loss of mitochondria in *Entamoeba histolytica*. *Proceedings of the National Academy of Sciences, USA* **92**, 6518–6521.

Clemens, D.L. and Johnson, P.J. (2000). Failure to detect DNA in hydrogenosomes of *Trichomonas vaginalis* by nick translation and immunomicroscopy. *Molecular and Biochemical Parasitology* **106**, 307–313.

Cline, K. and Henry, R. (1996). Import and routing of nucleus-encoded chloroplast proteins. *Annual Review of Cell and Developmental Biology* **12**, 1–26.

Davidson, E., van der Giezen, M., Horner, D., Embley, T.M. and Howe, C. (2002). An [Fe] hydrogenase from the anaerobic hydrogenosome-containing fungus *Neocallimastix frontalis* L2. *Gene* **296**, 45–52.

Dodge, J.D. and Lawes, G.B. (1974). Plastid ultrastructure in some parasitic and semi-parasitic plants. *Cytobiology* **9**, 1–9.

Downie, S.R. and Palmer, J.D. (1992). Use of chloroplast DNA rearrangements in reconstructing plant phylogeny. In: *Molecular Systematics of Plants* (P.S. Soltis, D.E. Soltis and J.J. Doyle, eds). pp. 14–35. New York: Chapman and Hall.

Dyall, S.D., Koehler, C.M., Delgadillo-Correa, M.G., Bradley, P.J., Plumper, E., Leuenberger, D., Turck, C.W. and Johnson, P.J. (2000). Presence of a member of the mitochondrial carrier family in hydrogenosomes: conservation of membrane-targeting pathways between hydrogenosomes and mitochondria. *Molecular Cell Biology* **20**, 2488–2497.

Edlind, T.D., Li, J., Visversvara, G.S., Vodkin, M.H., McLaughlin, G.L. and Katiyar, S.K. (1996). Phylogenetic analysis of the β-tubulin sequences from amitochondriate protozoa. *Molecular Phylogenetics and Evolution* **5**, 359–367.

El-Ani, A.S. (1967). Life cycle and variation of *Prototheca wickerhamii*. *Science* **156**, 1501–1503.

Embley, T.M. and Hirt, R.P. (1998). Early branching eukaryotes? *Current Opinion in Genetics and Development* **8**, 624–629.

Embley, T.M., Finlay, B.J., Dyal, P.L., Hirt, R.P., Wilkinson, M. and Williams, A.G. (1995). Multiple origins of anaerobic ciliates with hydrogenosomes within the radiation of aerobic ciliates. *Proceedings of the Royal Society of London. Series B, Biological Sciences* **262**, 87–93.

Embley, T.M., Horner, D.S. and Hirt, R.P. (1997). Anaerobic eukaryote evolution: hydrogenosomes as modified mitochondria? *Trends in Ecology and Evolution* **12**, 437–441.

Embley, T.M., van der Giezen, M., Horner, D.S., Dyal, P.L. and Foster, P. (2003). Mitochondria and hydrogenosomes are two forms of the same fundamental organelle. *Philosophical Transactions of the Royal Society of London. B* **358**, 191–204.

Fagan, T.M. and Hastings, J.W. (2002). Phylogenetic analysis indicates multiple origins of chloroplast glyceraldehyde-3-phosphate dehydrogenase genes in dinoflagellates. *Molecular Biology and Evolution* **19**, 1203–1207.

Fast, N.M. and Keeling, P.J. (2001). Alpha and beta subunits of pyruvate dehydrogenase E1 from the microsporidian *Nosema locustae*: Mitochondrion-derived carbon metabolism in microsporidia. *Molecular and Biochemical Parasitology* **117**, 201–209.

Fast, N.M., Logsdon, J.M., Jr. and Doolittle, W.F. (1999). Phylogenetic analysis of the TATA box binding protein (TBP) gene from *Nosema locustae*: evidence for a microsporidia-fungi relationship and spliceosomal intron loss. *Molecular Biology and Evolution* **16**, 1415–1419.

Fast, N.M., Kissinger, J.C., Roos, D.S. and Keeling, P.J. (2001). Nuclear-encoded, plastid-targeted genes suggest a single common origin for apicomplexan and dinoflagellate plastids. *Molecular Biology and Evolution* **18**, 418–426.

Fast, N.M., Xue, L., Bingham, S. and Keeling, P.J. (2002). Re-examining alveolate evolution using multiple protein molecular phylogenies. *Journal of Eukaryotic Microbiology* **49**, 30–37.

Feagin, J.E., Gardner, M.J., Williamson, D.H. and Wilson, R.J.M. (1991). The putative mitochondrial genome of *Plasmodium falciparum*. *Journal of Protozoology* **38**, 243–245.

Fenchel, T. and Finlay, B.J. (1995). *Ecology and Evolution in Anoxic Worlds*. New York: Oxford University Press.

Fichera, M.E. and Roos, D.S. (1997). A plastid organelle as a drug target in apicomplexan parasites. *Nature* **390**, 407–409.

Fitzpatrick, T., Ricken, S., Lanzer, M., Amrhein, N., Macheroux, P. and Kappes, B. (2001). Subcellular localization and characterization of chorismate synthase in the apicomplexan *Plasmodium falciparum*. *Molecular Microbiology* **40**, 65–75.

Foth, B.J., Ralph, S.A., Tonkin, C.J., Struck, N.S., Fraunholz, M., Roos, D.S., Cowman, A.F. and McFadden, G.I. (2003). Dissecting apicoplast targeting in the malaria parasite *Plasmodium falciparum*. *Science* **299**, 705–708.

Funes, S., Davidson, E., Reyes-Prieto, A., Magallón, S., Herion, P., King, M.P. and Gonzalez-Halphen, D. (2002). A green algal apicoplast ancestor. *Science* **298**, 2155.

Gajadhar, A.A., Marquardt, W.C., Hall, R., Gunderson, J., Ariztia-Carmona, E.V. and Sogin, M.L. (1991). Ribosomal RNA sequences of *Sarcocystis muris*, *Theileria annulata* and *Crypthecodinium cohnii* reveal evolutionary relationships among apicomplexans, dinoflagellates, and ciliates. *Molecular and Biochemical Parasitology* **45**, 147–154.

Gardner, M.J., Bates, P.A., Ling, I.T., Moore, D.J., McCready, S., Gunasekera, M.B., Wilson, R.J.M. and Williamson, D.H. (1988). Mitochondrial DNA of the human malarial parasite *Plasmodium falciparum*. *Molecular and Biochemical Parasitology* **31**, 11–17.

Gardner, M.J., Williamson, D.H. and Wilson, R.J.M. (1991). A circular DNA in malaria parasites encodes an RNA polymerase like that of prokaryotes and chloroplasts. *Molecular and Biochemical Parasitology* **44**, 115–123.

Gardner, M.J., Feagin, J.E., Moore, D.J., Rangachari, K., Williamson, D.H. and Wilson, R.J.M. (1993). Sequence and organization of large subunit rRNA genes from the extrachromosomal 35 kb circular DNA of the malaria parasite *Plasmodium falciparum*. *Nucleic Acids Research* **21**, 1067–1071.

Gardner, M.J., Goldman, N., Barnett, P., Moore, P.W., Rangachari, K., Strath, M., Whyte, A., Williamson, D.H. and Wilson, R.J.M. (1994). Phylogenetic analysis of the *rpo*B gene from the plastid-like DNA of *Plasmodium falciparum*. *Molecular and Biochemical Parasitology* **66**, 221–231.

Gardner, M.J., Hall, N., Fung, E., White, O., Berriman, M., Hyman, R.W., Carlton, J.M., Pain, A., Nelson, K.E., Bowman, S., Paulsen, I.T., James, K., Eisen, J.A., Rutherford, K., Salzberg, S.L., Craig, A., Kyes, S., Chan, M.S., Nene, V., Shallom, S.J., Suh, B., Peterson, J., Angiuoli, S., Pertea, M., Allen, J., Selengut, J., Haft, D., Mather, M.W., Vaidya, A.B., Martin, D.M., Fairlamb, A.H., Fraunholz, M.J., Roos, D.S., Ralph, S.A., McFadden, G.I., Cummings, L.M., Subramanian, G.M., Mungall, C., Venter, J.C., Carucci, D.J., Hoffman, S.L., Newbold, C., Davis, R.W., Fraser, C.M. and Barrell, B. (2002). Genome sequence of the human malaria parasite *Plasmodium falciparum*. *Nature* **419**, 498–511.

Germot, A., Philippe, H. and Le Guyader, H. (1996). Presence of a mitochondrial-type 70-kDa heat shock protein in *Trichomonas vaginalis* suggests a very early mitochondrial endosymbiosis in eukaryotes. *Proceedings of the National Academy of Sciences, USA* **93**, 14614–14617.

Germot, A., Philippe, H. and Le Guyader, H. (1997). Evidence for loss of mitochondria in Microsporidia from a mitochondrial-type HSP70 in *Nosema locustae*. *Molecular and Biochemical Parasitology* **87**, 159–168.

Ghosh, S., Field, J., Rogers, R., Hickman, M. and Samuelson, J. (2000). The *Entamoeba histolytica* mitochondrion-derived organelle (crypton) contains double-stranded DNA and appears to be bound by a double membrane. *Infection and Immunity* **68**, 4319–4322.

Gockel, G. and Hachtel, W. (2000). Complete gene map of the plastid genome of the nonphotosynthetic euglenoid flagellate *Astasia longa*. *Protist* **151**, 347–351.

Gray, M.W. (1999). Evolution of organellar genomes. *Current Opinion in Genetics and Development* **9**, 678–687.

Gray, M.W. and Doolittle, W.F. (1982). Has the endosymbiont hypothesis been proven? *Microbiological Reviews* **46**, 1–42.

Gray, M.W. and Spencer, D.F. (1996). Organellar evolution. In: *Evolution of Microbial Life* (D.M. Roberts, P. Sharp, G. Alderson and M.A. Collins, eds). pp. 109–126. Cambridge: Cambridge University Press.

Gray, M.W., Burger, G. and Lang, B.F. (1999). Mitochondrial evolution. *Science* **283**, 1476–1481.

Gray, M.W., Burger, G. and Lang, B.F. (2001). The origin and early evolution of mitochondria. *Genome Biology* **2**, Reviews 1018.

Gupta, R.S. (1999). Origin of eukaryotic cells: was metabolic symbiosis based on hydrogen the driving force? *Trends Biochemical Sciences* **24**, 423–424.

Hackstein, J.H., Akhmanova, A., Boxma, B., Harhangi, H.R. and Voncken, F.G. (1999). Hydrogenosomes: eukaryotic adaptations to anaerobic environments. *Trends in Microbiology* **7**, 441–447.

Hannaert, V., Saavedra, E., Duffieux, F., Szikora, J.-P., Rigden, D.J., Michels, P.A.M. and Opperdoes, F.R. (2003). Plant-like traits associated with metabolism of *Trypanosoma* parasites. *Proceedings of the National Academy of Sciences, USA* **100**, 1067–1071.

Harlow, D.R., Weinbach, E.C. and Diamond, L.S. (1976). Nicotinamide nucleotide transhydrogenase in *Entamoeba histolytica*, a protozoan lacking mitochondria. *Comparative Biochemistry and Physiology – Part B: Biochemistry and Molecular Biology* **53**, 141–144.

Harper, J.T. and Keeling, P.J. (2003). Nucleus-encoded, plastid-targeted glyceraldehyde-3-phosphate dehydrogenase (GAPDH) indicates a single origin for chromalveolate plastids. *Molecular Biology and Evolution* **28**, in press.

Harwood, J.L. (1996). Recent advances in the biosynthesis of plant fatty acids. *Biochimica et Biophysica Acta (BBA)/Lipids and Lipid Metabolism* **1301**, 7–56.
Hasegawa, M., Hashimoto, T., Adachi, J., Iwabe, N. and Miyata, T. (1993). Early branchings in the evolution of eukaryotes: ancient divergence of *Entamoeba* that lacks mitochondria revealed by protein sequence data. *Journal of Molecular Evolution* **36**, 380–388.
Hashimoto, T., Nakamura, Y., Nakamura, F., Shirakura, T., Adachi, J., Goto, N., Okamoto, K. and Hasegawa, M. (1994). Protein phylogeny gives a robust estimation for early divergences of eukaryotes: phylogenetic place of a mitochondria-lacking protozoan, *Giardia lamblia*. *Molecular Biology and Evolution* **11**, 65–71.
Hashimoto, T., Nakamura, Y., Kamaishi, T., Nakamura, F., Adachi, J., Okamoto, K. and Hasegawa, M. (1995). Phylogenetic place of mitochondrion-lacking protozoan, *Giardia lamblia*, inferred from amino acid sequences of elongation factor 2. *Molecular Biology and Evolution* **12**, 782–793.
Hashimoto, T., Sanchez, L.B., Shirakura, T., Müller, M. and Hasegawa, M. (1998). Secondary absence of mitochondria in *Giardia lamblia* and *Trichomonas vaginalis* revealed by valyl-tRNA synthetase phylogeny. *Proceedings of the National Academy of Sciences, USA* **95**, 6860–6865.
He, C.Y., Shaw, M.K., Pletcher, C.H., Striepen, B., Tilney, L.G. and Roos, D.S. (2001). A plastid segregation defect in the protozoan parasite *Toxoplasma gondii*. *EMBO Journal* **20**, 330–339.
Herrmann, K.M. (1995). The shikimate pathway as an entry to aromatic secondary metabolism. *Plant Physiology* **107**, 7–12.
Hiller, R.G. (2001). 'Empty' minicircles and *pet*B/*atp*A and *psb*D/*psb*E (cytb559 alpha) genes in tandem in *Amphidinium carterae* plastid DNA. *FEBS Letters* **505**, 449–452.
Hiltbrunner, A., Bauer, J., Alvarez-Huerta, M. and Kessler, F. (2001). Protein translocon at the *Arabidopsis* outer chloroplast membrane. *Biochemistry and Cell Biology* **79**, 629–635.
Hirt, R.P., Healy, B., Vossbrinck, C.R., Canning, E.U. and Embley, T.M. (1997). A mitochondrial Hsp70 orthologue in *Vairimorpha necatrix*: molecular evidence that microsporidia once contained mitochondria. *Current Biology* **7**, 995–998.
Hirt, R.P., Logsdon, J.M., Jr., Healy, B., Dorey, M.W., Doolittle, W.F. and Embley, T.M. (1999). Microsporidia are related to Fungi: evidence from the largest subunit of RNA polymerase II and other proteins. *Proceedings of the National Academy of Sciences, USA* **96**, 580–585.
Hopkins, J., Fowler, R., Krishna, S., Wilson, I., Mitchell, G. and Bannister, L. (1999). The plastid in *Plasmodium falciparum* asexual blood stages: a three-dimensional ultrastructural analysis. *Protist* **150**, 283–295.
Horner, D.S. and Embley, T.M. (2001). Chaperonin 60 phylogeny provides further evidence for secondary loss of mitochondria among putative early-branching eukaryotes. *Molecular Biology and Evolution* **18**, 1970–1975.
Horner, D.S., Hirt, R.P., Kilvington, S., Lloyd, D. and Embley, T.M. (1996). Molecular data suggest an early acquisition of the mitochondrion endosymbiont. *Proceedings of the Royal Society of London. Series B, Biological Sciences* **263**, 1053–1059.
Huss, V.A. and Sogin, M.L. (1990). Phylogenetic position of some *Chlorella* species within the chlorococcales based upon complete small-subunit ribosomal RNA sequences. *Journal of Molecular Evolution* **31**, 432–442.

Jomaa, H., Wiesner, J., Sanderbrand, S., Altincicek, B., Weidemeyer, C., Hintz, M., Turbachova, I., Eberl, M., Zeidler, J., Lichtenthaler, H.K., Soldati, D. and Beck, E. (1999). Inhibitors of the nonmevalonate pathway of isoprenoid biosynthesis as antimalarial drugs. *Science* **285**, 1573–1576.

Kamaishi, T., Hashimoto, T., Nakamura, Y., Masuda, Y., Nakamura, F., Okamoto, K., Shimizu, M. and Hasegawa, M. (1996a). Complete nucleotide sequences of the genes encoding translation elongation factors 1 alpha and 2 from a microsporidian parasite, *Glugea plecoglossi*: implications for the deepest branching of eukaryotes. *Journal of Biochemistry (Tokyo)* **120**, 1095–1103.

Kamaishi, T., Hashimoto, T., Nakamura, Y., Nakamura, F., Murata, S., Okada, N., Okamoto, K.-I., Shimizu, M. and Hasegawa, M. (1996b). Protein phylogeny of translation elongation factor EF-1α suggests microsporidians are extremely ancient eukaryotes. *Journal of Molecular Evolution* **42**, 257–263.

Katinka, M.D., Duprat, S., Cornillot, E., Méténier, G., Thomarat, F., Prenier, G., Barbe, V., Peyretaillade, E., Brottier, P., Wincker, P., Delbac, F., El Alaoui, H., Peyret, P., Saurin, W., Gouy, M., Weissenbach, J. and Vivarès, C.P. (2001). Genome sequence and gene compaction of the eukaryote parasite *Encephalitozoon cuniculi*. *Nature* **414**, 450–453.

Keeling, P.J. (1998). A kingdom's progress: Archezoa and the origin of eukaryotes. *BioEssays* **20**, 87–95.

Keeling, P.J. (2001). Parasites go the Full Monty. *Nature* **414**, 401–402.

Keeling, P.J. (2003). Congruent evidence from alpha-tubulin and beta-tubulin gene phylogenies for a zygomycete origin of microsporidia. *Fungal Genetics and Biology* **38**, 298–309.

Keeling, P.J. and Doolittle, W.F. (1996). Alpha-tubulin from early-diverging eukaryotic lineages and the evolution of the tubulin family. *Molecular Biology and Evolution* **13**, 1297–1305.

Keeling, P.J. and Doolittle, W.F. (1997). Evidence that eukaryotic triosephosphate isomerase is of alpha-proteobacterial origin. *Proceedings of the National Academy of Sciences, USA* **94**, 1270–1275.

Keeling, P.J. and Fast, N.M. (2002). Microsporidia: biology and evolution of highly reduced intracellular parasites. *Annual Review of Microbiology* **56**, 93–116.

Keeling, P.J. and Leander, B.S. (2003). Characterisation of a non-canonical genetic code in the oxymonad *Streblomastix strix*. *Journal of Molecular Biology* **326**, 1337–1349.

Keeling, P.J., Palmer, J.D., Donald, R.G., Roos, D.S., Waller, R.F. and McFadden, G.I. (1999). Shikimate pathway in apicomplexan parasites. *Nature* **397**, 219–220.

Keeling, P.J., Luker, M.A. and Palmer, J.D. (2000). Evidence from beta-tubulin phylogeny that microsporidia evolved from within the fungi. *Molecular Biology and Evolution* **17**, 23–31.

Keilin, D. (1921). On the life history of *Helicosporidium parasiticum* n. g., n. sp., a new species of protist parasite in the larvae of *Dashelaea obscura* Winn (Diptera: Ceratopogonidae) and in some other arthropods. *Parasitology* **13**, 97–113.

Kellen, W.R. and Lindegren, J.E. (1974). Life cycle of *Helicosporidium parasiticum* in the navel orangeworm, *Paramyelois transitella*. *Journal of Invertebrate Pathology* **23**, 202–208.

Kennedy, E.P. and Lehninger, A.L. (1949). Oxidation of fatty acids and tricarboxylic acid cycle intermediates by isolated liver mitochondria. *Journal of Biological Chemistry* **179**, 957–972.

Kies, L. (1974). Elektronenmikroskopische Untersuchungen an *Paulinella chromatophora* Lauterborn, einer Thekamöbe mit blau-grünen Endosymbionten (Cyanellen). *Protoplasma* **80**, 69–89.
Kilejian, A. (1975). Circular mitochondrial DNA from the avian malarial parasite *Plasmodium lophurae*. *Biochimica et Biophysica Acta* **390**, 267–284.
Knauf, U. and Hachtel, W. (2002). The genes encoding subunits of ATP synthase are conserved in the reduced plastid genome of the heterotrophic alga *Prototheca wickerhamii*. *Molecular Genetics and Genomics* **267**, 492–497.
Köhler, S., Delwiche, C.F., Denny, P.W., Tilney, L.G., Webster, P., Wilson, R.J.M., Palmer, J.D. and Roos, D.S. (1997). A plastid of probable green algal origin in apicomplexan parasites. *Science* **275**, 1485–1489.
Kroemer, G., Petit, P., Zamzami, N., Vayssiere, J.L. and Mignotte, B. (1995). The biochemistry of programmed cell death. *FASEB Journal* **9**, 1277–1287.
Kudo, R.R. (1947). *Protozoology*. Springfield, Il.: Charles C. Thomas.
Kuijt, J. (1969). *The Biology of Parasitic Flowering Plants*. Berkley, Ca.: University of California Press.
Kuijt, J., Bray, D. and Olson, A.R. (1985). Anatomy and ultrastructure of the endophytic system of *Pilostyles thurberi* (Rafflesiaceae). *Canadian Journal of Botany* **63**, 1231–1240.
Kurland, C.G. and Andersson, S.G. (2000). Origin and evolution of the mitochondrial proteome. *Microbiology and Molecular Biology Reviews* **64**, 786–820.
Kuvardina, O.N., Leander, B.S., Aleshin, V.V., Myl'nikov, A.P., Keeling, P.J. and Simdyanov, T.G. (2002). The phylogeny of colpodellids (Alveolata) using small subunit rRNA gene sequences suggests they are the free-living sister group to apicomplexans. *Journal of Eukaryotic Microbiology* **49**, 498–504.
Lake, J.A. and Rivera, M.C. (1994). Was the nucleus the first endosymbiont? *Proceedings of the National Academy of Sciences, USA* **91**, 2880–2881.
Lang, B.F., Gray, M.W. and Burger, G. (1999). Mitochondrial genome evolution and the origin of eukaryotes. *Annual Review of Genetics* **33**, 351–397.
Lang-Unnasch, N., Reith, M.E., Munholland, J. and Barta, J.R. (1998). Plastids are widespread and ancient in parasites of the phylum Apicomplexa. *International Journal of Parasitology* **28**, 1743–1754.
Lazarow, P.B. and Fujiki, Y. (1985). Biogenesis of peroxisomes. *Annual Review of Cell Biology* **1**, 489–530.
Leipe, D.D., Gunderson, J.H., Nerad, T.A. and Sogin, M.L. (1993). Small subunit ribosomal RNA of *Hexamita inflata* and the quest for the first branch of the eukaryotic tree. *Molecular and Biochemical Parasitology* **59**, 41–48.
Liaud, M.F., Lichtle, C., Apt, K., Martin, W. and Cerff, R. (2000). Compartment-specific isoforms of TPI and GAPDH are imported into diatom mitochondria as a fusion protein: evidence in favor of a mitochondrial origin of the eukaryotic glycolytic pathway. *Molecular Biology and Evolution* **17**, 213–223.
Lill, R., Diekert, K., Kaut, A., Lange, H., Pelzer, W., Prohl, C. and Kispal, G. (1999). The essential role of mitochondria in the biogenesis of cellular iron-sulfur proteins. *Biological Chemistry* **380**, 1157–1166.
Lindmark, D.G. and Müller, M. (1973). Hydrogenosome, a cytoplasmic organelle of the anaerobic flagellate *Tritrichomonas foetus*, and its role in pyruvate metabolism. *Journal of Biological Chemistry* **248**, 7724–7728.
Lloyd, D., Harris, J.C., Maroulis, S., Wadley, R., Ralphs, J.R., Hann, A.C., Turner, M.P. and Edwards, M.R. (2002). The "primitive" microaerophile *Giardia intestinalis* (syn. *lamblia, duodenalis*) has specialized membranes with electron

transport and membrane-potential-generating functions. *Microbiology* **148**, 1349–1354.

Lopez-Garcia, P. and Moreira, D. (1999). Metabolic symbiosis at the origin of eukaryotes. *Trends in Biochemical Sciences* **24**, 88–93.

Machado, M.A. and Zetsche, K. (1990). A structural, functional and molecular analysis of plastids of the holoparasites *Cuscuta reflexa* and *Cuscuta europaea*. *Planta* **181**, 91–96.

Madigan, M.M., Parker, J. and Madigan, M.T. (2002). *Brock's Biology of Microorganisms*. River, New Jersey: Prentice Hall.

Mai, Z., Ghosh, S., Frisardi, M., Rosenthal, B., Rogers, R. and Samuelson, J. (1999). Hsp60 is targeted to a cryptic mitochondrion-derived organelle ("crypton") in the microaerophilic protozoan parasite *Entamoeba histolytica*. *Molecular Cell Biology* **19**, 2198–2205.

Mangenot, G. and Mangenot, S. (1968). Sur la présence de leucoplastes chez les végétaux vasculaires mycotrophes ou parasites. *Comptes Rendus de l'Académie des Sciences* **267**, 1193–1195.

Martin, W. (1999). A briefly argued case that mitochondria and plastids are descendants of endosymbionts, but that the nuclear compartment is not. *Proceedings of the Royal Society of London. Series B, Biological Sciences* **266**, 1387–1395.

Martin, W. and Müller, M. (1998). The hydrogen hypothesis for the first eukaryote. *Nature* **392**, 37–41.

Martin, W., Brinkmann, H., Savonna, C. and Cerff, R. (1993). Evidence for a chimeric nature of nuclear genomes: eubacterial origin of eukaryotic glyceraldehyde-3-phosphate dehydrogenase genes. *Proceedings of the National Academy of Sciences, USA* **90**, 8692–8696.

Marvin-Sikkema, F.D., Kraak, M.N., Veenhuis, M., Gottschal, J.C. and Prins, R.A. (1993). The hydrogenosomal enzyme hydrogenase from the anaerobic fungus *Neocallimastix* sp. L2 is recognized by antibodies, directed against the C-terminal microbody protein targeting signal SKL. *European Journal of Cell Biology* **61**, 86–91.

McArthur, A.G., Morrison, H.G., Nixon, J.E., Passamaneck, N.Q., Kim, U., Hinkle, G., Crocker, M.K., Holder, M.E., Farr, R., Reich, C.I., Olsen, G.E., Aley, S.B., Adam, R.D., Gillin, F.D. and Sogin, M.L. (2000). The *Giardia* genome project database. *FEMS Microbiology Letters* **189**, 271–273.

McFadden, G.I. (1999). Plastids and protein targeting. *Journal of Eukaryotic Microbiology* **46**, 339–346.

McFadden, G.I. (2001). Primary and secondary endosymbiosis and the origin of plastids. *Journal of Phycology* **37**, 951–959.

McFadden, G.I. and Waller, R.F. (1997). Plastids in parasites of humans. *BioEssays* **19**, 1033–1040.

McFadden, G.I. and Roos, D.S. (1999). Apicomplexan plastids as drug targets. *Trends in Microbiology* **7**, 328–333.

McFadden, G.I., Reith, M., Munholland, J. and Lang-Unnasch, N. (1996). Plastid in human parasites. *Nature* **381**, 482.

McFadden, G.I., Gilson, P.R., Douglas, S.E., Cavalier-Smith, T., Hofmann, C.J. and Maier, U.G. (1997). Bonsai genomics: sequencing the smallest eukaryotic genomes. *Trends in Genetics* **13**, 46–49.

Mereschkowsky, C. (1905). Über Natur und Ursprung der Chromatophoren im Pflanzenreiche. *Biologisches Zentralblatt* **25**, 593–604.

Morrison, H.G., Roger, A.J., Nystul, T.G., Gillin, F.D. and Sogin, M.L. (2001). *Giardia lamblia* expresses a proteobacterial-like DnaK homolog. *Molecular Biology and Evolution* **18**, 530–541.
Müller, M. (1973). Biochemical cytology of trichomonad flagellates. I. Subcellular localization of hydrolases, dehydrogenases, and catalase in *Tritrichomonas foetus*. *Journal of Cell Biology* **57**, 453–474.
Müller, M. (1992). Energy metabolism of ancestral eukaryotes: a hypothesis based on the biochemistry of amitochondriate parasitic protists. *Biosystems* **28**, 33–40.
Müller, M. (1993). The hydrogenosome. *Journal of General Microbiology* **139**, 2879–2889.
Müller, M. (1997). Evolutionary origins of trichomonad hydrogenosomes. *Parasitology Today* **13**, 166–167.
Müller, M. and Lindmark, D.G. (1976). Uptake of metronidazole and its effect on viability in trichomonads and *Entamoeba invadens* under anaerobic and aerobic conditions. *Antimicrobial Agents and Chemotherapy* **9**, 696–700.
Nakamura, Y., Hashimoto, T., Yoshikawa, H., Kamaishi, T., Nakamura, F., Okamoto, K. and Hasegawa, M. (1996). Phylogenetic position of *Blastocystis hominis* that contains cytochrome-free mitochondria, inferred from the protein phylogeny of elongation factor 1 alpha. *Molecular and Biochemical Parasitology* **77**, 241–245.
Nickrent, D.L., Ouyang, Y., Duff, R.J. and dePamphilis, C.W. (1997a). Do nonasterid holoparasitic flowering plants have plastid genomes? *Plant Molecular Biology* **34**, 717–729.
Nickrent, D.L., Duff, R.J. and Konings, D.A.M. (1997b). Structural analysis of plastid-encoded 16S rRNAs in holoparasitic angiosperms. *Plant Molecular Biology* **74**, 731–743.
Nickrent, D.L., Duff, R.J., Colwell, A.E., Wolfe, A.D., Young, N.D., Steiner, K.E. and dePamphilis, C.W. (1998). Molecular phylogenetic and evolutionary studies of parasitic plants. In: *Molecular Systematics of Plants II. DNA Sequencing* (D.E. Soltis, P.S. Soltis and J.J. Doyle, eds). pp. 211–241. Boston: Kluwer.
O'Kelly, C.J., Farmer, M.A. and Nerad, T.A. (1999). Ultrastructure of *Trimastix pyriformis* (Klebs) Bernard *et al.*: similarities of *Trimastix* species with retortamonad and jakobid flagellates. *Protist* **150**, 149–162.
Palmer, J.D. (1997). Organelle genomes: going, going, gone! *Science* **275**, 790–791.
Palmer, J.D. (2003). The symbiotic birth and spread of plastids: how many times and whodunit? *Journal of Phycology* **39**, 1–9.
Palmer, J.D. and Delwiche, C.F. (1998). The origin and evolution of plastids and their genomes. In: *Molecular Systematics of Plants II. DNA Sequencing* (D.E. Soltis, P.S. Soltis and J.J. Doyle, eds). pp. 375–409. Boston: Kluwer.
Parsons, M., Furuya, T., Pal, S. and Kessler, P. (2001). Biogenesis and function of peroxisomes and glycosomes. *Molecular and Biochemical Parasitology* **115**, 19–28.
Perez-Martinez, X., Antaramian, A., Vazquez-Acevedo, M., Funes, S., Tolkunova, E., d'Alayer, J., Claros, M.G., Davidson, E., King, M.P. and Gonzalez-Halphen, D. (2001). Subunit II of cytochrome c oxidase in Chlamydomonad algae is a heterodimer encoded by two independent nuclear genes. *Journal of Biological Chemistry* **276**, 11302–11309.
Petrin, D., Delgaty, K., Bhatt, R. and Garber, G. (1998). Clinical and microbiological aspects of *Trichomonas vaginalis*. *Clinical Microbiology Reviews* **11**, 300–317.

Peyretaillade, E., Broussolle, V., Peyret, P., Metenier, G., Gouy, M. and Vivares, C.P. (1998). Microsporidia, amitochondrial protists, possess a 70-kDa heat shock protein gene of mitochondrial evolutionary origin. *Molecular Biology and Evolution* **15**, 683–689.

Pfanner, N. and Geissler, A. (2001). Versatility of the mitochondrial protein import machinery. *Nature Reviews Molecular Cell Biology* **2**, 339–349.

Piyophirapong, S., Linpiyawan, R., Mahaisavariya, P., Muanprasat, C., Chaiprasert, A. and Suthipinittharm, P. (2002). Cutaneous protothecosis in an AIDS patient. *British Journal of Dermatology* **146**, 713–715.

Pronk, J.T., Steensma, H.Y. and Van Dijken, J.P. (1996). Pyruvate metabolism in *Saccharomyces cerevisiae*. *Yeast* **12**, 1607–1633.

Ralph, S.A., D'Ombrain, M.C. and McFadden, G.I. (2001). The apicoplast as an antimalarial drug target. *Drug Resistance Updates* **4**, 145–151.

Reeves, R.E. (1984). Metabolism of *Entameoba histolytica* Schaudinn, 1903. *Advances in Parasitology* **23**, 105–142.

Riordan, C.E., Langreth, S.G., Sanchez, L.B., Kayser, O. and Keithly, J.S. (1999). Preliminary evidence for a mitochondrion in *Cryptosporidium parvum*: phylogenetic and therapeutic implications. *Journal of Eukaryotic Microbiology* **46**, 52S–55S.

Roberts, F., Roberts, C.W., Johnson, J.J., Kyle, D.E., Krell, T., Coggins, J.R., Coombs, G.H., Milhous, W.K., Tzipori, S., Ferguson, D.J., Chakrabarti, D. and McLeod, R. (1998). Evidence for the shikimate pathway in apicomplexan parasites. *Nature* **393**, 801–805.

Rodriguez, M.A., Garcia-Perez, R.M., Mendoza, L., Sanchez, T., Guillen, N. and Orozco, E. (1998). The pyruvate : ferredoxin oxidoreductase enzyme is located in the plasma membrane and in a cytoplasmic structure in *Entamoeba*. *Microbial Pathogenesis* **25**, 1–10.

Roger, A.J., Svard, S.G., Tovar, J., Clark, C.G., Smith, M.W., Gillin, F.D. and Sogin, M.L. (1998). A mitochondrial-like chaperonin 60 gene in *Giardia lamblia*: evidence that diplomonads once harbored an endosymbiont related to the progenitor of mitochondria. *Proceedings of the National Academy of Sciences, USA* **95**, 229–234.

Rohdich, F., Kis, K., Bacher, A. and Eisenreich, W. (2001). The non-mevalonate pathway of isoprenoids: genes, enzymes and intermediates. *Current Opinion in Chemical Biology* **5**, 535–540.

Rotte, C., Stejskal, F., Zhu, G., Keithly, J.S. and Martin, W. (2001). Pyruvate : NADP+ oxidoreductase from the mitochondrion of *Euglena gracilis* and from the apicomplexan *Cryptosporidium parvum*: a biochemical relic linking pyruvate metabolism in mitochondriate and amitochondriate protists. *Molecular Biology and Evolution* **18**, 710–720.

Saldarriaga, J.F., Taylor, F.J.R., Keeling, P.J. and Cavalier-Smith, T. (2001). Dinoflagellate nuclear SSU rRNA phylogeny suggests multiple plastid losses and replacements. *Journal of Molecular Evolution* **53**, 204–213.

Saldarriaga, J.F., McEwan, M.L., Fast, N.M., Taylor, F.J.R. and Keeling, P.J. (2003). Multiple protein phylogenies show that *Oxyhrris marina* and *Perkinsus marinus* are early branches of the dinoflagellate lineage. *International Journal of Systematic and Evolutionary Microbiology* **53**, 355–365.

Sato, S. and Wilson, R.J.M. (2002). The genome of *Plasmodium falciparum* encodes an active delta-aminolevulinic acid dehydratase. *Current Genetics* **40**, 391–398.

Schimper, A.F.W. (1883). Ueber die Entwickelung der Chlorophyllkörner und Farbkörper. *Botanische Zeitung* **41**, 105–114, 121–131, 137–146, 153–162.
Schwartz, R.M. and Dayhoff, M.O. (1978). Origins of prokaryotes, eukaryotes, mitochondria, and chloroplasts. *Science* **199**, 395–403.
Sekiguchi, H., Moriya, M., Nakayama, T. and Inouye, I. (2002). Vestigial chloroplasts in heterotrophic stramenopiles *Pteridomonas danica* and *Ciliophrys infusionum* (Dictyochophyceae). *Protist* **153**, 157–167.
Sepsenwol, S. (1973). Leucoplast of the cryptomonad *Chilomonas paramecium*. Evidence for presence of a true plastid in a colorless flagellate. *Experimental Cell Research* **76**, 395–409.
Shirakura, T., Hashimoto, T., Nakamura, Y., Kamaishi, T., Cao, Y., Adachi, J., Hasegawa, M., Yamamoto, A. and Goto, N. (1994). Phylogenetic place of a mitochondria-lacking protozoan, *Entamoeba histolytica*, inferred from amino acid sequences of elongation factor 2. *Japanese Journal of Genetics* **69**, 119–135.
Simpson, A.G.B. and Patterson, D.J. (1999). The ultrastructure of *Carpediemonas membranifera* (Eukaryota) with reference to the "Excavate hypothesis". *European Journal of Protistology* **35**, 353–370.
Simpson, A.G.B., van den Hoff, J., Bernard, C., Burton, H.R. and Patterson, D.J. (1997). The ultrastructure and systematic position of the euglenozoon *Postgaardi mariagerensis*, Fenchel et al. *Archiv für Protistenkunde* **147**, 213–225.
Simpson, A.G., Roger, A.J., Silberman, J.D., Leipe, D.D., Edgcomb, V.P., Jermiin, L.S., Patterson, D.J. and Sogin, M.L. (2002). Evolutionary history of "early-diverging" eukaryotes: the excavate taxon *Carpediemonas* is a close relative of *Giardia*. *Molecular Biology and Evolution* **19**, 1782–1791.
Siu, C., Swift, H. and Chiang, K. (1976). Characterization of cytoplasmic and nuclear genomes in the colorless alga *Polytoma*. I. Ultrastructural analysis of organelles. *Journal of Cell Biology* **69**, 352–370.
Sogin, M.L. (1989). Evolution of eukaryotic microorganisms and their small subunit ribosomal-RNAs. *American Zoologist* **29**, 487–499.
Sogin, M.L. and Silberman, J.D. (1998). Evolution of the protists and protistan parasites from the perspective of molecular systematics. *International Journal of Parasitology* **28**, 11–20.
Sogin, M.L., Gunderson, J.H., Elwood, H.J., Alonso, R.A. and Peattie, D.A. (1989). Phylogenetic meaning of the kingdom concept: an unusual ribosomal RNA from *Giardia lamblia*. *Science* **243**, 75–77.
Soltys, B.J. and Gupta, R.S. (1994). Presence and cellular distribution of a 60-kDa protein related to mitochondrial hsp60 in *Giardia lamblia*. *Journal of Parasitology* **80**, 580–590.
Stanier, R.Y. (1970). Some aspects of the biology of cells and their possible evolutionary significance. *Symposium of the Society for General Microbiology* **20**, 1–38.
Stechmann, A. and Cavalier-Smith, T. (2002). Rooting the eukaryote tree by using a derived gene fusion. *Science* **297**, 89–91.
Stiller, J.W., Reel, D.C. and Johnson, J.C. (2003). A single origin of plastids revisited: convergent evolution in organellar genome content. *Journal of Phycology* **39**, 95–105.
Stocking, C. and Gifford, E. (1959). Incorporation of thymadine into chloroplasts of *Spirogyra*. *Biochemical & Biophysical Research Communications* **1**, 159–164.

Strimmer, K. and von Haeseler, A. (1996). Quartet puzzling: a quartet maximum-likelihood method for reconstructing tree topologies. *Molecular Biology and Evolution* **13**, 964–969.

Tachezy, J., Sanchez, L.B. and Müller, M. (2001). Mitochondrial type iron-sulfur cluster assembly in the amitochondriate eukaryotes *Trichomonas vaginalis* and *Giardia intestinalis*, as indicated by the phylogeny of IscS. *Molecular Biology and Evolution* **18**, 1919–1928.

Tartar, A., Boucias, D.G., Adams, B.J. and Becnel, J.J. (2002). Phylogenetic analysis identifies the invertebrate pathogen *Helicosporidium* sp. as a green alga (Chlorophyta). *International Journal of Systematic and Evolutionary Microbiology* **52**, 273–279.

Tartar, A., Boucias, D.G., Adams, B.J. and Becnel, J.J. (2003). Comparison of plastid 16S rDNA (*rrn16*) from *Helicosporidium* spp.: evidence supporting the reclassification of Helicosporidia as green algae (Chlorophyta). *International Journal of Systematic and Evolutionary Microbiology* **in press**.

Taylor, F.J.R. (1987). *The biology of Dinoflagellates*. Oxford: Blackwell Scientific Publications.

Taylor, F.J.R. (1999). Morphology (Tabulation) and molecular evidence for dinoflagellate phylogeny reinforce each other. *Journal of Phycology* **35**, 1–3.

Tengs, T., Dahlberg, O.J., Shalchian-Tabrizi, K., Klaveness, D., Rudi, K., Delwiche, C.F. and Jakobsen, K.S. (2000). Phylogenetic analyses indicate that the 19′hexanoyloxy-fucoxanthin-containing dinoflagellates have tertiary plastids of haptophyte origin. *Molecular Biology and Evolution* **17**, 718–729.

Thiele, D. and Bergmann, A. (2002). Protothecosis in human medicine. *International Journal of Hygiene and Environmental Health* **204**, 297–302.

Titorenko, V.I., Ogrydziak, D.M. and Rachubinski, R.A. (1997). Four distinct secretory pathways serve protein secretion, cell surface growth, and peroxisome biogenesis in the yeast *Yarrowia lipolytica*. *Molecular Cell Biology* **17**, 5210–5226.

Tovar, J., Fischer, A. and Clark, C.G. (1999). The mitosome, a novel organelle related to mitochondria in the amitochondrial parasite *Entamoeba histolytica*. *Molecular Microbiology* **32**, 1013–1021.

Tudge, C. (2000). *The Variety of Life*. Oxford: Oxford University Press.

Vaidya, A.B., Akella, R. and Suplick, K. (1989). Sequences similar to genes for two mitochondrial proteins and portions of ribosomal RNA in tandemly arrayed 6-kilobase-pair DNA of a malarial parasite. *Molecular and Biochemical Parasitology* **35**, 97–107.

Van de Peer, Y., Ben Ali, A. and Meyer, A. (2000). Microsporidia: accumulating molecular evidence that a group of amitochondriate and suspectedly primitive eukaryotes are just curious fungi. *Gene* **246**, 1–8.

van der Giezen, M., Sjollema, K.A., Artz, R.R., Alkema, W. and Prins, R.A. (1997). Hydrogenosomes in the anaerobic fungus *Neocallimastix frontalis* have a double membrane but lack an associated organelle genome. *FEBS Letters* **408**, 147–150.

van der Giezen, M., Kiel, J.A., Sjollema, K.A. and Prins, R.A. (1998). The hydrogenosomal malic enzyme from the anaerobic fungus *Neocallimastix frontalis* is targeted to mitochondria of the methylotrophic yeast *Hansenula polymorpha*. *Current Genetics* **33**, 131–135.

van der Giezen, M., Slotboom, D.J., Horner, D.S., Dyal, P.L., Harding, M., Xue, G.P., Embley, T.M. and Kunji, E.R. (2002). Conserved properties of hydrogenosomal and mitochondrial ADP/ATP carriers: a common origin for both organelles. *EMBO Journal* **21**, 572–579.

Vávra, J. and Larsson, J.I.R. (1999). Structure of the Microsporidia. In: *The Microsporidia and Microsporidiosis* (M. Wittner and L.M. Weiss, eds). pp. 7–84. Washington B.C.: American Society for Microbiology.

Vossbrinck, C.R. and Woese, C.R. (1986). Eukaryotic ribosomes that lack a 5.8S RNA. *Nature* **320**, 287–288.

Vossbrinck, C.R., Maddox, J.V., Friedman, S., Debrunner-Vossbrinck, B.A. and Woese, C.R. (1987). Ribosomal RNA sequence suggests microsporidia are extremely ancient eukaryotes. *Nature* **326**, 411–414.

Waller, R.F., Keeling, P.J., Donald, R.G., Striepen, B., Handman, E., Lang-Unnasch, N., Cowman, A.F., Besra, G.S., Roos, D.S. and McFadden, G.I. (1998). Nuclear-encoded proteins target to the plastid in *Toxoplasma gondii* and *Plasmodium falciparum*. *Proceedings of the National Academy of Sciences, USA* **95**, 12352–12357.

Waller, R.F., Reed, M.B., Cowman, A.F. and McFadden, G.I. (2000). Protein trafficking to the plastid of *Plasmodium falciparum* is via the secretory pathway. *EMBO Journal* **19**, 1794–1802.

Waller, R.F., Ralph, S.A., Reed, M.B., Su, V., Douglas, J.D., Minnikin, D.E., Cowman, A.F., Besra, G.S. and McFadden, G.I. (2003). A type II pathway for fatty acid biosynthesis presents drug targets in *Plasmodium falciparum*. *Antimicrobial Agents and Chemotherapy* **47**, 297–301.

Waller, R.F., Keeling, P.J., van Dooren, G.G. and McFadden, G.I. (2003). Comment on "A green algal apicoplast ancestor". *Science* **301**, 49.

Weber, R., Bryan, R.T., Schwartz, D.A. and Owen, R.L. (1994). Human microsporidial infections. *Clinical Microbiology Reviews* **7**, 426–461.

Weidner, E., Findley, A.M., Dolgikh, V. and Sokolova, J. (1999). Microsporidian biochemistry and physiology. In: *The Microsporidia and Microsporidiosis* (M. Wittner and L.M. Weiss, eds.). Washington, D.C.: ASM Press.

Whatley, F.R. (1981). The establishment of mitochondria: *Paracoccus* and *Rhodopseudomonas*. *Annals of the New York Academy of Sciences* **361**, 330–340.

Whatley, J.M., John, P. and Whatley, F.R. (1979). From extracellular to intracellular: the establishment of mitochondria and chloroplasts. *Proceedings of the Royal Society of London. Series B, Biological Sciences* **204**, 165–187.

Williams, B.A.P., Hirt, R.P., Lucocq, J.M. and Embley, T.M. (2002). A mitochondrial remnant in the microsporidian *Trachipleistophora hominis*. *Nature* **418**, 865–869.

Williamson, D.H., Gardner, M.J., Preiser, P., Moore, D.J., Rangachari, K. and Wilson, R.J.M. (1994). The evolutionary origin of the 35 kb circular DNA of *Plasmodium falciparum*: new evidence supports a possible rhodophyte ancestry. *Molecular and General Genetics* **243**, 249–252.

Wilson, R.J.M., Fry, M., Gardner, M.J., Feagin, J.E. and Williamson, D.H. (1992). Subcellular fractionation of the two organelle DNAs of malaria parasites. *Current Genetics* **21**, 405–408.

Wilson, R.J.M., Denny, P.W., Preiser, D.J., Rangachari, K., Roberts, K., Roy, A., Whyte, A., Strath, M., Moore, D.J., Moore, P.W. and Williamson, D.H. (1996). Complete gene map of the plastid-like DNA of the malaria parasite *Plasmodium falciparum*. *Journal of Molecular Biology* **261**, 155–172.

Wolfe, K.H., Morden, C.W. and Palmer, J.D. (1992). Function and evolution of a minimal plastid genome from a nonphotosynthetic parasitic plant. *Proceedings of the National Academy of Sciences, USA* **89**, 10648–10652.

Wolters, J. (1991). The troublesome parasites: molecular and morphological evidence that Apicomplexa belong to the dinoflagellate-ciliate clade. *BioSystems* **25**, 75–84.

Yamin, M.A. (1979). Flagellates of the orders Trichomonadida Kirby, Oxymonadida Grassé, and Hypermastigida Grassi & Foà reported from lower termites (Isoptera Falilies Mastotermitidae, Kalotermitidae, Hodotermitidae, Termopsidae, Rhinotermitidae, and Serritermididae) and from the wood-feeding roach *Cryptocercus* (Dictyoptera: Ceyptocercidae). *Sociobiology* **4**, 1–120.

Yang, D., Oyaizu, Y., Oyaizu, H., Olsen, G.J. and Woese, C.R. (1985). Mitochondrial origins. *Proceedings of the National Academy of Sciences, USA* **82**, 4443–4447.

Yoon, H.S., Hackett, J.D., Pinto, G. and Bhattacharya, D. (2002). A single, ancient origin of the plastid in the Chromista. *Proceedings of the National Academy of Sciences, USA* **99**, 15507–15512.

Zhang, Z., Green, B.R. and Cavalier-Smith, T. (1999). Single gene circles in dinoflagellate chloroplast genomes. *Nature* **400**, 155–159.

Zhang, Z., Green, B.R. and Cavalier-Smith, T. (2000). Phylogeny of ultra-rapidly evolving dinoflagellate chloroplast genes: A possible common origin for sporozoan and dinoflagellate plastids. *Journal of Molecular Evolution* **51**, 26–40.

Zhu, G., Marchewka, M.J. and Keithly, J.S. (2000). *Cryptosporidium parvum* appears to lack a plastid genome. *Microbiology* **146**, 315–321.

Phylogenetic Insights into the Evolution of Parasitism in Hymenoptera

James B. Whitfield

Department of Entomology, 320 Morrill Hall, 505 S. Goodwin Avenue, University of Illinois, Urbana, IL 61801, USA

Abstact	69
1. Introduction	70
1.1. Parasitism and Phylogeny	70
1.2. Parasitoid Hymenoptera	71
2. Some Questions about Hymenopteran Parasitoid Evolution Addressed Using Phylogeny	74
2.1. Is the Parasitoid Lifestyle an Evolutionary 'Dead End'?	74
2.2. How Many Times did Parasitism Arise in Hymenoptera?	76
2.3. Evolution of Endoparasitism and Koinobiosis from Ectoparasitism	79
2.4. Do Parasitoids Exhibit Cophylogeny with their Hosts?	82
2.5. Cophylogeny Found in Associations with Viruses	82
3. Evolution from Parasitism to other Lifestyles	84
3.1. Gall-forming in Cynipoidea	84
3.2. Fig-pollinating in Chalcidoidea	87
3.3. Nest-building and Sociality in Aculeata	88
4. The Comparative Method and Parasitoids: Future Prospects	89
5. Conclusion	91
Acknowledgements	91
References	91

ABSTRACT

The Hymenoptera are one of the four megadiverse orders of insects, with over 100 000 described species and several times this number still waiting to be described. A major part of this diverse group is formed of large lineages of parasitoid wasps. Some of these lineages have in turn given rise to subgroups that have gone on to diversify into other lifestyles, such as gall-forming on, and pollination of, plants, as well as a broad array of food-collecting behaviors associated with social living in colonies. Thus, the Hymenoptera demonstrate the large evolutionary potential of parasitism as a lifestyle, in contrast to early assertions that parasitism tends to lead to evolutionary 'dead ends' driven by overspecialization. Phylogenetic approaches have already led to a number of important insights into the evolution of parasitism in Hymenoptera. A series of examples are discussed in this review, including the origin of parasitism in the order, the development of koinobiosis in some groups, coevolution with symbiotic viruses, and the evolution in some groups away from parasitism and into such habits as gall formation, pollination of figs, nest building and sociality. The potential for comparative analysis of hymenopteran habits is large, but progress is still in its early stages due to the paucity of available well-supported phylogenies, and the still limited accumulation of basic biological data for many taxa.

1. INTRODUCTION

1.1. Parasitism and Phylogeny

Since the very early days of parasitological inquiry, phylogeny has played an integral part in the interpretation of the diversity of parasites and their relationships with host organisms. Long before the introduction of explicit phylogenetic argumentation by Hennig and Wagner in the early 1950s, Kellogg (1896a, b) and Fahrenholz (1913) were interpreting the evolution of host–parasite relationships in terms of parallel phylogeny, where the evolution and diversification of parasites were driven by those of their hosts. This tradition culminated in the 'parasitological rules' made famous by Eichler (1942) and Manter (1940, 1955, 1966, 1967). The extreme emphasis on the influence of the host on parasite evolution, and on tight phylogenetic congruence, led to a number of generalizations that colored parasitological inference for many years, among them the views that parasites in general are simplified and degenerate compared to free-living organisms, and that parasite phylogeny should reflect that of their hosts ('Fahrenholz's Rule', erroneously attributed to Fahrenholz by Eichler (1942) despite earlier inferences of the same kind by Kellogg (1896a, b)). Excellent introductions to this early conceptual history are provided by Klassen (1992) and Brooks and McLennan (1993).

A further implication of presumed host-driven evolution is the idea that the tight host specificity exhibited by many parasites, while apparently often leading to increased species-level diversification (Price, 1980), is in the long run a sort of evolutionary 'dead end' (e.g. Rensch, 1960). Despite considerable debate over these issues (see Mayr, 1963; Futuyma and Moreno, 1988; Moran, 1988), and the absence of broad quantitative assessment of these generalizations (see Wiegmann *et al.*, 1993), they have persisted as assumptions until quite recently. Brooks and McLennan (1993) have presented abundant evidence that, at least in the case of true parasites of vertebrates, such generalizations do not hold, and may impede rather than enhance understanding of the evolution of parasitism.

A brief examination of the evidence for increased diversification in the parasitic groups of Hymenoptera is presented below in Section 2.1. Much of the rest of this review will also illustrate that parasitism has been anything but a 'dead end' for the Hymenoptera. Instead, the parasitic lifestyle has formed the basic groundplan for the diversification of most of the major lineages, eventually including such divergent habits as pollination of figs, formation of galls on plants, and nest-making and the evolution of sociality in the aculeate ants, wasps, and bees.

1.2. Parasitoid Hymenoptera

Parasites come in many forms, as should be clear from this volume. The different forms of parasitism have profoundly different implications for host specificity, as well as for impact upon the host organisms at both the individual and population level. This review will focus on members of a special class of parasite, indeed one that might well be argued to fall outside the concept of parasitism in its strict sense—the parasitoids.

What is a parasitoid? The term was coined by Reuter (1913) to refer to those parasites that invariably kill their hosts as part of the process of exploiting them. He had in mind specifically the Hymenoptera (the focus of this review) and Diptera that are important natural enemies of other insects, especially of their larval stages. These parasitoid insects, dependent upon their hosts as larvae but free-living as adults, have also been referred to as protelean parasites by Askew (1971). Typically, adult females seek out the host organisms, and deposit eggs in, on or near them; in a few species, the first instar larvae actually locate the host, after their parent female has placed them in a microhabitat in which it is likely that they will encounter hosts. The developing larvae then feed inside or attached to the outside of the host, and either emerge from the moribund host before pupating, or alter the host corpse to pupate within it (host mummification). A vast array of strategies for altering host immune responses, physiology and development are employed by the parasitoids, mostly mediated by biochemical venoms or symbiotic viruses injected by the parent female at oviposition (Stoltz, 1986).

Eggleton and Gaston (1990) have, quite rightly, suggested that the term parasitoid properly applies to a wide taxonomic spectrum of organisms, including nematodes, all of which share the characteristic of killing their host organism. Parasitoids differ from true predators in that they do not feed upon more than one host organism during their parasitic stages. This distinction is, however, blurred in some groups, such as those that attack egg cases of spiders or cockroaches, and those that may consume the food provisions in bee or wasp nest cells in addition to the host itself.

What are the evolutionary implications of the parasitoid lifestyle? For one thing, the habit of killing their hosts would appear, on the surface, to preclude the sorts of coevolution, in the reciprocal evolutionary response sense (Janzen, 1980; Thompson, 1989), one might otherwise expect to develop in parasitic associations. After all, how can a host respond evolutionarily when it is invariably killed by the parasitoid? Although we do not know of any cases where parasitoids have evolved to become less virulent to their hosts at the individual level (they either kill their host or are killed themselves), at the population level the story is more complex. For instance, Yves Carton, Charles Godfray and coworkers have elegantly demonstrated the presence of geographic variation in the ability of parasitoids to kill host *Drosophila*, as well as the ability of the host fly larvae to fend off the parasitoids; each of these abilities also responds to selection (Carton and Nappi, 1991, 1997, 2001; Kraaijeveld and Godfray, 1997, 1999; Kraaijeveld et al., 1998; Dupas and Boscaro, 1999; Castro et al., 2002). Thus, the battle between an individual host and its parasitoid(s) is a life-or-death struggle, with only one or the other eventually surviving, but the average outcome of this struggle is not always constant at the population level, especially over evolutionary time. There is thus still a broad scope for evolution to shape the interaction.

Eggleton and Belshaw (1992, 1993) have provided excellent comparative summaries of the evolution of parasitoids in the broad sense. I focus here upon the evolution of the parasitoid Hymenoptera, the largest group of parasitoids, with the primary aim of showing the insights that phylogenetic inference has brought to our understanding of the evolution of parasitism in the group. Figure 1 shows a scheme published by Eggleton and Belshaw (1992) indicating some of the evolutionary transitions in lifestyle among parasitoid Hymenoptera, and roughly how common each kind of transition is. This diagram is obviously not explicitly phylogenetic, although it was made with some considerable understanding of the phylogeny of

EVOLUTION OF PARASITISM IN HYMENOPTERA

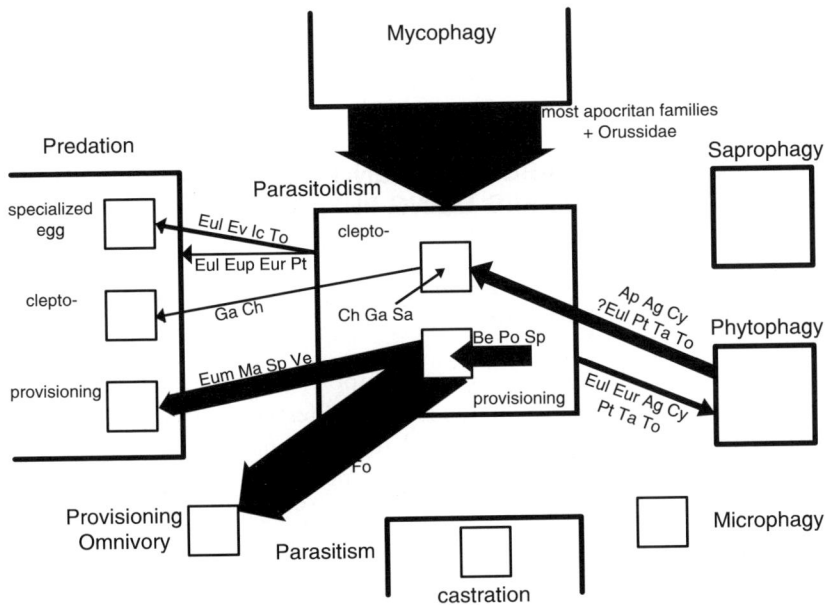

Figure 1 Flow chart of evolutionary transitions to and from parasitoidism in the Hymenoptera. The thickness of the arrows represents the relative numbers of species in the groups following the transition, not the number of independent times the transition has evolved. (Note: bethylids are unlikely to be truly provisioning, as is indicated as a possibility in the figure.) *Abbreviations:* Ag, Agaonidae; Ap, Apidae; Be, Bethylidae; Ch, Chrysididae; Cy, Cynipidae; Eul, Eulophidae; Eum, Eumenidae; Eup, Eupelmidae; Eur, Eurytomidae; Ev, Evaniidae; Ic, Ichneumonidae; Ga, Gasteruptiidae; Ma, Masaridae; Po, Pompilidae; Pt, Pteromalidae; Sa, Sapygidae; Sp, Sphecidae; Ta, Taeniostigmatidae; To, Torymidae; Ve, Vespidae. Reproduced with permission from Eggleton and Belshaw (1992).

Hymenoptera. My goal in this review is to show how several recent studies of phylogeny have (and in some cases have not) provided more specific insight into key issues in parasitoid Hymenopteran evolution, rather than to provide a detailed review of the phylogeny of the order and its full complexity of implications. For more comprehensive, albeit generally less phylogenetic, reviews of the biology of hymenopteran parasitoids, the reader is referred to the treatments by Askew (1971), Eggleton and Belshaw (1992, 1993), Godfray (1994), Hanson and Gauld (1995), Quicke (1997), and Whitfield (1998).

The remainder of this review will be organized into a series of issues or questions, each of which has been approached recently using phylogenetic analysis. Many other similar questions could be and are being asked, but these should provide a taste of how a phylogenetic approach can provide key insights into the evolution of parasitism.

2. SOME QUESTIONS ABOUT HYMENOPTERAN PARASITOID EVOLUTION ADDRESSED USING PHYLOGENY

2.1. Is the Parasitoid Lifestyle an Evolutionary 'Dead End'?

Wiegmann et al. (1993) attempted to supply the first real quantitative assessment of whether adoption of a parasitic lifestyle appears to have led to increased diversification in insects, and to represent a sort of evolutionary 'dead end'. Their analysis was based on a method of multiple sister-group comparisons, as originally used by Mitter et al. (1988) and Farrell et al. (1991) to examine whether herbivory led to increased diversification in insects.

Their conclusion was that parasitism in insects did not lead to increased diversification, or at least that there is no evidence that it did. The analysis was conducted across all insect groups, but included little detailed examination of the patterns within the Hymenoptera, by far the largest group of parasitic insects (in the broad sense including parasitoids) with approximately 250 000 parasitoid species among a total worldwide diversity of over 300 000 species. Table 1 provides a breakdown of both described and estimated total species for hymenopteran superfamilies gleaned from Gaston (1993) and Goulet and Huber (1993), along with a short synopsis of their biologies. With such numbers on the side of the parasites, how could parasitism not have led to increased diversification in this spectacular animal radiation (nearly rivaling that of the beetles)?

There are several reasons why their analysis would not have revealed higher diversification in parasitic lineages. First of all, it now appears, from current phylogenetic evidence, that parasitism may have had essentially a single origin early in hymenopteran evolution (see Section 2.2 below for details). Thus, the use of a multiple sister-group approach reduces the number of relevant comparisons to one. No matter how large the contrast in diversity between the two sister groups involved (and it is huge in this case), it all adds up to only one of many comparisons in their data set. Secondly, most of the other examples of parasitism in their study are of true parasites, rather than parasitoids, and the two lifestyles may be quite different with respect to diversification rate.

Finally, in stark contrast to the 'evolutionary dead end' view, parasitoid hymenopteran lineages have subsequently radiated into a wide variety of

Table 1 Species richness of Hymenoptera, by superfamily, with a brief synopsis of biology.

Superfamily	No. of described species[a]	Est. total no. of species[b]	Biology
Xyeloidea	56	60	Phytophagous
Megalodontoidea	302	300	Phytophagous
Tenthredinoidea	5328	7000	Phytophagous, some gall-formers, some predaceous as adults
Cephoidea	79	100	Phytophagous (bore in stems of plants)
Siricoidea	182	200	Mycophagous/phytophagous (introduce fungi into wood)
Orussoidea	69	70	Parasitoids
Stephanoidea	85	100	Parasitoids
Megalyroidea	80	100	Parasitoids
Ceraphronoidea	802	2000	Parasitoids
Evanoidea	1050	1200	Parasitoids/egg predators
Ichneumonoidea	25 000	100 000	Parasitoids, very few gall-formers
Chrysidoidea	4894	16 000	Parasitoids or cleptoparasites in insect nests
Vespoidea	21 000	48 000	Parasitoids, also predators, scavengers, mycophages, pollen-feeders
Apoidea	27 000	28 000	Parasitoids, many predaceous, most pollen-feeders (bees)
Cynipoidea	3290	4000	Most parasitoids, some gall-forming
Proctotrupoidea	2597	6000	Parasitoids
Platygastroidea	4022	10 000	Parasitoids
Mymarommatoidea	9	10	Not known
Chalcidoidea	18 600	60 000–100 000	Most parasitoids, few gall-formers, seed-eaters
Totals	114–445	283–323 000	Approximately 75% are parasitoids

[a]Data from Gaston (1993).
[b]Data extracted from Goulet and Huber (1993).

other lifestyles, as will be illustrated by examples in detail below. Two of these other lifestyles, herbivory (as exemplified by the bees, gall-forming wasps and pollinating fig wasps), and insect sociality (ants, many bees, and social wasps), have themselves been implicated as major spurs to either diversification (in the case of herbivory) or ecological dominance (in the case of insect sociality: Wilson, 1987). Thus, Hymenoptera seem, largely, to have become winners in their various evolutionary lifestyle 'choices', parasitism representing only one of these choices. They just did not happen to 'change their minds' often enough to provide very many independent tests. (Alternatively, perhaps only the highly successful diversifications have survived.)

2.2. How Many Times did Parasitism Arise in Hymenoptera?

As mentioned briefly above, most phylogenetic studies now suggest a single early origin of parasitism in the Hymenoptera. Throughout much of the twentieth century, several competing views coexisted on the origin of parasitism from phytophagy in Hymenoptera. An especially influential one was that of Malyshev (1968), who noted the apparent connections between gall-forming, inquilinism and parasitism in some families of wasps. He proposed a transition series from hypothetical ancestral 'protocephoids' through gall-forming and inquilinism to parasitism of gall-formers and other inhabitants of enclosed plant cavities. His ideas on these and some other transitions between biological habits in the Hymenoptera, while intriguing and biologically realistic, have mostly not been supported by subsequent studies using phylogenetic comparison. As we will see in Section 3.1, there are indeed some transitional stages between gall-forming and parasitism, but the trend is in the opposite direction—the gall-forming Apocrita are all relatively derived lineages with a parasitic heritage.

An earlier hypothesis on the origin of parasitism in Hymenoptera by Handlirsch (1906–8) seems to have hit the mark, however. Handlirsch noted the apparent morphological similarity between Siricoidea (horntails and woodwasps) and the putatively most primitive parasitic Apocrita, and the fact that the latter were often parasites of the former or of other wood-boring insects. He proposed a series of stages by which a gradual transition from wood-boring to ectoparasitism of wood-borers might have taken place. It is of interest in this regard to note that one relatively uncommon group of woodwasps, the Orussoidea, appears to have an ectoparasitic mode of life while still retaining much of the sawfly/woodwasp morphology.

In a series of papers richly informed by fossils and comparative morphology, Rasnitsyn (1968, 1975, 1980) produced the first comprehensive and well-resolved phylogenetic hypothesis (Rasnitsyn, 1988) for the Hymenoptera that fully developed the implications of Handlirsch's hypothesis. Rasnitsyn formally recognized the close relationship between the parasitic Orussoidea and the similarly parasitic Apocrita, clearly rejecting the Cephoidea/Apocrita sister group relationship that would have supported Malyshev's views. While Rasnitsyn's hypothesis was inferred without use of formal phylogenetic methodology or a fully explicit data set, he provided enough detail to set the stage for a series of subsequent studies by other workers. Many subsequent comparative morphological analyses (Gibson, 1985; Johnson, 1988; Whitfield *et al.*, 1989; Ronquist *et al.*, 1999; Sharkey and Roy, 2001), especially the superb series of monographs by Vilhelmsen and colleagues (Vilhelmsen, 1996, 1997, 2000a, b, c, 2001; Vilhelmsen *et al.*, 2001) have largely corroborated Rasnitsyn's views, at least for the evolutionary lineages relevant to the origin of parasitism.

By the 1990s, molecular data, especially DNA sequence data, began to be applied to hymenopteran phylogeny. While these studies have provided major insights into the evolution of many hymenopteran groups, sequence data have so far failed to provide much of a test of basal hymenopteran relationships (Dowton and Austin, 1999, 2001). Most of the genes so far used extensively for hymenopteran phylogenetics are either too rapidly evolving, or require more extensive investigation of non-phylogenetic biases, to track these Mesozoic divergences. At this point, analyses of both morphological and combined morphological + molecular data sets tend to confirm the (Siricoidea + (Orussoidea + Apocrita)) relationship that is consistent with the hypotheses of Handlirsch and Rasnitsyn. The most recent such analysis, at the time of writing, is Dowton and Austin (2001); their figure showing the basal transition to parasitism is reproduced in Figure 2, along with the earlier, and in many regards similar, 'synthetic consensus' phylogeny from Whitfield (1998).

The implications of these relationships seem fairly clear. First, the earliest form of parasitism found in the Hymenoptera (ectoparasitism of wood-boring insects) had its origin in the common ancestor of Orussoidea + Apocrita. Thus, the parasitoid habit is likely to have had a single early origin in Hymenoptera. This ancestral parasitoid would have had, as its precursors, wood-boring woodwasps that fed upon a rich diet of fungi whose growth they enhanced by an ovipositional mucous secretion from the adult female, as still occurs today in Siricoidea (Spradbery and Kirk, 1978; Gauld and Bolton, 1988). Modern hymenopteran parasitoids similarly use a variety of venoms and symbiotic organisms to modify their

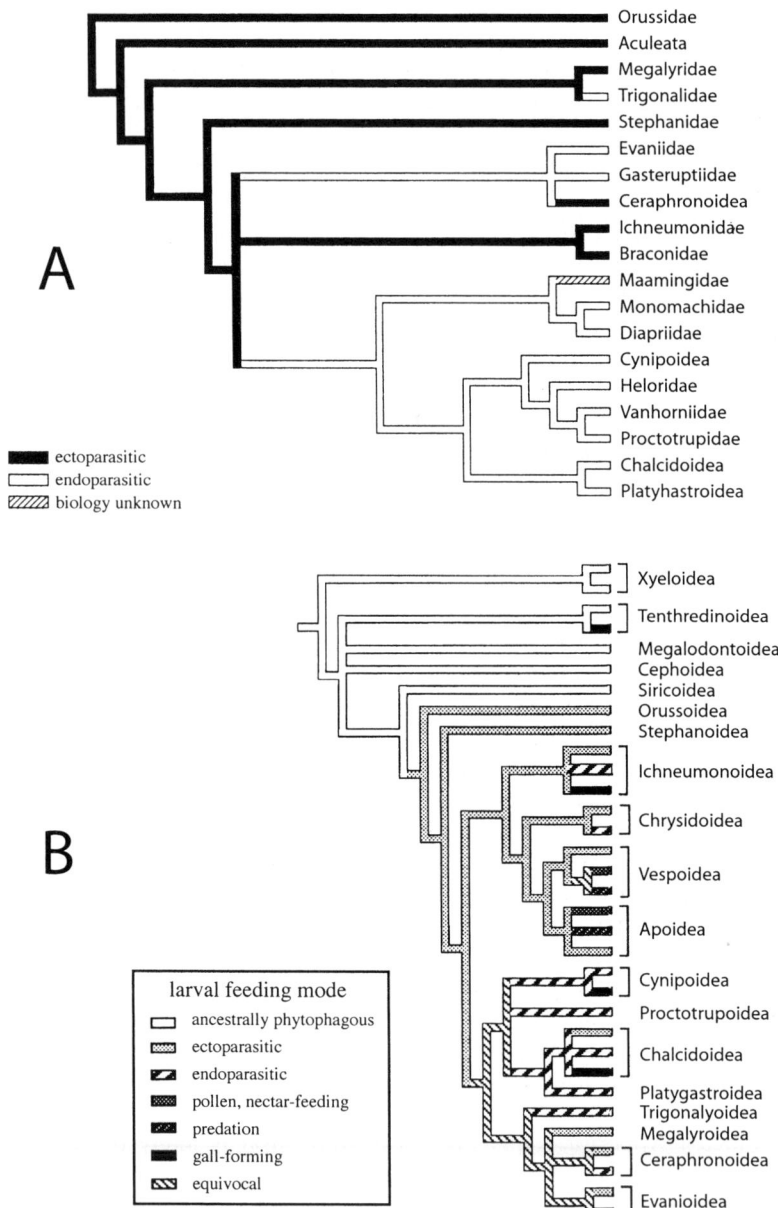

Figure 2 **A.** Partial phylogenetic tree of Hymenoptera, based on analysis of combined morphological and molecular data, with transitions to ectoparasitism and endoparasitism indicated. (Reproduced with permission from Dowton and Austin

host substrate or organism (Piek and Spanjer, 1986; Steiner, 1986; Stoltz, 1986; Coudron, 1990; Stoltz and Whitfield, 1992; Jones and Coudron, 1993). Thus, there is a rich, but still mostly undeveloped, field of inquiry opening up in the area of comparative analysis of the venoms and viruses Hymenoptera are known to employ. Progress in this area is slow, in part due to the minute quantities and the variety of these substances (many of them likely to be undescribed peptides) parasitoids inject into their hosts. Clearly, as the phylogeny of Hymenoptera becomes better understood, comparative study of parasitoid biology at the mechanistic level will lead to a fuller elucidation of the evolutionary pathways parasitoids have taken.

2.3. Evolution of Endoparasitism and Koinobiosis from Ectoparasitism

Parasitoids can be easily divided functionally into internal (endoparasites) and external (ectoparasitoids). As already mentioned (Section 2.2), it has been known for a long time that the earliest lineages of parasitoids in Hymenoptera were ectoparasitoids, typically of wood-boring or otherwise concealed immature insect hosts.

From a series of field studies of forest parasitoids and their hosts, Haeselbarth (1979) provided another distinction that seemed to correlate more closely with patterns in host specificity of parasitoids, that between parasitoids which quickly paralyze or kill their host organisms and feed upon them quickly (idiophytes), and those that allow the host to regain or maintain activity and develop further while parasitized (koinophytes). Askew and Shaw (1986) developed this biological distinction in further detail, renaming these modes idiobiosis and koinobiosis to avoid plant connotations. As koinobiosis implies rather more sophisticated and long-term interactions with host organisms, it was expected that koinobionts would tend to show more restricted host ranges than would idiobionts, and this pattern largely held in their and many other subsequent analyses, albeit most clearly so when closely related groups are compared (Sheehan and Hawkins, 1991; Eggleton and Gaston, 1992; Hawkins *et al.*, 1992; Shaw, 1994).

2001.) **B.** Summary tree for Hymenoptera, showing major biological transitions within the order. This tree was intended to depict consensus between studies at the time, and was not based on an explicit phylogenetic analysis. (Reproduced with permission from Whitfield, 1998.) Both trees show a single basal origin of ectoparasitism in the Orussoidea, a heritage held in common with Apocrita.

In general, ectoparasitoids are idiobionts, and most endoparasitoids are koinobionts, but exceptions in both directions occur. It was thus of interest to know which transition arose earlier (and perhaps made the other possible)—to endoparasitism from ectoparasitism, or to koinobiosis from idiobiosis. How do these transitions relate to the apparent evolution toward parasitization of exposed hosts from the more ancestral habit of attacking hosts in concealed locations? Belshaw and Quicke (2002) reviewed a number of hypotheses on this topic, and attempted to test them using a phylogeny of Ichneumonoidea based on 28S rDNA sequence data. Their conclusions, which are still limited due to the need for further testing of their phylogeny using other data sets, include the observation, unsurprising in retrospect, that the answer one gets depends on the lineage one is studying. At this point it would not be productive to be more critical of the most general patterns; we need not only better supported and more comprehensive phylogenies, but also more complete biological data for many taxa to advance further. Instead, I will review in more detail results from one of the more intensively studied lineages, the cyclostome Braconidae. It is already becoming clear from preliminary studies that other lineages will provide equally fascinating insights into the evolution of parasitic strategy as their phylogenies become better known (S.R. Shaw, 1988; Gauld and Wahl, 2000; Smith and Kambhampati, 2000).

The cyclostome Braconidae are so called because they have a more or less circular excavation above their mandibles that appears to enhance the ability to chew their way out of enclosed sites such as tunnels in wood. They are a large, putatively monophyletic lineage of wasps that are predominantly ectoparasitoids of concealed hosts. Several subfamilies within this complex are, however, composed of endoparasitoids: the Opiinae and Alysiinae, which parasitize the larvae and puparia of Diptera, and the Rogadinae, which mummify caterpillars of butterflies and moths (Shaw and Huddleston, 1991). The forms of endoparasitism in these groups of parasitoids are different enough that it seemed reasonable that endoparasitism might have arisen separately in these lineages.

Mark Shaw (1983), noting some remarkable similarities in oviposition between the koinobiont ectoparasitoid *Rhysipolis* and the koinobiont endoparasitoid *Clinocentrus*, proposed a scenario for how endoparasitism might have developed from ectoparasitic koinobiosis in the rogadine lineage. In his scenario, the transition from idiobiosis to koinobiosis occurred before that from ectoparasitism to endoparasitism; in this way the koinobiont ectoparasitoid *Rhysipolis* is a logical transitional form. Whitfield (1992) tested Shaw's hypothesis by inferring a phylogeny of rogadine and related cyclostome braconid lineages using morphological data. He found (Figure 3) that there appeared to be at least two

EVOLUTION OF PARASITISM IN HYMENOPTERA

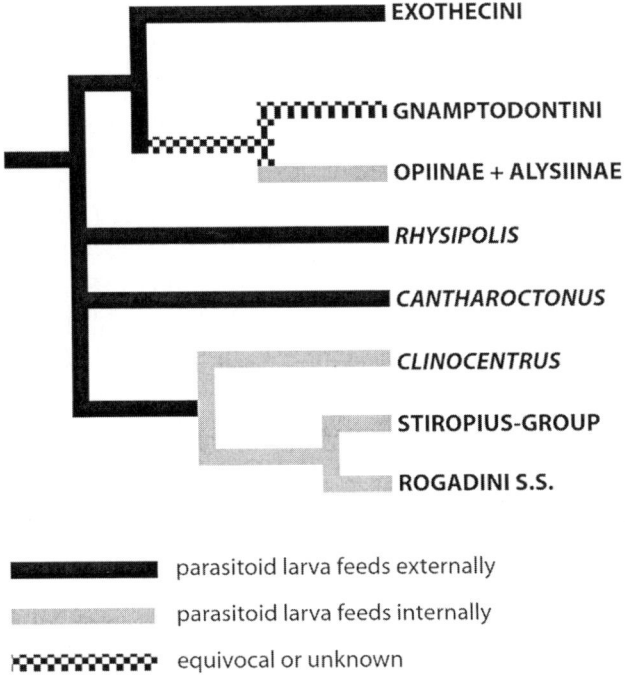

Figure 3 Partial phylogenetic tree of Rogadinae and related cyclostome braconid wasp groups, based on analysis of morphological data, and showing hypothesized transitions between biological lifestyles. Note the dual origin of endoparasitism even within this restricted group. Reproduced with permission from Whitfield (1992).

independent origins of endoparasitism within the cyclostome braconids, one exactly as Shaw (1983) proposed with the ectoparasitoid *Rhysipolis* the sister-group to the endoparasitoid mummifier *Clinocentrus*, and a second in the clade including the Opiinae/Alysiinae lineage. Quicke (1993) challenged a few character interpretations from Whitfield's treatment, but his reanalysis found much the same overall pattern (with only the position of the Gnamptodontinae different), and this pattern was later confirmed using molecular data from 28S rRNA gene sequences (Belshaw et al., 1998). Thus all evidence at this stage points to at least two independent origins of koinobiont endoparasitism within just the cyclostome braconid wasps; there must be many others, especially in the Ichneumonidae and Chalcidoidea (Whitfield, 1998). Once endoparasitism has been established in hymenopteran lineages, the tendency to diversify its mode of action appears to be strong.

2.4. Do Parasitoids Exhibit Cophylogeny with their Hosts?

It appears from the general parasitological literature that an obvious field of study among parasitoid Hymenoptera would be direct comparison of host and parasitoid phylogenies. However, this has not generally been the case. Early suggestions from Griffiths (1964) and Mackauer (1965) that fly and aphid parasitoids, respectively, showed correspondence between the parasitoid and host phylogenies have not been supported by further work. Indeed, detailed studies of host ranges in specific parasitoid groups (see Askew and Shaw, 1974, 1986; Shaw and Askew, 1976; Gross and Price, 1988; Whitfield and Wagner, 1988; Askew, 1994; Shaw, 1994) show that limits to host ranges tend to follow host life-history, host locations on plants, or other ecological patterns more closely than strictly historical or taxonomic ones. Ecological/behavioral determination of host relationships has appeared to be so common that few recent studies have even attempted to compare host and parasitoid phylogenies. It would be interesting in future to examine this lack of correspondence in more depth and quantitative rigor, as it appears likely that at least some host range patterns would be phylogenetically correlated. One would expect that the patterns would just tend to be much more complex than simple correspondence between single phylogenies, and thus would not be obvious from superficial comparison of phylogenetic topologies.

2.5. Cophylogeny Found in Associations with Viruses

If cophylogenetic patterns (parallel phylogenies) have been difficult to find in relationships between parasitoids and their hosts, they are beginning to appear in another context—the relationships between parasitoids and the symbiotic entities they employ against their hosts. Vertically transmitted endosymbionts have proven to be some of the best examples of cophylogeny yet discovered (e.g., see Moran and Baumann, 1994). The most fully studied case of this with respect to parasitoid wasps is the association of polydnaviruses with braconid wasps. These unusual viruses (actually proviruses) are fully integrated into the chromosomal DNA of the wasps (Fleming and Summers, 1986, 1991; Stoltz et al., 1986; Xu and Stoltz, 1991; Belle et al., 2002) and thus are inherited in Mendelian fashion (Stoltz, 1990). After replication and packaging of the viruses in the wasp ovarian calyx tissue, the viral particles are injected into the host caterpillars together with the wasp eggs. A variety of effects of the viral coat proteins, as well as of polydnaviral genes expressed in the hosts, has now been documented (Edson *et al.*, 1981; Stoltz, 1986; Fleming, 1992; Stoltz and Whitfield, 1992; Asgari *et al.*, 1998; Shelby and Webb, 1999; Schmidt *et al.*, 2001). These include impacts on the host endocrine and immune function that allow the wasp eggs and larvae to survive successfully within the host. How and when this remarkable partnership between wasp and virus evolved thus became of considerable interest for the understanding of host/parasitoid relationships.

From the early days of recognition that viruses were involved in parasitism by braconid wasps, it was noticed that the viruses were found in related groups of wasps, and in all examined individuals of them (Stoltz and Vinson, 1979). Subsequent studies showed that viruses from more closely related wasps cross-reacted more strongly with each other in Southern blots and immunoblots than those from more distantly related wasps (Cook and Stoltz, 1983; Stoltz and Whitfield, 1992). The combined evidence, together with the mode of inheritance of the viruses, led to explicit predictions (Whitfield, 1990; Stoltz and Whitfield, 1992) that the phylogenetic relationships among the viruses would reflect those of the wasps.

These predictions proved more difficult to test than anticipated. First of all, few polydnaviruses had been studied in any detail, so comparative information concerning shared genes was lacking. Thus, a considerable effort in testing and redesigning polymerase chain reaction primers, and then comparing sequences from multiple wasp species, was required to accumulate any data on viral relationships. Second, the genera within the microgastroid assemblage of braconids that carry polydnaviruses appear to have radiated quite quickly and extensively, complicating phylogenetic analysis at this taxonomic level (Mardulyn and Whitfield, 1999). So a two-pronged approach was used to address the cophylogeny question, with the hope that the two prongs would eventually converge to provide a coherent picture. The first prong was testing the monophyly of the 'microgastroid' assemblage of braconid subfamilies known to carry the polydnaviruses. Whitfield and Mason (1994), using morphological data, showed that this group of subfamilies appeared to form a true lineage; the addition of DNA sequence data by Whitfield (1997) and later by Dowton and Austin (1998) corroborated the monophyly of the polydnavirus-bearing wasp lineage.

The second prong of attack worked at the species level, to test true cophylogeny between wasp species and the polydnaviruses they carry. The genus *Cotesia* was selected as a starting point since the viruses from several species in this genus were already under investigation and some viral genes had been characterized. Whitfield (2000) showed that the phylogeny of six species of *Cotesia* (based on sequence data from the 16S rDNA and ND1 genes) is essentially identical to that of the polydnaviruses they carry (based on sequence data from the viral CrV1 gene). More recent studies (A. Michel-Salzat and J.B. Whitfield, unpublished observations) are extending these findings through many more species of *Cotesia* as well as related microgastrine-group genera. At this point, it does seem likely that the cophylogeny between wasps and viruses extends through the entire polydnavirus-bearing lineage of braconids, but this has not yet been fully demonstrated. The origin of this lineage of associated partners has been dated at roughly 74 ± 11 million years ago (mya), using molecular clock techniques calibrated with wasp fossils (Whitfield, 2002). A summary of what we know at this point (extracted from Whitfield and Asgari, 2003) is presented in Figure 4.

3. EVOLUTION FROM PARASITISM TO OTHER LIFESTYLES

3.1. Gall-forming in Cynipoidea

Several groups of ancestrally parasitoid Hymenoptera have secondarily radiated into gall-forming habits on different plant groups. These include the Braconidae (de Macedo and Monteiro, 1989; Infante *et al.*, 1995; Austin and Dangerfield, 1998), several families of Chalcidoidea (Hanson and Gauld, 1995) and, most familiar to Northern Hemisphere biologists, the Cynipidae (Kinsey, 1920; Cornell, 1983; Ronquist and Liljeblad, 2001). The last group has been most extensively studied from both biological and evolutionary perspectives.

The Cynipidae in its current restricted sense (Ronquist, 1995) consists entirely of wasps that feed in and/or form galls in plant tissue, or are inquilines of gall-makers. All of the close relatives of the Cynipidae, including their immediate sister lineage the Figitidae, are parasitoids of other insects (Ronquist, 1995; Liljeblad and Ronquist, 1998). Thus we have no obvious intermediate stages that would provide insight into the evolutionary transition between parasitism and gall-making, except that several representative sister taxa (Thrasorinae and Parnipinae) are parasitoids of insects that are themselves gall-formers. Thus, Malyshev's transition between gall-making and parasitism might appear to have some truth in it, although the direction of change is of course backwards. We do, in contrast, have detailed analyses of the subsequent evolution of gall-making itself in the Cynipidae by Ronquist and Liljeblad (2001). They mapped such traits as gall structure, position on the plant, mode of attachment, host plant growth form, and host plant families onto a wasp phylogeny based on a large morphological data set. The resulting picture

EVOLUTION OF PARASITISM IN HYMENOPTERA

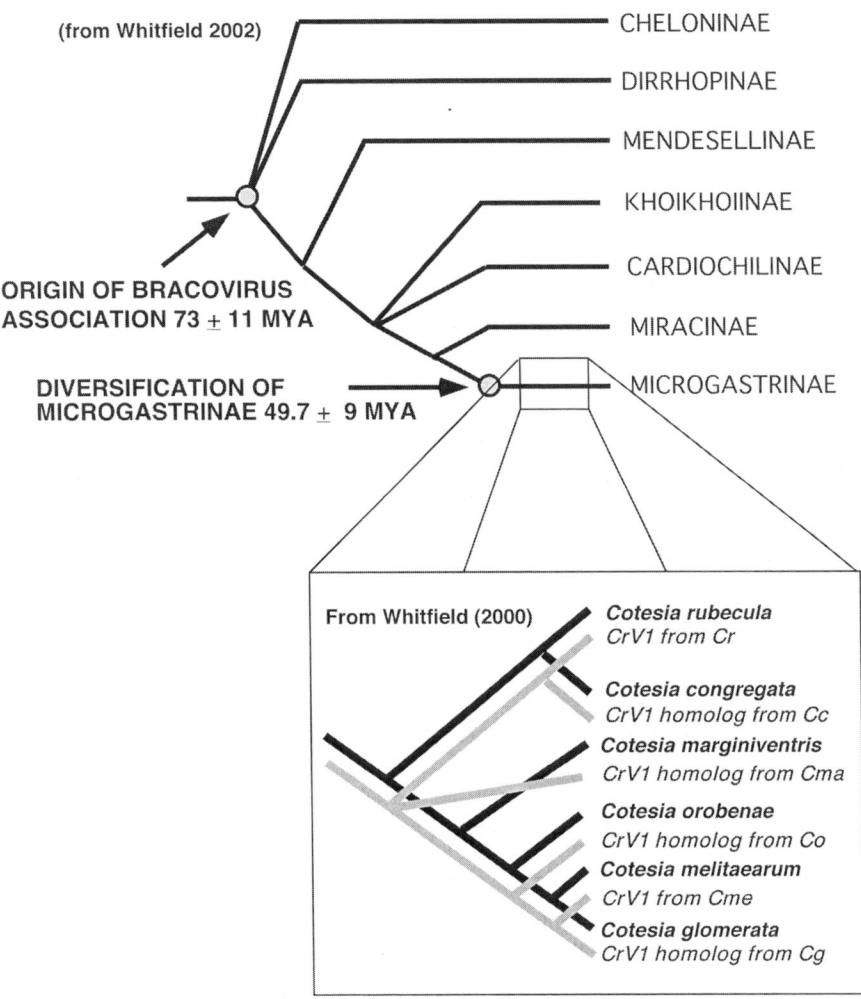

Figure 4 Summary of findings on coevolution of polydnaviruses with braconid parasitoid wasps. Note the very early origin of the virus/wasp association, and the co-phylogeny (at least in one recent group) between the two associates. Modified from Whitfield and Asgari (2003).

is generally that of basal taxa tending to be more generalized gallers of herbaceous plants, then the subsequent radiation of gall types and independent origins of galling of woody plants and inquilinism in woody galls (Figure 5). They found no evidence of strict cophylogeny between

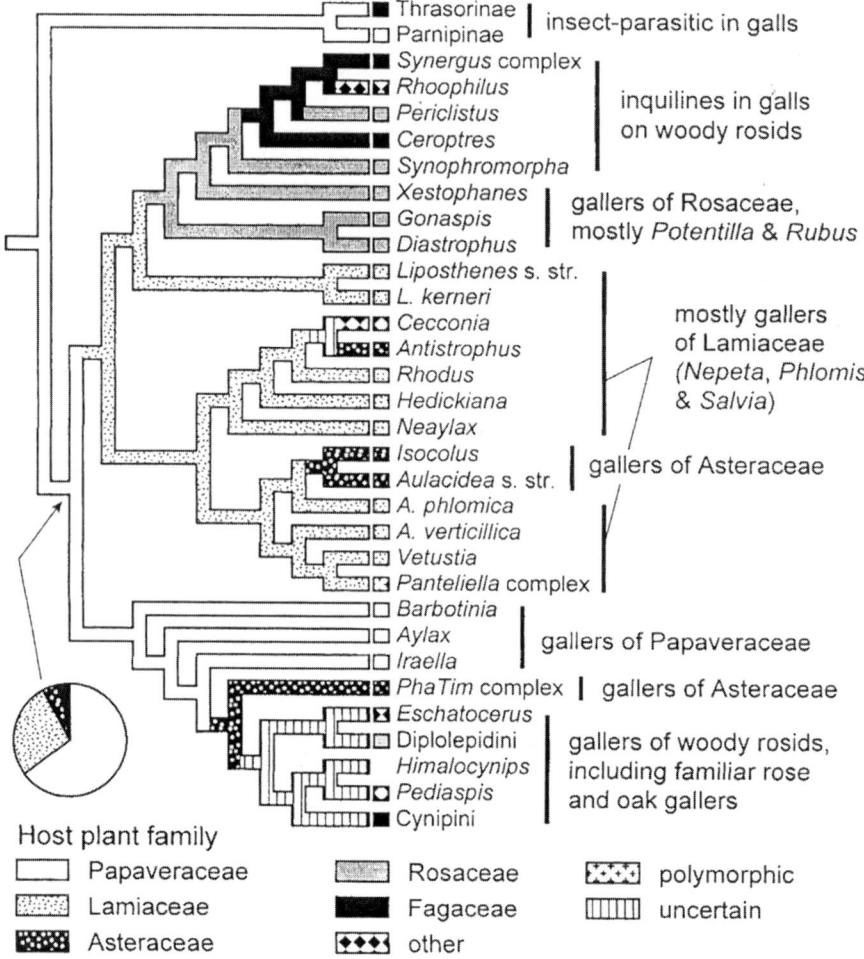

Figure 5 Evolution of host–plant associations in the gall wasps (Cynipidae). The two taxa at the top of the tree are non-galling outgroups. Note the dual origins of galling on woody plants. Reproduced with permission from Ronquist and Liljeblad (2001).

gall wasps and their host plants, although there is some clear tendency for wasp lineages to be restricted to particular host higher taxa. In their results, inquilinism was found to be a derived state relative to gall-making, with no apparent relationship to the origin of parasitism in cynipoids, as Malyshev (1968) would have predicted.

3.2. Fig-pollinating in Chalcidoidea

As with the origin of gall-making in Cynipoidea, the fig-pollinating wasps in the family Agaonidae appear to have a single origin from parasitoid ancestors (Machado et al., 1996). These wasps have long been considered a classic example of animal/plant coevolution. Unlike the cynipids, there are clear patterns of cophylogeny with host plants both in higher taxon comparisons (Figure 6) and in species-by-host-species associations (Herre et al., 1996; Cook and Lopez-Vaamonde, 2001; Machado et al., 2001; Weiblen and Bush, 2002). There are some known exceptions to strict host specificity and strict cophylogeny, which appear to be explainable by host shifts during secondary sympatry of wasps originally isolated on allopatric host plants (Kerdelhue et al., 1999; Machado et al., 2001). On the whole, the evolutionary patterns suggest that the kinds of very close association we see now between fig wasps and their host plants may have operated for quite

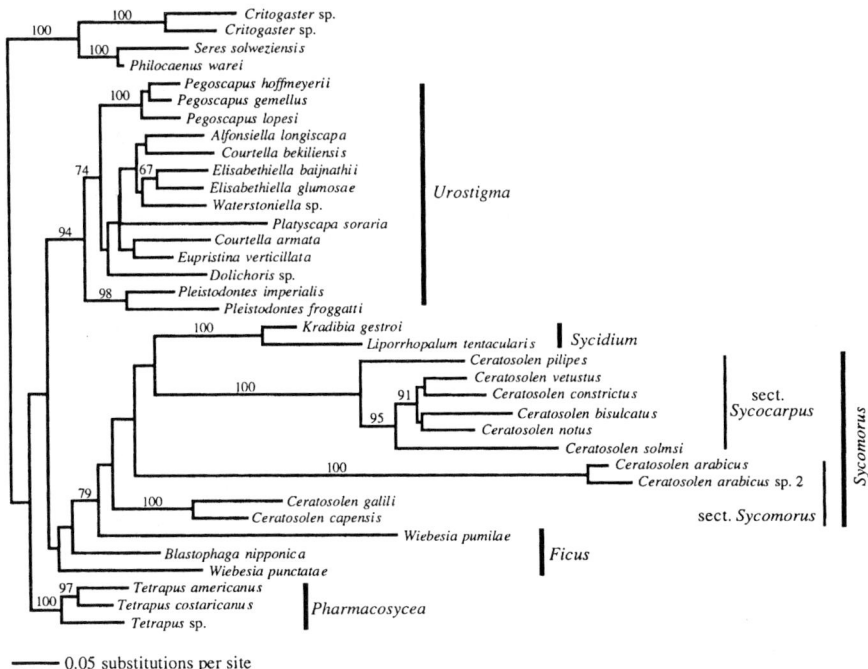

Figure 6 Evolutionary association between fig wasps (on left) and the major groupings of their fig hosts (on right). At finer scales some host-switching is known to occur. Reproduced with permission from Machado et al. (2001).

some time; the origin of the mutualism was dated at approximately 87.5 mya by Machado *et al.* (2001).

An interesting aspect of the fig wasp system is that the figs are also sometimes inhabited by 'parasitic' (i.e. non-pollinating) wasps living in the figs as well as by parasitoids on these fig non-pollinators. Surprisingly, there appears to be not only significant phylogenetic association between the non-pollinators and their host figs, but also extensive shared cophylogeny between pollinating and non-pollinating wasps in the same figs (Machado *et al.*, 1996; Lopez-Vaamonde *et al.*, 2001). The overall fig-fig wasp system is highly complex and will continue to produce fascinating phylogenetic interpretations for some time to come.

3.3. Nest-building and Sociality in Aculeata

The Aculeata (or 'stinging Hymenoptera') are a demonstrably monophyletic group of Hymenoptera, in which the ancestral ovipositor structure no longer functions directly in egg-laying. Instead, eggs are laid through a separate opening at the base of the ovipositor, and the 'ovipositor' itself has been modified into a specialized structure for delivering venoms into host, prey or enemy animals (Oeser, 1961). In the non-aculeate apocritan Hymenoptera ('Parasitica'), the ovipositor typically has this stinging function, but it also serves to deliver the eggs onto or into host organisms (either as part of the same stinging action or separately; Steiner, 1986).

Thus, the distinctive aspect of the Aculeata in the broad sense is not the stinging activity, but the loss of egg-laying function by the ovipositor structure. The basal groups of Aculeata are biologically similar to the ectoparasitoid representatives of other apocritan Hymenoptera, paralyzing host organisms, typically insects or other arthropods, and depositing their eggs upon them. Nevertheless, most aculeate wasps have been referred to traditionally as 'solitary wasps' or 'predatory wasps', and not explicitly as ectoparasitoids (Evans and West-Eberhard, 1970; Iwata, 1976; Yeo and Corbet, 1983; O'Neill, 2001). This distinction appears to have been made in order to compare them with the social wasps, bees, and ants, which are derived from some of them, but it tends to obscure the fact that the behavior of ancestral Aculeata is biologically very much like the groundplan of ectoparasitoid Hymenoptera.

Rather than the huge diversification in specialized parasitoid mode of life, we see in such groups as the Chalcidoidea and Ichneumonoidea (Whitfield, 1998) that some lineages of the aculeate Hymenoptera accumulated quite different evolutionary innovations such as the tendency to move hosts (prey) to nests or burrows of their own making, in order to progressively provision these nests for their young (not necessarily with animal prey), to care for their young, and ultimately, in a few highly successful groups, to form societies of close relatives for care of offspring (Evans and West Eberhard, 1970; Iwata, 1976; O'Neill, 2001). It is the lineages leading to the social wasps, ants, and bees that are most familiar to most non-entomologists. While these represent only a relatively small slice of the phylogenetic 'pie' of Hymenoptera, they are often ubiquitous and ecologically dominant animals. Thus, they represent one of the most spectacular ways in which parasitism in Hymenoptera has been anything but a 'dead end'.

4. THE COMPARATIVE METHOD AND PARASITOIDS: FUTURE PROSPECTS

Most of the examples cited above demonstrate the utility of optimizing various traits onto phylogenies in order to study the number and relative timing of the origin(s) of each trait. This basic approach has been widely used, and is discussed in detail by Brooks and McLennan (1991, 1993). An additional form of 'comparative method' uses a phylogeny to develop a series of phylogenetically independent contrasts (Felsenstein, 1985) or other kinds of comparison that can then be studied statistically without being confounded by historical ancestry (Felsenstein, 1985; Harvey and Pagel, 1991); exemplified above by the studies of relative diversification of parasitic versus non-parasitic groups discussed in Section 2.1. This general approach is especially useful for testing whether the evolution of a specific biological trait tends to be associated with a common ecological background, or to be correlated with some other trait(s) (Maddison, 1990; Harvey and Pagel, 1991; Martins, 1996). The independent contrasts type of analysis has not been used very extensively in Hymenoptera until recently, especially in parasitoid groups (Mayhew, 1998; but see the examples from Mayhew and colleagues included in Table 2). The reasons for this appear to be twofold.

First, for most groups of parasitoids, we still are lacking comprehensive and well-supported phylogenies. Thus, for many traits, more traditional classifications must be used as a substitute for phylogeny, resulting in an obvious loss of power and accuracy in inference. This might still be reasonably successful if it were not for the second problem with parasitoid Hymenoptera: the biological (behavioral and ecological) data we have are often too widely scattered and unrepresentative to make clear generalizations about traits of larger lineages (Whitfield, 1998).

Thus, we still need much more basic information for many lineages, as well as greater taxon representation and data sources in phylogenies, in

Table 2 Some comparative traits in parasitoid Hymenoptera requiring further phylogenetic information.

Trait	References
Complementary sex determination	Godfray (1994)
Sex ratio, sexual dimorphism and maturing structure in bethylids	Hardy and Mayhew (1998)
Constraints in sex ratio adjustment	West and Sheldon (2002)
Nonsiblicidal behavior and clutch size in bethylids	Mayhew and Hardy (1998)
Development mode vs. life-history pattern	Mayhew and Blackburn (1999)
Number of ovarioles per ovary vs. host stage attacked	Price (1973)
Egg production strategies	Price (1974)
Gregrarious vs. solitary parasitism	Le Masurier (1987); Mayhew (1998)
Heteronomous parasitism in Aphelinidae	Hunter and Woolley (2001)
Host-feeding behavior	Heimpel and Collier (1996)
Ovigeny	Jervis et al. (2001)
Ovipositor morphology—lineage vs. functional aspects	Le Ralec et al. (1996); Quicke et al. (1999)
Polyembrony	Strand and Grbic (1997)
Syncytial vs. holobastic cleavage with respect to life history	Grbic and Strand (1998)
Venom composition and effects on host insects	Piek (1986); Steiner (1986); Coudron (1990); Jones and Coudron (1993)
Forewing size, ovipositor length and ovarian structure	Price (1972)

order to advance with the comparative method in parasitoid evolution. Of the two problems, the need for phylogenetic rigor through greater taxon representation and greater clade support is clearly the more easily soluble. In fact, ongoing work in a number or laboratories on many groups is likely to make this assessment seem altogether too conservative in the very near future.

Accumulating more biological data, and for a wider array of taxa, is more of a long-term issue. Nevertheless, when combined with strongly supported phylogenies, this should help to provide a predictive framework for interpretation, as well as for targeting taxa for future field and laboratory study. Table 2 provides a list of some biological traits already under study that will clearly benefit from the comparative approach.

5. CONCLUSION

Our understanding of the evolution of the parasitoid lifestyle in the wasps has benefited enormously from the adoption of phylogenetic approaches. We can already see that this mode of life has not only been an ecological and evolutionary success in its own right, but has also provided the raw material for the evolution of additional successful lifestyles, and contributed mightily to the huge radiation of animals known as the Hymenoptera.

We are still placed very early on the learning curve, however. The development and refinement of phylogenetic methodology came to the study of Hymenoptera at a time when many of the parasitoid groups were still relatively poorly known, both taxonomically and biologically. It is clear that the species richness of Hymenoptera, while already large based only on described species, is truly enormous, composing one of the largest groups of organisms on Earth (Gaston, 1993; Dolphin and Quicke, 2001). The parasitoid groups are likely to richly repay study, even at the purely descriptive taxonomic and biological level, for a long time. The development of a rigorous phylogenetic framework for the Hymenoptera will play an important role in the direction and interpretation of this work in progress.

ACKNOWLEDGMENTS

I especially wish to thank Tim Littlewood for inviting me to contribute this article to the volume; it has proven rewarding (to me at least) to survey the work of many other fellow hymenopterists with a general theme in mind. Discussions with Andy Austin, Rob Belshaw, Mark Dowton, Charles Godfray, George Heimpel, Allen Herre, Paul Ode, Donald Quicke, Fredrik Ronquist, Mark Shaw, and Lars Vilhelmsen were all immensely helpful in keeping abreast of the pieces of their work presented here. My own work in this area was funded by several grants from the National Science Foundation of the USA and the US Department of Agriculture; their support is much appreciated.

REFERENCES

Asgari, S., Theopold, U., Wellby, C., Schmidt, O. (1998). A protein with protective properties against the cellular defence reactions in insects. *Proceedings of the National Academy of Sciences of the USA* **95**, 3690–3695.

Askew, R.R. (1971). *Parasitic Insects*. New York: American Elsevier.

Askew, R.R. (1994). Parasitoids of leaf-mining Lepidoptera: what determines their host ranges? In: *Parasitoid Community Ecology* (B. A. Hawkins and W. Sheehan, eds), pp. 177–202. Oxford: Oxford University Press.

Askew, R.R. and Shaw, M.R. (1974). An account of the Chalcidoidea (Hymenoptera) parasitising leaf-mining insects of deciduous trees in Britain. *Biological Journal of the Linnean Society* **6**, 289–335.

Askew, R.R. and Shaw, M.R. (1986). Parasitoid communities: their size, structure, and development. In: *Insect Parasitoids* (J. Waage and D. Greathead, ed.). pp. 225–264. London: Academic Press.

Austin, A.D. and Dangerfield, P.C. (1998). Biology of *Mesostoa kerri* (Insecta: Hymenoptera: Braconidae: Mesostoinae), an endemic Australian wasp that causes stem galls on *Banksia marginata*. *Australian Journal of Botany* **46**, 559–569.

Belle, E., Beckage, N.E., Rousselet, J., Poirie, M., Lemeunier, F. and Drezen, J.-M. (2002). Visualization of polydnavirus sequences in a parasitoid wasp chromosome. *Journal of Virology* **76**, 5793–5796.

Belshaw, R. and Quicke, D.L.J. (2002). Robustness of ancestral state estimates: evolution of life history strategy in ichneumonoid parasitoids. *Systematic Biology* **51**, 450–477.

Belshaw, R., Fitton, M., Herniou, E., Gimeno, C. and Quicke, D.L.J. (1998). A phylogenetic reconstruction of the Ichenumonoidea (Hymenoptera) based on the D2 variable region of 28S ribosomal RNA. *Systematic Entomology* **23**, 109–123.

Brooks, S.R. and McLennan, D.A. (1991). *Phylogeny, Ecology and Behavior: A Research Program in Evolutionary Biology*. Chicago: University of Chicago Press.

Brooks, D.R. and McLennan, D.A. (1993). *Parascript: Parasites and the Language of Evolution*. Washington, DC: Smithsonian University Press.

Carton, Y. and Nappi, A.J. (1991). The *Drosophila* immune reaction and the parasitoid capacity to evade it: genetic and coevolutionary aspects. *Acta Oecologia* **12**, 89–104.

Carton, Y. and Nappi, A.J. (1997). *Drosophila* cellular immunity against parasitoids. *Parasitology Today* **13**, 218–227.

Carton, Y. and Nappi, A.J. (2001). Immunogenetic aspects of the cellular immune response of *Drosophila* against parasitoids. *Immunogenetics* **52**, 157–164.

Castro, L.R., Austin, A.D. and Dowton, M. (2002). Contrasting rates of mitochondrial molecular evolution in parasitic Diptera and Hymenoptera. *Molecular Biology and Evolution* **19**, 1100–1113.

Cook, D.I. and Stoltz, D.B. (1983). Comparative serology of viruses isolated from ichneumonid parasitoids. *Virology* **130**, 215–220.

Cook, J.M. and Lopez-Vaamonde, C. (2001). Fig biology: turning over new leaves. *Trends in Ecology and Evolution* **16**, 11–13.

Cornell, H.V. (1983). The secondary chemistry and complex morphology of galls formed by the Cynipinae (Hymenoptera): why and how? *American Midland Naturalist* **110**, 225–234.

Coudron, T.A. (1990). Host regulating factors associated with parasitic Hymenoptera. In: *Naturally Occurring Pest Bioregulators* (P. A. Hedin, ed), pp. 41–65. New York: ACS Books.

Dolphin, K. and Quicke, D.L.J. (2001). Estimating the global species richness of an incompletely described taxon: an example using parasitoid wasps (Hymenoptera: Braconidae). *Biological Journal of the Linnean Society* **73**, 279–286.

Dowton, M. and Austin, A.D. (1999). Models of analysis for molecular datasets for the reconstruction of basal hymenopteran relationships. *Zoologica Scripta* **28**, 69–74.
Dowton, M. and Austin, A.D. (2001). Simultaneous analysis of 16S, 28S, CO1 and morphology in the Hymenoptera: Apocrita—evolutionary transitions among parasitic wasps. *Biological Journal of the Linnean Society* **74**, 87–111.
Dupas, S. and Boscaro, M. (1999). Geographic variation and evolution of immune suppressive genes in a *Drosophila* parasitoid. *Ecography* **22**, 284–291.
Edson, K.M., Vinson, S.B., Stoltz, D.B. and Summers, M.D. (1981). Virus in a parasitoid wasp: suppression of the cellular immune response in the parasitoid's host. *Science* **211**, 582–583.
Eggleton, P. and Belshaw, R. (1992). Insect parasitoids: an evolutionary overview. *Philosophical Transactions of the Royal Society of London, Series B* **337**, 1–20.
Eggleton, P. and Belshaw, R. (1993). Comparisons of dipteran, hymenopteran and coleopteran parasitoids: provisional phylogenetic explanations. *Biological Journal of the Linnean Society* **48**, 213–226.
Eggleton, P. and Gaston, K.J. (1990). 'Parasitoid' species and assemblages: convenient definitions or misleading compromises? *Oikos* **59**, 417–421.
Eggleton, P. and Gaston, K.J. (1992). Tachinid host ranges: a reappraisal (Diptera: Tachinidae). *Entomologist's Gazette* **43**, 139–143.
Eichler, W. (1942). Die Entfaltungsregel und andere Gesetzmässigkeiten in den parasitologishen Beziehungen der Mallophagen und anderer ständigen Parasiten zu ihren Wirten. *Zoologische Anzeiger* **137**, 77–83.
Evans, H.E. and West Eberhard, M.J. (1970). *The Wasps*. Ann Arbor: University of Michigan Press.
Fahrenholz, H. (1913). Ectoparasiten und Abstammungslehre. *Zoologische Anzeiger* **41**, 371–374.
Farrell, B., Mitter, C. and Doussard, D. (1991). Macroevolution of plant defense: do latex/resin secretory canals spur diversification? *American Naturalist* **138**, 881–900.
Felsenstein, J. (1985). Phylogenies and the comparative method. *American Naturalist* **125**, 1–15.
Fleming, J.G.W. (1992). Polydnaviruses: mutualists and pathogens. *Annual Review of Entomology* **37**, 401–423.
Fleming, J.G.W. and Summers, M.D. (1986). *Campoletis sonorensis* endoparasitic wasps contain forms of *C. sonorensis* virus DNA suggestive of integrated and extrachromosomal polydnavirus DNAs. *Journal of Virology* **57**, 552–562.
Fleming, J.G.W. and Summers, M.D. (1991). Polydnavirus DNA is integrated in the DNA of its parasitoid wasp host. *Proceedings of the National Academy of Sciences of the USA* **88**, 9770–9774.
Futuyma, D.J. and Moreno, G. (1988). The evolution of ecological specialization. *Annual Review of Ecology and Systematics* **19**, 207–233.
Gaston, K.J. (1993). Spatial patterns in the description and richness of the Hymenoptera. In: *Hymenoptera and Biodiversity* (J. LaSalle and I. D. Gauld, eds), pp. 277–293. Wallingford, UK: CAB International.
Gauld, I.D. and Bolton, B. (1988). *The Hymenoptera*. Oxford: Oxford University Press and The Natural History Museum.
Gauld, I.D. and Wahl, D.B. (2000). The Labeninae (Hymenoptera: Ichneumonidae): a study in phylogenetic reconstruction and evolutionary biology. *Zoological Journal of the Linnean Society* **129**, 271–347.

Gibson, G.A.P. (1985). Some pro- and mesothoracic characters important for phylogenetic analysis of Hymenoptera, with a review of terms used for structures. *Canadian Entomologist* **117**, 1395–1443.

Godfray, H.C.J. (1994). *Parasitoids: Behavioral and Evolutionary Ecology*. Princeton, NJ: Princeton University Press.

Gokhman, V. E. (1995). Trends of biological evolution in the subfamily Ichneumoninae and related groups (Hymenoptera Ichneumonidae): an attempt of phylogenetic reconstruction. *Russian Entomological Journal* **4**, 91–103.

Goulet, H. and Huber, J.T. (1993). *Hymenoptera of the World: an Identification Guide to Families*. Ottawa: Research Branch, Agriculture Canada.

Grbic, M. and Strand, M.R. (1998). Shifts in the life history of parasitic wasps correlate with pronounced alterations in early development. *Proceedings of the National Academy of Sciences of the USA* **95**, 1097–1101.

Griffiths, G.C.D. (1964). The Alysiinae parasites of the Agromyzidae. I. General questions of taxonomy, biology and evolution. *Beiträge zur Entomologie* **14**, 823–914.

Gross, P. and Price, P.W. (1988). Plant influences on parasitism of two leafminers: a test of enemy-free space. *Ecology* **69**, 1506–1516.

Haeselbarth, E. (1979). Zur Parasitierung der Puppen von Forleule (*Panolis flammea* (Schiff.)), Kiefernspanner (*Bupalus piniarius* (L.)) und Heidelbeerspanner (*Boarmia bistortata* (Goeze)) in bayerischen Kiefernwäldern. *Zeitschrift für angewandte Entomologie* **87**, 186–202.

Handlirsch, A. (1906–8). *Die fossilen Insekten und die Phylogenie der rezenten Formen*. Leipzig: W. Engelmann.

Hanson, P.E. and Gauld, I.D. (1995). The biology of Hymenoptera. In: *The Hymenoptera of Costa Rica* (P. E. Hanson and I. D. Gauld, eds), pp. 20–88. Oxford: Oxford University Press.

Hardy, I.C.W. and Mayhew, P.J. (1998). Sex ratio, sexual dimorphism and mating structure in bethylid wasps. *Behavioral Ecology and Sociobiology* **42**, 383–397.

Harvey, P.H. and Pagel, M.D. (1991). *The Comparative Method in Evolutionary Biology*. Oxford: Oxford University Press.

Hawkins, B.A., Shaw, M.R. and Askew, R.R. (1992). Relations among assemblage size, host specialization, and climatic variability in North American parasitoid communities. *American Naturalist* **139**, 58–79.

Heimpel, G.E. and Collier, T.R. (1996). The evolution of host-feeding behavior in insect parasitoids. *Biological Review* **71**, 373–400.

Herre, E.A., Machado, C.A., Bermingham, E., Nason, J.D., Windsor, D.M., McCafferty, S.S., Houten, W.V. and Bachmann, K. (1996). Molecular phylogenies of figs and their pollinator wasps. *Journal of Biogeography* **23**, 521–530.

Hunter, M.S. and Woolley, J.B. (2001). Evolution and behavioral ecology of heteronomous aphelinid parasitoids. *Annual Review of Entomology* **46**, 251–290.

Infante, F., Hanson, P.E., and Wharton, R. (1995). Phytophagy in the genus *Monitoriella* (Hymenoptera: Braconidae), with a description of a new species. *Annals of the Entomological Society of America* **88**, 406–415.

Iwata, K. (1976) *Evolution of Instinct: Comparative Ethology of the Hymenoptera*. New Delhi: Amerind Publishing.

Janzen, D.H. (1980). When is it coevolution? *Evolution* **34**, 611–612.

Jervis, M.A., Heimpel, G.E., Ferns, P.N., Harvey, J.A. and Kidd, N.A.C. (2001). Life-history strategies in parasitoid wasps: a comparative analysis of 'ovigeny'. *Journal of Animal Ecology* **70**, 442–458.

Johnson, N.F. (1988). Midcoxal articulations and the phylogeny of the Order Hymenoptera. *Annals of the Entomological Society of America* **81**, 870–881.

Jones, D. and Coudron, T. (1993). Venoms of parasitic Hymenoptera as investigatory tools. In: *Parasites and Pathogens of Insects* (N. E. Beckage, S. N. Thompson and B. A. Federici, eds), pp. 227–244. San Diego: Academic Press.

Kellogg, V.L. (1896a). New Mallophaga I, with special reference to a collection from maritime birds of the Bay of Monterey, California. *Proceedings of the California Academy of Sciences, 2nd Series* **6**, 31–168.

Kellogg, V.A. (1896b). New Mallophaga II, from land birds; together with an account of the mallophagous mouthparts. *Proceedings of the California Academy of Sciences, 2nd Series* **6**, 431–548.

Kerdelhue, C., Clainche, I.L. and Rasplus, J.Y. (1999). Molecular phylogeny of the *Ceratosolen* species pollinating *Ficus* of the subgenus *Sycomorus sensu stricto*: biogeographical history and the origins of the species-specificity breakdown cases. *Molecular Phylogenetics and Evolution* **11**, 401–414.

Kinsey, A.C. (1920). Phylogeny of cynipid genera and biological characteristics. *Bulletin of the American Museum of Natural History* **42**, 357–402.

Klassen, G.J. (1992). Coevolution: a history of the macroevolutionary approach to studying host–parasite associations. *Journal of Parasitology* **78**, 573–587.

Kraaijeveld, A.R. and Godfray, H.C.J. (1997). Trade-off between parasitoid resistance and larval competitive ability in *Drosophila melanogaster*. *Nature* **389**, 278–280.

Kraaijeveld, A.R. and Godfray, H.C.J. (1999). Geographic patterns in the evolution of resistance and virulence in *Drosophila* and its parasitoids. *American Naturalist* **153**, S61–S74.

Kraaijeveld, A.R., Alphen, J.J.M.v. and Godfray, H.C.J. (1998). The coevolution of host resistance and parasitoid virulence. *Parasitology* **116**, S29–S45.

Liljeblad, J. and Ronquist, F. (1998). A phylogenetic analysis of higher-level gall wasp relationships (Hymenoptera: Cynipidae). *Systematic Entomology* **23**, 229–252.

Lopez-Vaamonde, C., Rasplus, J.Y., Weiblen, G.D. and Cook, J.M. (2001). Molecular phylogenies of fig wasps: partial cocladogenesis of pollinators and parasites. *Molecular Phylogenetics and Evolution* **21**, 55–71.

Macedo, M.V. de, and Monteiro, R. (1989). Seed predation by a braconid wasp, *Allorhogas* sp. (Hymenoptera). *Journal of the New York Entomological Society* **97**, 358–362.

Machado, C.A., Herre, E.A., McCafferty, S. and Bermingham, E. (1996). Molecular phylogenies of fig pollinating and non-pollinating wasps and the implications for the origin and evolution of the fig–fig wasp mutualism. *Journal of Biogeography* **23**, 531–542.

Machado, C.A., Jousselin, E., Kjellberg, F., Compton, S.G. and Herre, E.A. (2001). Phylogenetic relationships, historical biogeography and character evolution of fig-pollinating wasps. *Proceedings of the Royal Society of London, Series B* **268**, 685–694.

Mackauer, M. (1965). Parasitological data as an aid in aphid classification. *Canadian Entomologist* **97**, 1016–1024.

Maddison, W.P. (1990). A method for testing the correlated evolution of two binary characters: are gains or losses concentrated on certain branches of a phylogenetic tree? *Evolution* **44**, 539–557.

Malyshev, S.I. (1968). *Genesis of the Hymenoptera, and the Phases of their Evolution*. London: Methuen.

Manter, H.W. (1940). The geographical distribution of digenetic trematodes of marine fishes of the tropical American Pacific. *Allan Hancock Pacific Expedition Reports* **2**, 531–547.
Manter, H.W. (1955). The zoogeography of trematodes of marine fishes. *Experimental Parasitology* **4**, 62–86.
Manter, H.W. (1966). Parasites of fishes as biological indicators of recent and ancient conditions. In: *Host–Parasite Relationships* (J. E. McCauley, ed.), pp. 59–71. Corvallis: Oregon State University Press.
Manter, H.W. (1967). Some aspects of the geographical distribution of parasites. *Journal of Parasitology* **53**, 3–9.
Mardulyn, P. and Whitfield, J.B. (1999). Phylogenetic signal in the COI, 16S and 28S genes for inferring relationships among genera of Microgastrinae (Hymenoptera: Braconidae): evidence of a high diversification rate in this group of parasitoids. *Molecular Phylogenetics and Evolution* **12**, 282–294.
Martins, E.P., ed. (1996). *Phylogenies and the Comparative Method in Animal Behavior*. Oxford: Oxford University Press.
Masurier, A.D.L. (1987). A comparative study of the relationship between host size and brood size in *Apanteles* spp. (Hymenoptera: Braconidae). *Ecological Entomology* **12**, 383–393.
Mayhew, P.J. (1998). The evolution of gregariousness in parasitoid wasps. *Proceedings of the Royal Society of London, Series B* **265**, 383–389.
Mayhew, P.J. (2001). Shifts in hexapod diversification and what Haldane could have said. *Proceedings of the Royal Society of London, Series B* **269**, 969–974.
Mayhew, P.J. and Blackburn, T.M. (1999). Does development mode organize life-history traits in the parasitoid Hymenoptera? *Journal of Animal Ecology* **68**, 906–916.
Mayhew, P.J. and Hardy, I.C.W. (1998). Nonsiblicidal behavior and the evolution of clutch size in bethylid wasps. *American Naturalist* **151**, 409–424.
Mayr, E. (1963). *Animal Species and Evolution*. Cambridge, MA: Belknap.
Mitter, C., Farrell, B. and Wiegmann, B. (1988). The phylogenetic study of adaptive zones: has phytophagy promoted insect diversification? *American Naturalist* **132**, 107–128.
Moran, N. (1988). The evolution of host-plant alternation in aphids: evidence for specialization as a dead end. *American Naturalist* **132**, 681–706.
Moran, N. and Baumann, P. (1994). Phylogenetics of cytoplasmically inherited microorganisms of arthropods. *Trends in Ecology and Evolution* **9**, 15–20.
Oeser, R. (1961). Vergleichend-morphologische Untersuchungen über den Ovipositor der Hymenopteren. *Mitteilungen des Zoologisches Museums Berlin* **37**, 1–119.
O'Neill, K.M. (2001). *Solitary Wasps: Behavior and Natural History*. Ithaca: Cornell University Press.
Piek, T. and Spanjer, W. (1986). Chemistry and pharmacology of solitary wasp venoms. In: *Venoms of the Hymenoptera: Biochemical, Pharmacological and Behavioural Aspects* (T. Piek, ed.), pp. 161–307. London: Academic Press.
Price, P.W. (1972). Parasitoids using the same host: adaptive nature of differences in size and form. *Ecology* **53**, 190–195.
Price, P.W. (1973). Reproductive strategies in parasitoid wasps. *American Naturalist* **107**, 684–693.
Price, P.W. (1974). Strategies for egg production. *Evolution* **28**, 76–84.
Price, P.W. (1980). *Evolutionary Biology of Parasites*. Princeton, NJ: Princeton University Press.

Quicke, D.L.J. (1993). The polyphyletic origin of endoparasitism in the cyclostome lineages of Braconidae (Hymenoptera): a reassessment. *Zoologische Mededelingen* **67**, 159–177.
Quicke, D.L.J. (1997). *Parasitic Wasps*. London: Chapman and Hall.
Quicke, D.L.J., LeRalec, A. and Vilhelmsen, L. (1999). Ovipositor structure and function in the parasitic Hymenoptera. *Rendiconti* **47**, 197–239.
Ralec, A.L., Rabasse, J.M. and Wajnberg, E. (1996). Comparative morphology of the ovipositor of some parasitic Hymenoptera in relation to characteristics of their hosts. *Canadian Entomologist* **128**, 413–433.
Rasnitsyn, A.P. (1968). Evolution of the function of the ovipositor in relation to the origin of parasitism in Hymenoptera. *Entomological Review* **47**, 35–40.
Rasnitsyn, A.P. (1975). [Early stages of evolution of Hymenoptera.] *Zoologicheskii Zhurnal* **54**, 848–860. [In Russian.]
Rasnitsyn, A.P. (1980). [Origin and evolution of hymenopterous insects.] *Transactions of the Paleontological Institute of the Academy of the Sciences of the USSR* **174**, 1–192. [In Russian.]
Rasnitsyn, A.P. (1988). An outline of evolution of the hymenopterous insects (Order Vespida). *Oriental Insects* **22**, 115–145.
Rensch, B. (1960). *Evolution Above the Species Level*. New York, NY: Columbia University Press.
Reuter, O.M. (1913). *Lebensgewohnheiten und Instinkte der Insekten*. Berlin: Friendlander.
Ronquist, F. (1994). Evolution of parasitism among closely related species: phylogenetic relationships and the origin of inquilinism in gall wasps (Hymenoptera: Cynipidae). *Evolution* **48**, 241–266.
Ronquist, F. (1995). Phylogeny and early evolution of the Cynipoidea (Hymenoptera). *Systematic Entomology* **20**, 309–335.
Ronquist, F. and Liljeblad, J. (2001). Evolution of the gall wasp–host plant association. *Evolution* **55**, 2503–2522.
Ronquist, F., Rasnitsyn, A.P., Roy, A., Eriksson, K. and Lindgren, M. (1999). Phylogeny of the Hymenoptera: a cladistic reanalysis of Rasnitsyn's (1988) data. *Zoologica Scripta* **28**, 13–50.
Schmidt, O., Theopold, U. and Strand, M.R. (2001). Innate immunity and its evasion and suppression by hymenopteran endoparasitoids. *BioEssays* **23**, 344–351.
Sharkey, M.J. and Roy, A. (2001). Phylogeny of the Hymenoptera: a reanalysis of the Ronquist et al. (1999) reanalysis, emphasizing wing venation and apocritan relationships. *Zoologica Scripta* **31**, 57–66.
Shaw, M.R. (1983). On(e) evolution of endoparasitism: the biology of some genera of Rogadinae (Braconidae). *Contributions of the American Entomological Institute* **20**, 307–328.
Shaw, M.R. (1994). Parasitoid host ranges. In: *Parasitoid Community Ecology* (B. A. Hawkins and W. Sheehan, eds), pp. 111–162. Oxford: Oxford University Press.
Shaw, M.R. and Askew, R.R. (1976). Ichneumonoidea (Hymenoptera) parasitic upon leaf-mining insects of the orders Lepidoptera, Hymenoptera and Coleoptera. *Ecological Entomology* **1**, 127–133.
Shaw, M.R. and Huddleston, T. (1991). *Classification and Biology of Braconid Wasps (Hymenoptera: Braconidae)*. London: Royal Entomological Society.
Shaw, S.R. (1988). Euphorine phylogeny: the evolution of diversity in host-utilization by parasitoid wasps (Hymenoptera: Braconidae). *Ecological Entomology* **13**, 323–335.

Sheehan, W. and Hawkins, B.A. (1991). Attack strategy as an indicator of host range in metopiine and pimpline Ichneumonidae (Hymenoptera). *Ecological Entomology* **16**, 129–131.

Shelby, K.S. and Webb, B.A. (1999). Polydnavirus-mediated suppression of insect immunity. *Journal of Insect Physiology* **45**, 507–514.

Smith, P.T. and Kambhampati, S. (2000). Evolutionary transitions in Aphidiinae (Hymenoptera: Braconidae). In: *Hymenoptera: Evolution, Biodiversity and Biological Control* (A. D. Austin and M. Dowton, eds), pp. 106–113. Collingwoood, Victoria: CSIRO.

Spradbery, J.P. and Kirk, A. (1978). Aspects of ecology of siricid woodwasps (Hymenoptera: Siricidae) in Europe, North Africa and Turkey with special reference to the biological control of *Sirex noctilio* F. in Australia. *Bulletin of Entomological Research* **68**, 341–359.

Stary, P. (1964). Food specificity in the Aphidiidae. *Entomophaga* **9**, 91–99.

Stary, P. (1970). *Biology of Aphid Parasites (Hymenoptera: Aphidiidae) with Respect to Integrated Control*. The Hague: Junk.

Stary, P. (1981). On the strategy, tactics and trends of host specificity evolution in aphid parasitoids (Hymenoptera: Aphidiidae). *Acta Entomologica Bohemoslava* **78**, 65–75.

Steiner, A.L. (1986). Stinging behaviour of solitary wasps. In: *Venoms of the Hymenoptera: Biochemical, Pharmacological and Behavioural Aspects* (T. Piek, ed.), pp. 63–160. London: Academic Press.

Stoltz, D.B. (1986). Interactions between parasitoid-derived products and host insects: an overview. *Journal of Insect Physiology* **32**, 347–350.

Stoltz, D.B. (1990). Evidence for chromosomal transmission of polydnavirus DNA. *Journal of General Virology* **71**, 1051–1056.

Stoltz, D.B. and Vinson, S.B. (1977). Baculovirus-like particles in the reproductive tracts of female parasitoid wasps: II. The genus *Apanteles*. *Canadian Journal of Microbiology* **22**, 1013–1023.

Stoltz, D.B. and Vinson, S.B. (1979). Viruses and parasitism in insects. *Advances in Virus Research* **24**, 125–171.

Stoltz, D.B. and Whitfield, J.B. (1992). Viruses and virus-like entities in the Parasitic Hymenoptera. *Journal of Hymenoptera Research* **1**(1), 125–139.

Stoltz, D.B., Guzo, D. and Cook, D. (1986). Studies on polydnavirus transmission. *Virology* **155**, 120–131.

Strand, M.R. and Grbic, M. (1997). The development and evolution of polyembryonic insects. *Current Topics in Developmental Biology* **35**, 121–160.

Thompson, J.N. (1989). Concepts of coevolution. *Trends in Ecology and Evolution* **4**, 179–183.

Vilhelmsen, L. (1996). The preoral cavity of lower Hymenoptera (Insecta): comparative morphology and phylogenetic significance. *Zoologica Scripta* **25**, 143–170.

Vilhelmsen, L. (1997). The phylogeny of lower Hymenoptera (Insecta), with a summary of the early evolutionary history of the order. *Zeitschrift für zoologische Systematik und Evolutionsforschung* **35**, 49–70.

Vilhelmsen, L. (2000a). Before the wasp-waist: comparative anatomy and phylogenetic implications of the skeleto-musculature of the thoraco-abdominal boundary region in basal Hymenoptera (Insecta). *Zoomorphology* **119**, 185–221.

Vilhelmsen, L. (2000b). Cervical and prothoracic skeleto-musculature in the basal Hymenoptera (Insecta): comparative anatomy and phylogenetic implications. *Zoologische Anzeiger* **239**, 103–136.

Vilhelmsen, L. (2000c). The ovipositor apparatus of basal Hymenoptera (Insecta): phylogenetic implications and functional morphology. *Zoologica Scripta* **29**, 319–345.

Vilhelmsen, L. (2001). Phylogeny and classification of the extant basal lineages of the Hymenoptera (Insecta). *Zoological Journal of the Linnean Society* **131**, 393–442.

Vilhelmsen, L., Isadoro, N., Romani, R., Basibuyuk, H. and Quicke, D.L.J. (2001). Host location and oviposition in a basal group of parasitic wasps: the subgenual organ, ovipositor apparatus and associated structures in the Orussidae (Hymenoptera, Insecta). *Zoomorphology* **121**, 63–84.

Weiblen, G.D and Bush, G.L. (2002). Speciation in fig pollinators and parasites. *Molecular Ecology* **11**, 1573–1578.

West, S.A. and Sheldon, B.C. (2002). Constraints in the evolution of sex ratio adjustment. *Science* **295**, 1685–1688.

Whitfield, J.B. (1990). Parasitoids, polydnaviruses and endosymbiosis. *Parasitology Today* **6**, 381–384.

Whitfield, J.B. (1992). The polyphyletic origin of endoparasitism in the cyclostome lineages of Braconidae (Hymenoptera). *Systematic Entomology* **17**, 273–286.

Whitfield, J.B. (1997). Molecular and morphological data suggest a single origin of the polydnaviruses among braconid wasps. *Naturwissenschaften* **84**, 502–507.

Whitfield, J.B. (1998). Phylogeny and evolution of host–parasitoid interactions in Hymenoptera. *Annual Review of Entomology* **43**, 129–151.

Whitfield, J.B. (2000). Phylogeny of microgastroid braconid wasps, and what it tells us about polydnavirus evolution. In: *Hymenoptera: Evolution, Biodiversity and Biological Control* (A. D. Austin and M. Dowton, ed.). pp. 97–105. Collingwood, Victoria: CSIRO.

Whitfield, J.B. (2002). Estimating the age of the polydnavirus/braconid wasp symbiosis. *Proceedings of the National Academy of Sciences of the USA* **99**, 7508–7513.

Whitfield, J.B. and Asgari, S. (2003). Virus or not?—phylogenetics of polydnaviruses and their wasp carriers. *Journal of Insect Physiology* **49**, 397–405.

Whitfield, J.B. and Mason, W.R.M. (1994). Mendesellinae, a new subfamily of braconid wasps (Hymenoptera, Braconidae), with a review of relationships within the microgastroid assemblage. *Systematic Entomology* **19**, 61–76.

Whitfield, J.B. and Wagner, D.L. (1988). Patterns in host ranges within the nearctic species of the parasitoid genus *Pholetesor* Mason (Hymenoptera: Braconidae). *Environmental Entomology* **17**, 608–615.

Whitfield, J.B., Johnson, N.F. and Hamerski, M.R. (1989). Identity and phylogenetic significance of the metapostnotum in nonaculeate Hymenoptera. *Annals of the Entomological Society of America* **82**, 663–673.

Wiegmann, B.M., Mitter, C. and Farrell, B. (1993). Diversification of carnivorous parasitic insects: extraordinary radiation or specialized dead end. *American Naturalist* **142**, 737–754.

Wilson, E.O. (1987). The little things that run the world. *Conservation Biology* **1**, 344–346.

Xu, D. and Stoltz, D.B. (1991). Evidence for a chromosomal location of polydnavirus DNA in the ichneumonid parasitoid, *Hyposoter fugitivus*. *Journal of Virology* **65**, 6693–6704.

Yeo, P.F. and Corbet, S.A. (1983). *Solitary Wasps. Naturalists' Handbooks no. 3.* Cambridge: Cambridge University Press.

Nematoda: Genes, Genomes and the Evolution of Parasitism

Mark L. Blaxter

Institute of Cell, Animal and Population Biology, University of Edinburgh, King's Buildings, Edinburgh, EH9 3JT, UK

Abstract	102
1. Nematode Genomes and the Evolution of Parasitism	103
1.1. What are nematodes? The phylum Nematoda in the kingdom of animals	103
1.2. Nematode phylogeny	106
1.2.1. Molecular analyses of nematode diversity and phylogeny	107
1.2.2. Completing the puzzle: limitations of the current molecular phylogeny of Nematoda and routes to overcoming them	108
2. Nematode Parasitism	111
2.1. What is parasitism? parasites as small, non-mutualist symbionts	111
2.2. The spectrum of nematode exploitative symbioses: association to parasitism	112
2.3. Phylogenetic placement of nematode parasites	116
2.3.1. Clade I: Dorylaimia	117
2.3.2. Clade II: Enoplia	117
2.3.3. Clade C&S (III, IV and V): Chromadoria	118
2.4. Patterns in nematode parasitism of animals	119
2.4.1. The L3 rule: the Stage that Invades the Definitive Host is the Third Stage Larva	119
2.4.2. Is there a Link Between Arthropod Association and Vertebrate Parasitism?	120
2.4.3. Life Cycle Evolution in the Strongyloidoidea	122
2.4.4. Coevolution	124
3. Nematode Genomes and Parasitism	126
3.1. Genome evolution in nematodes	129
3.1.1. The Genome of *Caenorhabditis elegans*	132

ADVANCES IN PARASITOLOGY VOL 54
0065-308X $35.00

Copyright © 2003 Elsevier Ltd
All rights of reproduction in any form reserved

3.1.2. Patterns of Genome Evolution in *Caenorhabditis* Species . 137
3.1.3. Patterns of Genomic Evolution Across the Nematoda 138
3.1.4. An Example of Rapid Genomic Change:
the Evolution of the Nematode HOX Gene Cluster 143
3.1.5. Nematode Mitochondria 144
3.1.6. Nematode Genome Evolution and Parasitism 148
3.2. ORFeome/proteome evolution and the
parasitic phenotype... 148
3.2.1. How Many Genes do Nematodes Have? 149
3.2.2. The 'Parasitome': Genes and Proteins Implicated in the
Parasitic Phenotype.. 156
3.2.3. Arms Races Between Hosts and Parasites 161
4. Summary 165
Acknowledgements 167
References 168

ABSTRACT

Nematodes are remarkably successful, both as free-living organisms and as parasites. The diversity of parasitic lifestyles displayed by nematodes, and the diversity of hosts used, reflects both a propensity towards parasitism in the phylum, and an adaptability to new and challenging environments. Parasitism of plants and animals has evolved many times independently within the Nematoda. Analysis of these origins of parasitism using a molecular phylogeny highlights the diversity underlying the parasitic mode of life. Many vertebrate parasites have arthropod-associated sister taxa, and most invade their hosts as third stage larvae: these features co-occur across the tree and thus suggest that this may have been a shared route to parasitism. Analysis of nematode genes and genomes has been greatly facilitated by the *Caenorhabditis elegans* project. However, the availability of the whole genome sequence from this free-living rhabditid does not simply permit definition of 'parasitism' genes; each nematode genome is a mosaic of conserved features and evolutionary novelties. The rapid progress of parasitic nematode genome projects focussing on species from across the diversity of the phylum has defined sets of genes that have patterns of evolution that suggest their involvement with various facets of parasitism, in particular the problems of acquisition of nutrients in new hosts and the evasion of host immune defences. With the advent of functional genomics techniques in parasites, and in particular the possibility of gene knockout using RNA interference, the roles of many putative parasitism genes can now be tested.

1. NEMATODE GENOMES AND THE EVOLUTION OF PARASITISM

The success of the phylum Nematoda as parasites of other animals and plants is striking. Parasitic species play a significant role in regulating the productivity of wild populations, and impact negatively on human agricultural endeavour as well as human health. The emergence of robust phylogenetic analyses of parasitic nematodes in relation to their hosts (Hugot et al., 1996; Hugot, 1999; Dorris et al., 2002), and of extensive genomic information on free-living and parasitic species (Consortium, 1998; Blaxter et al., 1999), permits for the first time an attempt at elucidating the predispositions and adaptations of nematodes as parasites, and the mode and tempo of evolution of a fascinating and crucially important part of planetary biological diversity. In this essay, I have summarised my understanding of the current status of nematode phylogenetics and genomics, attempting a synthesis that reveals the particular biology that underlies diverse parasitic phenotypes.

1.1. What are nematodes? The phylum Nematoda in the kingdom of animals

Nematodes are abundant, diverse and ubiquitous members of the meiofauna. They are found in astronomical numbers in all sorts of sediments (Platonova and Gal'tsova, 1976), where they play core ecological roles in nutrient recycling and link the producer/saprophytic levels of food nets with the upper, predatory guilds (Bongers, 1990; Ekschmitt et al., 2001). Nematodes have also evolved to be parasitic a number of times (Blaxter et al., 1998; Dorris et al., 2002), and have successfully adapted to use other non-vertebrates, vertebrates and plants as hosts. Most vertebrate species, especially terrestrial ones, have an associated nematode parasite fauna (Bundy, 1997).

The nematodes are ranked as a phylum (sometimes a class) Nematoda, and traditionally were placed within an unresolved assemblage of lower invertebrates, the 'Aschelminthes' (Brusca and Brusca, 1990). Nematodes lack a true coelom, and are not visibly segmented. In addition, their development, particularly in groups such as the rhabditid terrestrial bacteriovores (Sulston et al., 1983; Sulston et al., 1988) and ascaridid gut parasites (Strassen, 1896), was not simply classifiable as either protostome or deuterostome, or radial or spiral. The Aschelminthes hypothesis suggests that nematodes are a representative of a radiation from a pre-protostome-deuterostome split ancestor, and thus can be considered primitive (see Brusca and Brusca, 1990, for a 'classical' phylogenetic treatment). Evidence adduced to this view included the (usually) small size of nematodes, and the presence of eutely. However, these phenotypes are commonly associated with animals that inhabit the water film, or live within sediments, and may be a secondary adaptation to size reduction and rapid development (Nelson, 1982; Kinchin, 1994; Goldstein and Blaxter, 2002).

More recently, cladistic analyses of morphological datasets from metazoan phyla have prompted a reappraisal of the position of Nematoda (Nielsen, 1995). More significantly, the advent of molecular phylogenetic methods for the analysis of deep divergence has suggested that nematodes are not a basally derived metazoan, but a crown member of a new superphylum, Ecdysozoa, or the moulting animals (Aguinaldo et al., 1997). The Ecdysozoa hypothesis is a radical reworking of the structure of the Metazoa, and has been the subject of much debate: traditional morphological zoology and some molecular phylogeneticists find it unpalatable (Giribet and Wheeler, 1999; Wägele et al., 1999; Giribet et al., 2000; Nielsen, 2001; Blair et al., 2002; Hedges, 2002), while molecular biology and genetics has taken it to heart (Mushegian et al., 1998; Ruvkun and Hobert, 1998a; de Rosa et al., 1999; Haase et al., 2001). On balance, I conclude that the molecular evidence is compelling, though not without its problems. Under this new view of metazoan evolution (Figure 1), Nematoda are associated with other phyla, often obscure and less-speciose, including the Nematomorpha (hair worms), Priapulida, Kinorhyncha (mud dragons), Gnathostomulida, Loricifera, Tardigrada (water bears) and Onychophora (velvet worms) but also the well-known and speciose Arthropoda. The interrelationships of the Ecdysozoa have still to be firmly resolved (Peterson and Eernisse, 2001), but it is generally agreed that Nematomorpha are the sister taxon to Nematoda, with either Priapulida or a clade of (Priapulida + Kinorhyncha + Loricifera + Gnathostomulida) as sister taxon to (Nematoda + Nematomorpha). The chaetognaths (phylum Chaetognatha) have also been placed in Ecdysozoa, though the molecular evidence for this is qualified by methodological uncertainties (Telford and Holland, 1997; Giribet et al., 2000; Peterson and Eernisse, 2001).

With this new view of metazoan life, some of the phenotypes of nematodes can be reassessed. The arthropods, onychophorans and tardigrades are each segmented, and have paired ventral nerve cords. Nematodes do not display evident segmentation, but serially repeated patterns of cell division revealed by embryonic cell lineaging in *Caenorhabditis elegans* are suggestive

GENES, GENOMES AND PARASITISM IN THE NEMATODA

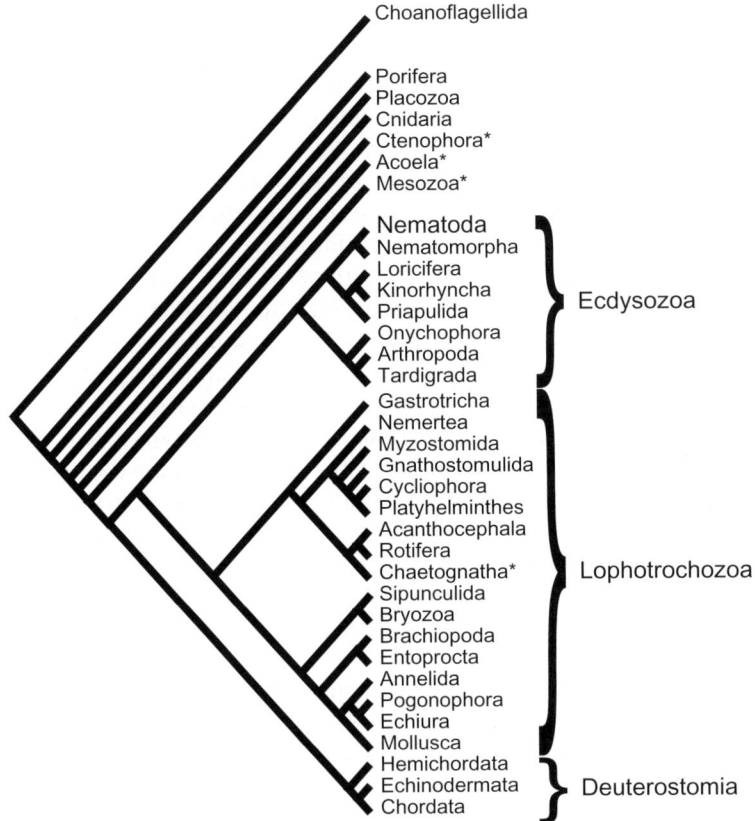

Figure 1 Nematoda in the tree of metazoan life. Interrelationships of the living phyla of Metazoa (based on the works of Nielsen, 1995; Winnepenninckx *et al.*, 1995; Aguinaldo *et al.*, 1997; Giribet *et al.*, 2000; Peterson and Eernisse, 2001). The placement (and status) of some phyla (marked with asterisks) is still the subject of much debate. Note also that the cladogenesis events leading to the living phyla are believed to have occurred in the Precambrian, and possibly in a very short time period (geologically speaking) and thus this cladogram may suggest greater resolution of ancestral divergences than is the case.

of some degree of anterior–posterior metamerism (Sulston and Horvitz, 1977; Sulston *et al.*, 1980). The ventral nerve cord in nematodes is always single, but again cell lineaging has revealed near-symmetrical left-right developmental contributions which could be a reduced remnant of paired cords. The nematode pharynx is a triradiate, myoepithelial pump, and similar structures are found in both tardigrades and onychophorans (phyla with obvious arthropod affinity) as well as other ecdysozoan phyla (Garey and Schmidt-Rhaesa, 1998). While the triradiate structure of such a pump is one that is optimal in terms of mechanics, the use of myoepithelial cells, rather than a lining epithelium acted upon by distinct, mesodermal muscles, is an ecdysozoan peculiarity.

The significance of the new view of the position of nematodes for parasitology is that the old adjective (and sometimes taxonomic group) 'helminths' is no longer valid (Winnepenninckx *et al.*, 1995). Rather than discussing nematode and flatworm parasites as if they were phylogenetically related, we should view them as independent adaptations to the same sorts of environment, much as marsupial and eutherian 'wolves' are unrelated as wolves, and the various reef communities evident through geological time are similar in gross structure but were formed by very different animals. Thus, if a platyhelminth and a nematode parasite display similar phenotypes in the face of a particular host and its immunity, this should tell us more about the host's immunity than any anciently shared biology of the parasites.

1.2. Nematode phylogeny

Recognition of the phylum status of Nematoda, and the probability that nematodes have had a separate evolutionary trajectory for over 550 million years (Adoutte *et al.*, 1999, 2000), raises a problem. Nematodes are notoriously lacking in morphology. This is not to say that they have no morphology; indeed, the trained nematologist can demonstrate all the diverse setae, annulae, papillae and teeth any morphologist might desire to see. It is rather that, as small, interstitial animals with a locomotory hydroskeleton, they have in general been constrained to a simple, vermiform appearance. The mode of locomotion also probably precluded appendage evolution (or even promoted appendage loss), and also reduced cephalisation. Thus the legs and bristles so useful to arthropod taxonomy are missing (though external genitalia, in the form of copulatory bursae and peri-genital sense organs are abundantly informative (Fitch, 1997, 2000)). So is the current nematode fauna the result of recent expansion of a relict or degenerate form, or is it a representation of a deep evolutionary divergence?

There are no informative nematode fossils (Conway Morris, 1981, 1993), and the size of the majority of species (and the even smaller size and fragility of diagnostic characters) suggests that in the absence of novel deposits or methods it is unlikely that any resolution of these questions will come from micropalaeontology. While stunning embryological fossils are being reported from some phosphatic deposits of lower Ordovician and

Cambrian age (Chen *et al.*, 2000), these are generally thought to be of taxa which used the water column to disperse propagules. Nematodes are conspicuously absent, as adults or larvae, from the zooplankton, and thus even these incredible windows on the palaeontology of the evolution of development are unlikely to open on nematode ancestry.

In addition, there is a daunting number of nematode taxa to analyse and understand. About 25 000 different species have been described, and a large proportion of these are parasitic (De Ley and Blaxter, 2002; de Meeus and Renaud, 2002). Estimates of the total number of nematode species on the planet vary from 100 000 to 100 000 000 (Lambshead, 1993), though the upper estimates may be too high (Lambshead *et al.*, 2003). If the ratio of parasitic to free-living taxa was maintained in this diversity, the problem for parasitology becomes mind-boggling. However, the majority of undescribed taxa are estimated to be meiobenthic marine forms (Lambshead, 1993). The actual total number of parasitic nematode species is likely to be several-fold higher than is currently known, but unlikely to approach the several million suggested by scaling current data.

1.2.1. Molecular analyses of nematode diversity and phylogeny

Morphology is continually selected and channelled into forms that permit successful reproduction, and the little stochastic leeway that is available is hard to quantify. The genome however is a mixture of stasis (conserved genes with essential function) and clock-like change (stochastic mutational events and fixation). Even within conserved genes, some residues (for example the third bases of codons encoding amino acids which can be encoded by four related codons; four-fold redundant sites) are free to vary. Thus while nematode bodies may have limited superficial diversity, nematode genomes will carry a history of the evolution of the phylum. Just as molecular phylogenetics has permitted an integrated and testable hypothesis of metazoan evolution (Peterson and Eernisse, 2001), the application of molecular methods to the Nematoda has yielded a remarkably stable, new phylogeny (Blaxter *et al.*, 1998; De Ley and Blaxter, 2002).

The gene selected for analysis of nematode phylogeny has in the main been that encoding small subunit ribosomal RNA gene (SSU rRNA), but other analyses have used globin genes (Vanfleteren *et al.*, 1994), mitochondrial genes (Thomas and Wilson, 1991), segments of the large subunit ribosomal RNA gene (Thomas *et al.*, 1997), partial elongation factor 1 alpha genes (Sidow and Thomas, 1994), the internal transcribed spacer regions of the ribosomal RNA repeats (Gasser and Newton, 2000) and other RNA-encoding genes (Xie *et al.*, 1994b). The SSU rRNA genes have proved most useful, as they are relatively easy to isolate, even from single specimens of sub-millimetre sized nematodes (Floyd et al., 2002), they are reliably orthologous between species, and a robust set of models of their evolution has been developed for nematodes and other taxa. They may not be ideal, as there are issues with differences in evolutionary rates of change between higher taxa, the reliability of alignment between distantly related taxa, and in the effects of overall genome base composition on the patterns of substitution in SSU rRNA genes, but they currently represent the best marker available (see Swofford et al., 1996; Peterson and Eernisse, 2001). An important additional criterion is that sequences are available for phyla that are credible outgroups for the Nematoda.

An initial analysis of nematode SSU rRNA genes (based on 53 sequences from mostly parasitic taxa) (Blaxter et al., 1998) has been confirmed and extended by subsequent analyses (Aleshin et al., 1998a, b; Kampfer et al., 1998; Dorris et al., 1999; De Ley and Blaxter, 2002). The current dataset of nematode SSU rRNA gene sequences is over 300 taxa, with many more free-living taxa represented, but the overall structure of the tree has not changed significantly since first publication. In addition, analyses of other genes are in agreement with the SSU rRNA tree where taxon datasets overlap (Vanfleteren et al., 1994; Burr et al., 2000). The phylogeny derived from the SSU rDNA appears to be robust in that most clades are supported by high bootstrap support and/or posterior probabilities.

The Nematoda can be split into three major clades, called Clades I, II and C&S in the original publication (Blaxter et al., 1998), but now given the names Dorylaimia, Enoplia and Chromadoria, respectively (Figure 2) (De Ley and Blaxter, 2002). The relative branching order between the base of the Nematoda and these clades is as yet unresolved: some datasets yielding a basal rooting of Enoplia while others suggest Chromadoria separated first. All three groups are biologically diverse, and include marine and terrestrial forms. Parasitic taxa are also found in all three.

1.2.2. Completing the puzzle: limitations of the current molecular phylogeny of Nematoda and routes to overcoming them

The global initiative to sequence SSU rRNA genes from nematodes has yielded very positive results, and new sequences are still being gathered. A recent survey suggests that sequences of taxa deriving from nearly half the described families of nematodes have been determined, though many are yet to be published or deposited in public databases (De Ley and Blaxter, 2002). Against this success must be placed the known severe undersampling

GENES, GENOMES AND PARASITISM IN THE NEMATODA

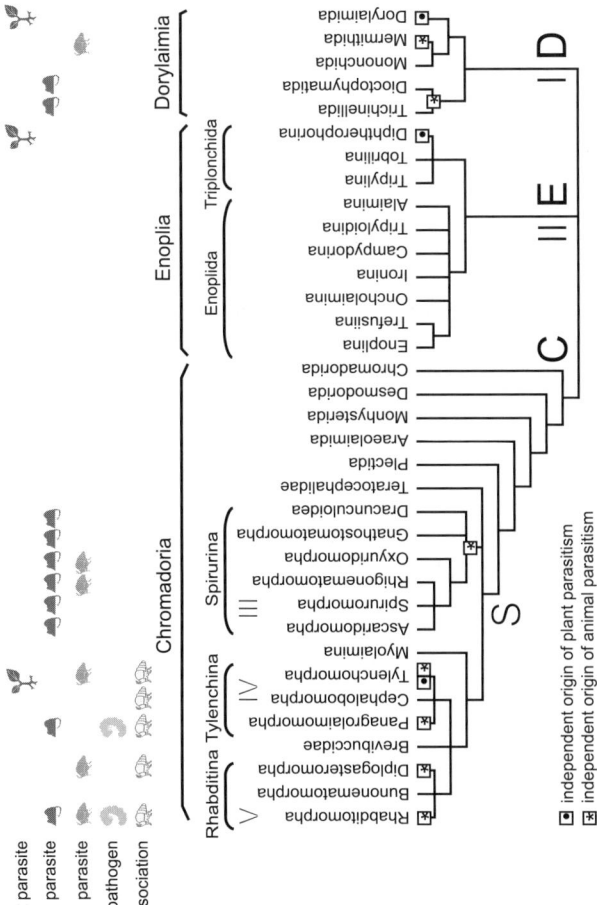

Figure 2 The new nematode phylogeny. The tree in this figure is derived from an analysis of over 300 nematode SSU sequences, complemented by a consideration of morphological characters. It is an attempt to put into a traditional, ranked classification the cladistic findings of molecular studies (De Ley and Blaxter, 2002). To retain relative ranks, the status of some taxa had to be changed, in particular the "downgrading" of several taxa of parasitic groups previously afforded ordinal status. Intellectual credit for this reconciliation with traditional naming is primarily due to my collaborator, Dr. P. De Ley. The trophic lifestyle of each taxon is given above and inferred independent origins of parasitic phenotypes marked on the tree. The major clades of nematodes, **C** (Chromadoria, called C&S in Blaxter *et al*., 1998 indicating a fusion of Chromadorida and Secernentea), **E** (Enoplia, Clade II of Blaxter *et al*. (1998)) and **D** (Dorylaimia, Clade I) are marked, as are clades III (Spirurina), IV (Tylenchina) and V (Rhabditina).

of diversity that described nematode species represents. It is difficult to estimate what grade of deficit is present. In bacterial community surveys using ribosomal RNA sequencing, the shocking discovery was that there are many bacterial groups (of the equivalent status to animal phyla) that were unknown to culturing-based bacteriological taxonomy (Pace, 1997). A similar conclusion has been derived from molecular surveys of protist communities (Lopez-Garcia et al., 2001; Moon-van der Staay et al., 2001; Amaral Zettler et al., 2002; Moreira and Lopez-Garcia, 2002). In contrast, while current morphological surveys have only covered a minute portion of the earth's surface, traditional methods for isolation and identification of nematodes are unlikely to have systematically missed major groups as most major ecosystems have been sampled (Lawton et al., 1996; Bloemers et al., 1997; Floyd et al., 2002; Blaxter, 2003). Thus, I would expect that the future of nematode systematics will be in filling in the current tree (with perhaps some groups at family rank still to be discovered) rather than in discovering large branches unsuspected by previous authors. Equally, this view may be insufficiently nematophilic.

Several parasitic groups are significantly underrepresented in current public datasets compared with their known diversity. For example, the Oxyuridomorpha (pinworms and relatives) are represented by a single sequence, despite their common occurrence as host-specific parasites of mammals and insects. Other groups, particularly those parasitising non-vertebrate animals, are also lacking. There are very few parasites of marine non-vertebrate hosts described, and none sampled. Often, the community has sequenced from parasites with significant economic impact on human, animal or plant health, sacrificing breadth of coverage for depth of resolution of a particular group (Xie et al., 1994a, b; Dorris and Blaxter, 2000; Dorris et al., 2002).

The advent of significantly cheaper sequencing technology, the promulgation of basic molecular biology skills to additional nematode systematics laboratories, and the development of techniques for rapid and reliable gene isolation using the polymerase chain reaction, even from fixed or archive specimens (Thomas et al., 1997; Herniou et al., 1998; Dorris and Blaxter, 2000; Dorris et al., 2002), will undoubtedly facilitate advances in coverage, even from the most obscure groups. The SSU rRNA gene has proved useful for most genus-level discriminations, and very useful for deeper analyses. However, other genes will need to be sampled and tested for resolution of the histories of species flocks, and other groups where very little SSU sequence change separates taxa (Powers et al., 1997; Adams et al., 1998). Similarly, the SSU rRNA gene may not retain enough phylogenetic signal to resolve the current trichotomy at the base of the Nematoda and other sequences, perhaps those encoding slowly evolving protein-coding genes will have to be investigated for this problem (Mushegian et al., 1998; Blair et al., 2002).

2. NEMATODE PARASITISM

In some older systematic treatments of the phylum Nematoda, parasites were grouped together as a monophyletic clade sharing a particular trophic mode. This *a priori* classificatory act made it relatively difficult to discuss the origins of and current relationships of the parasitic taxa to each other. The new molecular phylogeny intersperses free-living and parasitic taxa, suggesting multiple events of acquisition of the parasitic mode of life (see below) (Dorris et al., 1999). This in turn raises questions of what constitutes parasitism.

2.1. What is parasitism? parasites as small, non-mutualist symbionts

Simply put, symbiosis means living together, and most life on this planet could be classified as symbiotic. However, it seems incongruent that the cow browsing on grass should be thought as living in symbiosis with the grass: surely it is a herbivore? Discussion of symbiosis also distinguishes between mutualistic versus exploitative interactions, and obligate versus temporary or opportunistic interactions. Both these axes grade across the range of extremes. Mutualistic, obligate symbionts are always found in association with each other, and both symbionts derive fitness advantages from the interaction. Thus, aphids carry bacteria, *Buchnera* species, that assist by providing essential amino acid synthesis to supplement the aphid's poor sap diet. In return, the bacteria are protected from the vagaries of a hostile environment, and are passed from mother to daughter aphid (Baumann et al., 1998). Some symbioses are probably mutualistic, but not obligatory for one or both partners. For example, the acoel flatworm *Convoluta roscoffensis* carries photosynthetic algae. *C. roscoffensis* locomotes so as to ensure the algae are in the sunlight, and the algae provide the products of photosynthesis to the flatworm that does not feed otherwise. However, while *C. roscoffensis* cannot survive without the alga (*C. roscoffensis* is an obligate symbiont), the alga can survive without the flatworm, and is widely distributed and is hence a temporary or opportunistic symbiont (Keeble, 1910).

Exploitative symbioses are usually described as parasitic, though they include types of commensalism and association (where only one partner gains any benefit from the interaction, but the other does not necessarily suffer any cost). These symbioses are usually between organisms of similar size, or organisms where the larger one (the host) is exploited by the smaller (the parasite). When the organism suffering through the interaction is smaller than the exploiter, the interaction becomes more akin to a prey–predator relationship, even if the predator does not kill the prey while exploiting it. Thus, most herbivores prey on plants, but tend not to kill them. Because nematodes are small animals, and probably were ancestrally small animals, their relationships with other multicellular organisms (fungi, plants and animals) are more likely to be classified as parasitic than predatory or herbivorous.

Parasitism is usually taken to indicate an association where the parasitic partner harms but does not kill its host: parasites that kill hosts are sometimes called 'poorly adapted'. While it is true that a parasite in the wrong host can be more virulent than in its usual host, it is important to realise that relative pathogenicity and virulence of parasites will be an adaptive trait, selected depending on the biology of the system (Anderson and May, 1982). For example, it might make adaptive sense for a parasite to kill or disable its host if it is transmitted by cannibalism, as appears to be the case with *Trichinella spiralis* in mice. Similarly, causing pathology in a host, such as lethargy and fever, might aid in transmission of a species using a biting arthropod as a vector. Thus, we should expect to find all grades of effects on hosts, from apparent indifference to severe, even lethal pathology, depending on the life cycle dynamics of the parasite. Many associations between organisms are temporary and for the purposes of transport or protection of the smaller partner only. The cost to the larger is not that it loses nutrition to the smaller (i.e. it is not eaten) but that it expends energy transporting the smaller around. If the smaller subsequently invades and exploits the food source of the larger partner, the association may start to have significant costs for the exploited.

From the above discussion, I hope it is clear that I consider that the origins of parasitism, while based in free-living microbivory, must also be traced through associative, more-or-less exploitative interactions, and may include considerations of mutualistic symbiosis.

2.2. The spectrum of nematode exploitative symbioses: association to parasitism

Finding new food sources is a trial for any animal, but particularly acute for microfauna with no access to long-distance means of dispersal. Many nematodes use other animals, commonly arthropods and molluscs, as transport hosts. Thus, microbivorous nematodes exploiting mammal dung often form close associations with one or a few species of insects (dung flies and dung beetles) which transport them between food foci (Kühne, 1996). The association is specific and highly evolved, as the nematodes display developmental and behavioural adaptations to acquiring passage, including nictation on the substrate surface, and secretion of adhesive structures to cement their relationships (Bovien, 1937). Nematodes often lodge in specific locations on the host, for example under the elytra, or just within the cloaca. Even the archetypical free-living nematode, the much-studied *Caenorhabditis elegans*, may have specific interactions with transport hosts. The congeneric *Caenorhabditis remanei vulgaris* was first isolated from pulmonate gastropods and terrestrial isopods (Baird *et al.*, 1994).

No cost to the transporter has been measured in most cases. However, nematode associates can be costly for burying beetles exploiting the highly dispersed resource of small mammal carcasses (Richter, 1993). The nematode associates of burying beetles have a life cycle closely mapped onto that of their host, and emerge from their transport host (wherein they have been diapaused within the diverticula of the gut) when she lays eggs on a recently buried carcass. As the carcass rots, the nematodes multiply on the bacterial bloom. The beetle larvae also feed on the carcass, which is carefully guarded by the parent beetles, and thus nematodes and larvae are in direct food competition. As the carcass and bacterial flora is exhausted, and the larvae prepare to enter pupation, the nematodes invade through the rectum and diapause in the gut. They are thus carried by the larva into the soil, survive within the pupa, and are carried to the new food source within the next generation adult. In captive colonies of beetles, the nematodes can have a significant fitness effect, but whether this is also true in the wild is unresolved (Richter, 1993).

Some nematode associates of invertebrates are described as necromenic: that is they find a suitable 'host', which they either attach to or invade, and diapause. When the host dies, the nematode feasts on the decaying carcass. While this biology has frequently been described in the laboratory for hosts that die or are killed, and left to rot, it is not clear whether this is truly a life cycle strategy for the nematode species involved. A diapaused microbivore on a dead host will exit from diapause and feed on local food sources, just as it would exit if the host, still living, brought it to a food source. Perhaps this should usually be called opportunistic necromeny until proven otherwise.

Importantly, some nematodes are truly necromenic, as they hasten their host's death. The insect-pathogenic nematodes of the genera *Heterorhabditis*

Table 1. Nematode bacterial symbionts.

Nematode taxon	Nematode phylogenetic position*	Bacterial taxon	Bacterial phylogenetic position	Symbiosis type
Heterorhabditis	Rhabditomorpha	*Photorhabdus*	gamma proteobacteria	extracellular symbiont
Steinernema	Panagrolaimomorpha	*Xenorhabdus*	gamma proteobacteria	extracellular symbiont
several species	Tylenchomorpha	not known		intracellular symbiont
several species in the Onchocercinae	Spiruromorpha	*Wolbachia*	alpha proteobacteria	intracellular symbiont
Astomonema	Monhysterida	not known (more than one type)		extracellular (gut) symbionts
several species in the Stilbonematidae	Desmodorida	several different species	gamma proteobacteria	extracellular (cuticle) symbionts
Xiphinema	Dorylaimida	a verrucomicrobe	gamma proteobacteria	intracellular symbiont

* see Figure 2 for nematode phylogeny.

and *Steinernema* are phylogenetically very distant (see below) but share a very specific adaptation to obtain food (Gaugler and Kaya, 1990; Nealson, 1991). Both genera live in symbiosis with specific bacteria (a different bacterium in each case: see Table 1) carried by the invasive stage of the nematodes (third stage larvae or L3). These bacteria are lethal to insects, as they secrete a battery of toxins, and are also very efficient at maintaining single-isolate, 'clonal' cultures, as they also secrete effective wide-spectrum antibiotics (ffrench-Constant and Bowen, 1999; ffrench-Constant *et al.*, 2000). The L3 seek out insect larvae in the soil (by either active search or ambush strategies), and invade through the cuticle or through cuticular openings. Once inside the haemolymph, the nematode releases the bacterium, which kills the insect and multiplies rapidly. The nematode feeds on the pure culture of bacteria, reproduces, and the new L3 collect some of the bacteria to take with them into the soil, to look for new prey. This relationship between nematode and insect larva could be regarded as predatory, or as highly pathogenic parasitism. Some plant-root pathogens may similarly be associated with plant pathogenic bacteria and fungi: rather than feeding on the plant itself they exploit and promote the work of other pathogens.

Some species, such as *Pristionchus lheritieri*, have been described as facultative parasites, able to invade and kill an insect host when one is available, but also able to propagate on bacterial food alone (Geraert *et al.*, 1989). Some reports of facultative parasitism should perhaps be regarded with some caution, as often the 'parasite' becomes 'adapted' to free-living conditions and loses the ability to 'parasitise' with passage. While this could be a real biological phenomenon, it is possible that the initial 'parasitic' association was rather the use of a dead or severely compromised 'host' as a rich bacterial food source, and carry-over of pathogenic bacteria in the first rounds of culture. An important step in studies of these animals must be proof of infectivity and completion of the life cycle (essentially satisfying Koch's postulates).

Parasites are commonly divided into ectoparasites and endoparasites. For plant-attacking nematodes this division is relatively robust. While some plant parasitic nematodes are sedentary endoparasites, exploiting one or a very few local feeding sites, others behave more as browsers. The browsing nematodes, including those feeding outside the plant and those burrowing within without setting up a specialised feeding site, could perhaps be better described as herbivores: they have a similar relationship to the plant as do large herbivores but are just orders of magnitude smaller (Maggenti, 1981; Dropkin, 1989).

For parasites of metazoans, a division into ectoparasitic and endoparasitic is less useful. Very few nematodes exploit animal hosts by living or feeding on the outer surface. A more meaningful division is between gut-dwelling parasites and tissue-dwelling parasites. Gut-dwelling parasites either exploit the protected food resource of the commensal flora of the host gut lumen, and can thus perhaps be conceived as bacteriovores exploiting a rich food source in a peculiarly warm environment, or feed on and sometimes dwell within the gut wall. This second group have thus evolved to feed on animal tissue, an important change from the lumenal parasites. Tissue-dwelling parasites invade the host's body and take up reproductive residence in specific organ sites, usually linked to their modes of transmission. Many gut parasites also have a tissue-migratory, and sometimes tissue-diapause, stage in their life cycle before taking up reproductive residence in the gut. Thus, behaviour is associated with an increase in adult size and reproductive output of the migrating species compared to related, non-migrating ones.

Nematode parasites of vertebrates in general cause morbidity in, rather than mortality of, their hosts, though in situations of poor nutrition or immunosuppression, they can be fatal (Maizels *et al.*, 1993b). These virulence patterns reflect an adaptive balance between having the host live a long enough life for the parasite to complete its life cycle, and maximising egg production by food source exploitation (Read and Skorping, 1995a, b).

Many parasites of vertebrates, particularly the spirurids, use vector or intermediate hosts to effect transmission between final hosts (see Anderson (1992) for an encyclopaedic description of parasitic nematode life cycles). For some taxa, such as *Trichinella*, an optional paratenic host is used that acts as a reservoir of diapausal infective propagules in the environment. Many nematodes use arthropods as vectors, suggesting to some that the arthropod association is perhaps ancestral, and that the vertebrate has been added during evolution. As discussed below, while the pattern revealed by the new molecular phylogeny is not unequivocal, parsimony considerations argue against this view, and some vector hosts are likely to have been acquired by parasites already resident in vertebrates.

2.3. Phylogenetic placement of nematode parasites

In traditional phylogenies, then, parasitic taxa were usually classified in distinct higher groups, and this made investigation of the derivation of parasitism problematic. The new molecular phylogenetic framework is very revealing. In general, parasitic taxa are deeply nested within radiations of free-living taxa, and thus their origins can be modelled. There have been many independent events of the origin of both plant- and animal-parasitism in the Nematoda (Blaxter et al., 1998; Dorris et al., 1999) and all the major clades of nematodes defined by molecular analyses include parasites.

2.3.1. Clade I: Dorylaimia

Within Dorylaimia are placed the Dorylaimida (free-living microbivores and plant parasites), Mermithida (all insect parasites), Mononchida (free-living predators), Dioctophymatida (all vertebrate parasites) and Trichinellida (all animal parasites). The mermithids are not the sister-group of the vertebrate-parasitic lineages, and indeed have a parasitic life cycle distinct from all the other nematode parasites of animals. In mermithids, only the first three larval stages are parasitic. The L3 transitions to a free-living mode of life and the adult reproduces outside any host. Strictly, this mode of life renders the insect an 'intermediate' host, as it is not the locus of reproduction. Perhaps the origin of parasitism in mermithids lies in predation of small arthropods, transitioning to endotokous parasitism. The closest free-living nematode group, the mononchids, includes predators and is not known to have associations with other invertebrates. The plant-parasitic dorylaimids are microherbivores, using a stylet to pierce root tissue and extract the cell contents. Other major groups in the Dorylaimia also have a tooth or stylet, and thus this adaptation to herbivory has probably evolved from a piercing tooth. The tooth is also evident in larvae of trichinellids such as *Trichinella*. *Trichinella spiralis* is unusual in metazoan parasites in that it resides intracellularly (in paratenic and definitive host muscle tissue) during the diapausal L1 stage. The adult nematodes are also 'intracellular', but their relative size compared to the cells of the gut epithelium results in host cell death as the nematodes migrate through them (ManWarren et al., 1997; Romaris and Appleton, 2001).

2.3.2. Clade II: Enoplia

The Enoplia contains no (known) animal parasites, though the paucity of described parasites of marine invertebrates suggests that there are certainly hosts available to these mostly marine nematode groups. The plant-parasitic (or microherbivorous) enoplids, the triplonchids such as *Trichodorus*, feed using a piercing oral spear, which has had an independent origin from the stylets of tylenchs and the odontostyles of nematodes in the Dorylaimia.

2.3.3. Clade C&S (III, IV and V): Chromadoria

Chromadoria includes a vast diversity of free-living marine and terrestrial forms, as well as the many parasitic groups of the Secernentea (clades III, IV and V). The chromadorids excluding the Secernentea include a very numerous radiation of mainly marine benthic forms, and have been poorly sampled molecularly. This deficiency is being addressed.

(a) Clade III: Spiruria. Clade III comprises only animal parasites, including the ascaridid, spirurid, and oxyurid parasites of vertebrates, and oxyurid and rhigonematid parasites of arthropods. The evolutionary origins of this clade are still obscure. The free-living plectids are the sister group of the whole secernentean radiation (Clades III, IV and V), not just Clade III. Within Clade III, while Ascaridomorpha, Rhigonematomorpha and Oxyuridomorpha are robust taxa present in both traditional and molecular analyses, the classical 'Spirurida' is paraphyletic in molecular trees.

(b) Clade IV: Tylenchia. Within Clade IV, plant- and fungal-feeding groups are placed together suggesting (uncontroversially) that tylenchid fungivores may have given rise to or been derived from stylet-feeding herbivores. Parasitism of animals, as has been indicated above, takes many forms, from opportunistic to essential, complex interactions. The presence of true (endo)parasites within the tylenchids suggests that they are derived from ectoparasitic taxa, and also that this route may have been followed more than once. Thus this scenario would suggest that understanding of the parasitic adaptations of *Meloidogyne, Heterodera, Globodera et al.* should be underpinned by a firm grounding in the biology of ectoparasitic herbivory from which they have been derived in evolution. It will be very interesting to investigate the biology and molecular phylogenetic relationships of the Hexatylina, parasites of insects related to the tylenchid plant and fungal parasites (see Nickle, 1991): how does this animal-parasitic radiation link to the plant parasites?

The vertebrate-parasitic Strongyloidoidea (*Strongyloides, Parastrongyloides, Rhabdias* and relatives) are also placed in Tylenchia, as a sister clade to the free-living Panagrolaimomorpha and the entomopathogenic genus *Steinernema* (Dorris *et al.*, 1999, 2002). Additionally, nematodes in the Drilonematidae are parasites of the haemocoel of oligochaetes. The relationship of *Drilonema* and related taxa to other Clade IV parasites is unknown, and knowledge of their biology is rudimentary but deserving of attention. Opportunistic or accidental pathogens such as *H. gingivalis* are robustly placed in groups of free-living taxa, suggesting that their acquisition of animal tropism is a recent event (De Ley *et al.*, 1995; Felix *et al.*, 2000).

(c) Clade V: Rhabditia. There are a large number of nematode–arthropod associations that have not been studied extensively, and many of these may turn out to be recently acquired traits. However, it is possible that some of these less well-known associations may be ancient. For example, within the rhabditine nematodes, the use of other invertebrates as transport hosts is widespread, and the dauer larva has been proposed to be an adaptation to the use of these hosts amongst other factors (Bovien, 1937; Kiontke, 1996). *C. remanei vulgaris* was first described from pulmonate mollusc and terrestrial isopod transport hosts (Baird *et al.*, 1994), and may reveal the role of the dauer-transport host association in the biology of the genus *Caenorhabditis* as a whole, including *C. elegans*. Other rhabditids, such as *Rhabditis coarctata*, have specific and highly specialised associations with particular beetle species (Richter, 1993), and many species within the Diplogasterina are also truly parasitic (Geraert *et al.*, 1989). The speciose and economically important vertebrate-parasitic group Strongylida (Durette-Desset *et al.*, 1994; Gasser and Newton, 2000) is robustly placed in Clade V, as a sister taxon to the entomopathogenic *Heterorhabditis* (Blaxter *et al.*, 1998; Dorris *et al.*, 1999). Other non-strongylid vertebrate parasites are also found in Clade V, such as *Rhabditis orbitalis* and *Pelodera strongyloides*, parasites of the orbit and skin of wild and domesticated animals (Sudhaus and Asakawa, 1991). For *P. strongyloides* and congeners, molecular analyses reveal a remarkably divergent SSU sequence that affirms placement in Clade V, but makes robust assignment of affinity problematic (Fitch *et al.*, 1995; Fitch, 1997, 2000). It is likely that both *P. strongyloides* and *R. orbitalis* are additional independent events of acquisition of animal parasitism.

2.4. Patterns in nematode parasitism of animals

2.4.1. The L3 rule: the Stage that Invades the Definitive Host is the Third Stage Larva

It is striking that in the Spirurina, the Strongyloidoidea and the Strongylida, the stage of the nematode life cycle that transitions between the environment or the vector host is the third stage larva. It was believed that the larvae of *Ascaris suum* (and other ascaridids) emerged from the egg and infected their new hosts as first stage larvae (L1), but recent direct observation has shown that there are two moults within the egg, and that the larva emerges as a standard L3 (Geenen *et al.*, 1999). The third stage larva is also involved as a developmental switch or decision point in free-living nematodes, particularly free-living Rhabditia. The dauer larva is an alternative, diapausal L3 that is adapted to endure adverse conditions. Extensively studied in *C. elegans*, where the developmental decision to enter the dauer developmental pathway has been shown to be influenced by the concentration of food, the ambient temperature, the presence of congeneric nematodes and the genetic complement of the individual (Riddle *et al.*, 1981), the dauer larva is a common morph in many species. Many nematodes that associate with transport hosts do so as dauer L3. The common role in invasion of the parasitic L3 probably arose multiple times from this plesiomorphic life cycle pattern of free-living ancestors. In *Strongyloides ratti* the proportion of larvae entering the direct as opposed to the indirect (i.e. dauer-like as opposed to standard) pathways is affected, as in *C. elegans*, by environmental stress, ambient temperature and nematode genetics (Viney, 1996; Gemmil *et al.*, 1997; Viney, 1999; Harvey *et al.*, 2000).

However, in some nematodes, the stage that initiates symbiosis is not the L3. *Bursaphelenchus* sp. (Clade IV) associate with wood-boring beetles, exiting with the beetles from dying trees. In this case the associating stage is the L4. In parasites of the Dorylaimia, the parasitic stages also deviate from this pattern. Mermithid parasites exit from their hosts as L3, and are free-living thereafter, a reversal of the direction of life cycle change seen in the secernentean parasites. In *Trichinella spiralis* (Clade I), the infective stage is the L1, which is able to invade and infect the host direct from the egg. Thus the L3 as infective or transition stage is limited to secernentean parasites, and its origin may be sought in the biology of diapause in L3 of the secernentean ancestor.

2.4.2. Is there a Link Between Arthropod Association and Vertebrate Parasitism?

Given that nematodes have shared the environment with both vertebrates and non-vertebrates for over 550 million years, events of the origin of exploitative interactions such as parasitism could be obscured by deep time. However, it is notable that many of the parasitic radiations of nematodes are apparently significantly more recent than the origin of the phylum, and thus it may be possible to look at the antecedents to, and predispositions towards, the acquisition of parasitic lifestyles by comparing extant taxa. An anthropocentric view of life divides animals into lower and higher forms, with 'invertebrates' below vertebrates, and ourselves as a pinnacle (Haeckel, 1874). While this viewpoint is untenable today, it has left a legacy of ingrained perception in the consideration of the origins of parasitism. Thus, by association, vertebrate parasitic forms are more advanced than invertebrate parasites, and vertebrate parasitism's origins are to be sought within the parasites of invertebrates. Rationally, there is no reason to presuppose that there should be greater barriers to a parasite moving from vertebrate to non-vertebrate hosts than vice versa. However, within the Tylenchida and Rhabditida, the major vertebrate parasites have as sister groups nematodes that exploit arthropod hosts.

The closest relatives to the vertebrate parasitic strongylids and strongyloidoids are in each case insect pathogens. In *Steinernema* and *Heterorhabditis*, it is the L3 that invades the insect larva, just as in the strongyloidoids and strongyles. The similarity in life cycle pattern also extends in these cases to the anterior sensory anatomy (Ashton *et al.*, 1995; Ashton and Schad, 1996; Fine *et al.*, 1997; Li *et al.*, 2000) and the mechanics of invasion of the host, and a direction of evolution of vertebrate parasitism from insect pathogenicity can be robustly proposed (Dorris *et al.*, 1999, 2002).

In the Spirurida, the situation is unclear. Current molecular sampling of Spirurina, and Spiruromorpha in particular, is still very patchy, but current data yields a strongly supported tree that has striking implications. There is a wide variety of patterns of intermediate host use in the Spiruria. The rhigonematids and oxyurids usually directly infect the final host. The oxyurids parasitise both insects and vertebrates, but too few sequences have been obtained from these to derive any indication of which was ancestral. In Spiruromorpha, however, arthropod species are utilised as intermediate hosts, and sometimes additional intermediate hosts are exploited between a primary arthropod host and the definitive vertebrate host. *A priori* it might be expected that the 'more complex' life cycles should be observed in derived taxa, and the simpler life cycle should be plesiomorphic. However, SSU rRNA sequence from the gnathostomes (unpublished GenBank submission from Almeyda-Artigas *et al.*; see Almeyda-Artigas *et al.*, 2000), which utilise crustacean and vertebrate intermediate hosts in effecting transmission to the carnivore or pinniped definitive hosts, suggest that this group arose from the base of Spirurina. The ascarids, with both indirect and direct life cycles, show a dynamic pattern of change in this character (Nadler and Hudspeth, 2000).

Was the ancestor of the Spiruria an arthropod parasite? Within spirurids, a wide range of arthropods is used as intermediate hosts: from dipterans to mites to crustaceans (see Anderson, 1992). The pattern of use of these hosts suggests to me that their recruitment has been based on what was adaptive to the nematode, and not restricted by phylogenetic history. For example, the filarial nematodes use biting flies, mosquitoes and mites to effect transmission, and while there is within-genus conservation of vector taxa utilised, the pattern across the family is complex. The gnathostomes have a marine-based life cycle. Despite the fact that the vast majority of other Spiruria have terrestrially limited life cycles, this may indicate that the origins of spirurid parasitism may lie in the sea.

2.4.3. Life Cycle Evolution in the Strongyloidoidea

Strongyloides display a striking life cycle that has elements of both free-living and parasitic modes (Harvey *et al.*, 1999). Strongyloides parasitise a wide range of tetrapod vertebrates, including mammals, and squamate reptiles. A sister genus, *Parastrongyloides*, has been described from moles (in the UK) and possums (in Australia). In *Strongyloides*, a gut-lumenal parasitic generation of parthenogenetic females produces eggs that hatch as they are passed in the host faeces. The emerging L1 larvae are chromosomally determined to be male or female (Harvey and Viney, 2001), and the male larvae develop as functional, free-living males. The female larvae can undergo a free-living reproductive cycle, as amphimictic females, or develop directly to infective L3 (Viney, 1996; Harvey *et al.*, 2000). This dual reproductive mode is found in all *Strongyloides*, but there is variation between species in the proportions of larvae that follow each route, and in the ability to undergo more than one free-living adult generation (Figure 3A) (Yamada *et al.*, 1991; Viney *et al.*, 1992). Even in species that can have more than one free-living cycle, the number of additional cycles is limited to one or very few. *Parastrongyloides* spp. are also gut parasites, but here the parasitic generation is also sexually reproducing with males and females, and *Parastrongyloides trichosuri* can undergo multiple free-living cycles, at least in the laboratory, before passage through the next possum host. Another genus of vertebrate parasites, *Rhabdias* spp., found as adults in the lungs of amphibians, also has a free-living reproductive cycle, in this case self-fertilising (Smyth and Smyth, 1980; Spieler and Schierenberg, 1995).

Molecular phylogenetics using partial SSU rRNA places *Parastrongyloides* as the sister taxon to *Strongyloides*, and places *Rhabdias* as a sister to *Strongyloides* and *Parastrongyloides* (Dorris and Blaxter, 2000; Dorris *et al.*, 2002). All these three groups require passage through a vertebrate host to maintain themselves: they are obligate parasites that use a free-living

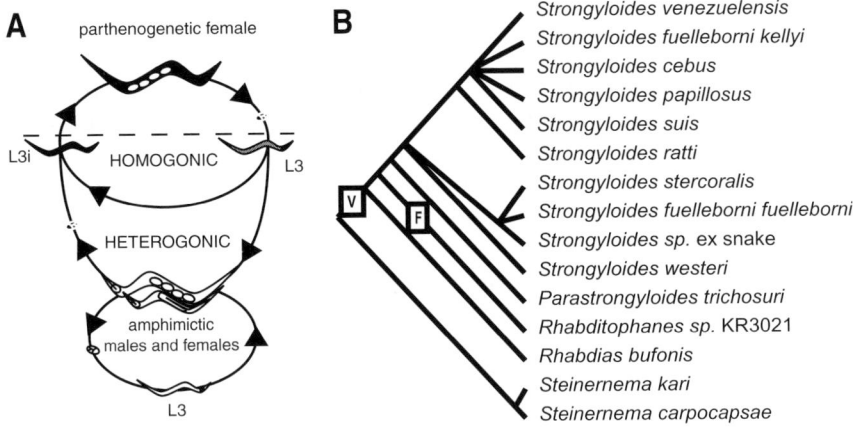

Figure 3 Life cycle evolution in the Strongyloidoidea. **A.** A cartoon of the life cycle of an idealised *Strongyloides*. The dashed line indicates transition between the host (above the line) and the environment (generally host faeces, below the line). Parasitic stages are filled in black, free-living in white, and stages in which an environmentally conditioned developmental choice is made are in grey. **L3**, third stage larva, **L3i**, infective L3. The additional free-living cycle has only been observed in some species. **B.** A cladogram illustrating the possible reversion to a free-living lifestyle in *Rhabditophanes* sp. KR3021. The tree was derived from Bayesian analysis of the 5' end of the SSU rRNA, and all resolved nodes were supported with >65% posterior probability. The lineage leading to *Rhabditophanes* arises after that leading to *Rhabdias*. The boxed **V** indicates acquisition of vertebrate parasitism, the boxed **F** a reversion. *Steinernema* sp. are insect pathogens. Adapted from Dorris, Viney and Blaxter (2002).

cycle to increase the number of infectious propagules in the environment (and to have sex, in the case of *Strongyloides*). *Strongyloides* have apparently evolved to have restricted free-living cycles and parthenogenetic females.

Surprisingly, the SSU rRNA sequence from a free-living nematode is robustly placed within this clade of vertebrate parasites (Felix *et al.*, 2000; Dorris *et al.*, 2002). *Rhabditophanes* are associates of gastropod molluscs, but are readily propagated in culture and are not known to be parasitic or pathogenic in any species. Other related nematodes (alloionematids) are associates or parasites of gastropods (Cabaret and Morand, 1990). Parsimony analysis of the parasitic phenotype suggests that *Rhabditophanes* are derived from a vertebrate parasitic lineage but have lost the need for a host. This reversion is the first to be demonstrated for any metazoan parasitic lineage, and is strongly supportive of a view that the acquisition of a parasitic phenotype is a gain in abilities, and does not necessarily result in an atrophy-related loss of free-living capacity (Figure 3B).

2.4.4. Coevolution

The long associations between nematodes and their hosts (the only fossil nematodes are insect parasites in 30 Mya amber, and L3 of gut parasites in coprolites of recent South American megafauna) leave ample time for mutual adaptation. The immune systems of arthropod and vertebrate hosts have evolved to specifically combat nematode (and other) infections, and one would expect the parasites to have similar evolutionary responses in a coevolving arms race (Burt and Bell, 1987). Coevolution can be examined in two ways: one is through analysis of the molecular and physiological interactions between the partners to uncover their mutual adaptations. Another is to examine the pattern of cladogenesis in groups of parasites and hosts to determine whether the parasite is tracking host speciation, or whether it is capturing new hosts without regard to their phylogenetic relatedness to the one previously parasitised (Page and Hafner, 1996).

In the strongylids (Clade V; Rhabditomorpha), it is clear that there has been much host capture during the evolution of this widely spread and successful group. All strongylids are very closely related genetically, based on SSU rDNA sequence (Blaxter et al., 1998; Dorris et al., 1999), and transfer of the molecular evolutionary rate derived for *C. elegans* and *C. briggsae* (Coghlan and Wolfe, 2002) (see below) to the strongylids suggests an origin at about 100–150 Mya. This time is congruent with the emergence and radiation of the mammals, their major hosts. Examination of phylogenetic analyses of the strongylids, derived for example from ribosomal internal transcribed spacer sequences (Gasser et al., 1993, 1997; Chilton et al., 1995, 1997a, b; Hoste et al., 1995; Romstad et al., 1998; Gasser and Newton, 2000), show that, except at a very local (subfamily) level there is little concordance between the mammalian and nematode trees. For example, the parasites of the grazing marsupials of Australia, kangaroos and wombats, appear to be derived from within radiations of parasites resident in eutherian mammals (Chilton et al., 1997a). Likewise, in *Strongyloides*, no evidence of cospeciation was found (Dorris et al., 2002). Indeed, humans are parasitised by three species of Strongyloides, and one of these, *Strongyloides kelleyi* (Kelly et al., 1976; Ashford and Barnish, 1989) is closely related to parasites of domesticated ungulates and suids rather than other primates. In the Onchocercinae, the human-parasitic *Onchocerca volvulus* is genetically extremely close to the cattle parasite *Onchocerca ochengi* (Xie et al., 1994b). Given the common occurrence of onchocercids in ungulates, it seems likely that onchocercal disease in humans is a naturalised zoonosis from domesticated cattle.

In contrast, cladistic analyses of morphology of pinworm parasites (Enterobiinae) of primates have revealed significant cospeciation between the two partners, with some discrepancies best explained by a limited number of host switching and lineage sorting events (Hugot et al., 1996; Sorci et al., 1997; Hugot, 1999). Overall therefore, the patterns of coevolution between nematode parasites and their hosts are variable, and are probably affected by the interaction between the persistence of environmental infective stages, the specialisation of the reproductive stage, and the pattern of vector use. In chewing lice, a close relationship was found between parasite and pocket gopher hosts, and this was proposed to have arisen through the coincidence of the louse breeding cycle with residence on a single host organism (Page and Hafner, 1996). A parasite that has a non-host associated dispersal stage will be more prone to finding itself in the 'wrong' host, and thus more able to speciate without reference to its original host.

Another coevolutionary feature of nematode biology that has been examined using molecular phylogenetic data is that of nematodes and their bacterial symbionts. Several groups of nematodes have been described as having intracellular bacterial symbionts (in the Tylenchomorpha, *Xiphinema* (Vandekerckhove et al., 2000, 2002) and Onchocercinae (Sironi et al., 1995)). Also, as described above, two genera of insect pathogens utilise toxic extracellular symbionts. In addition, stilbonematid chromadorids live (exosymbiotically) with sulphur-oxidising bacteria that specifically associate with the cuticle of these anoxic mud dweller, and *Astomonema* has a gut endosymbiont community (Kampfer et al., 1998). Molecular identification of most of these symbionts of nematodes has been achieved (Table 1).

Steinernema species have a very specific association with particular clones of *Xenorhabdus*, but there is no obvious congruence between the bacterial and nematode phylogenies (M. Dorris, W. Rogers and M.L. Blaxter, unpublished). *Heterorhabditis* species appear to be less closely tied physiologically with one particular clone of *Photorhabdus*. The interrelationships of stilbonematid symbionts and their nematode partners are, as yet, incompletely resolved.

Within the Onchocercinae (filarial parasites of vertebrates), most species harbour intracellular alpha-proteobacterial symbionts closely related to the *Wolbachia* reproductive parasites of arthropods (Sironi et al., 1995). In arthropods, *Wolbachia* are maternally transmitted parasites that can use a number of reproductive manipulations to favour their transmission to the next generation to the detriment of uninfected offspring (Werren, 1997). Comparison of the phylogenies of *Wolbachia* and their arthropod hosts reveals that there has been recent, massive, lateral transfer (host capture) by *Wolbachia*, and there is little evidence of cospeciation between the two groups of organisms. In contrast, in the onchocercine nematodes the bacterial trees mirror very closely those of their hosts (Bandi *et al.*, 1998; Casiraghi *et al.*, 2001). Transmission is maternal (Taylor *et al.*, 1999), but no reproductive modifications have been documented, and thus it would seem that the bacteria are at worst symbionts with little cost to their hosts. Two additional lines of evidence suggest that the association may be mutualistic. In species that are 'infected', all individuals carry the bacteria. Any cost of the bacterial load would result in a pattern of partial coverage. Secondly, treatment of the nematodes with anti-alpha-proteobacterial antibiotics, such as tetracycline and derivatives, results in stunting of growth, inhibition of moulting and severe interference with fecundity (Hoerauf *et al.*, 1999, 2000a, b, 2001). This suggests that the bacteria are in some way useful or essential to their filarial hosts.

Onchocercine nematodes are transmitted by a number of different vector insects and chelicerates, which, as arthropods, may carry *Wolbachia*. The nematode bacteria, however, are probably not derived from any (known) arthropod *Wolbachia* as the nematode bacterial groups are placed outside the clade of arthropod *Wolbachia*, and may even form a paraphyletic group that includes the ancestors of the arthropod *Wolbachia* (Bandi *et al.*, 1998). Not all onchocercine species carry these endosymbionts, suggesting that while the symbiosis may be mutualistic, it is not essential (in a long-term phylogenetic sense) to the nematode: for example *Onchocerca flexuosa* of deer, a close relative of the bacteria-positive *O. volvulus* of humans, is negative (Figure 4) (Bandi *et al.*, 2001).

3. NEMATODE GENOMES AND PARASITISM

The phenotype of parasitism within the Nematoda has arisen multiple times from within free-living groups. It is thus unlikely to be fruitful to seek a single genetic change or set of changes that have enabled parasitic species to exploit hosts: we should expect a multiplicity of different genotypic adaptations to parasitism. Present day nematode genomes will include ancient relics of their shared ancestry as nematodes, more recent mementoes of the local branches of the nematode tree they have traversed, and idiosyncratic, perhaps stochastic, changes that relate to present day conditions and adaptations. We might expect that some nematode groups, whether parasitic or not, will have genome compositions and architectures that are peculiar when compared to the remainder of the phylum. It would

GENES, GENOMES AND PARASITISM IN THE NEMATODA 127

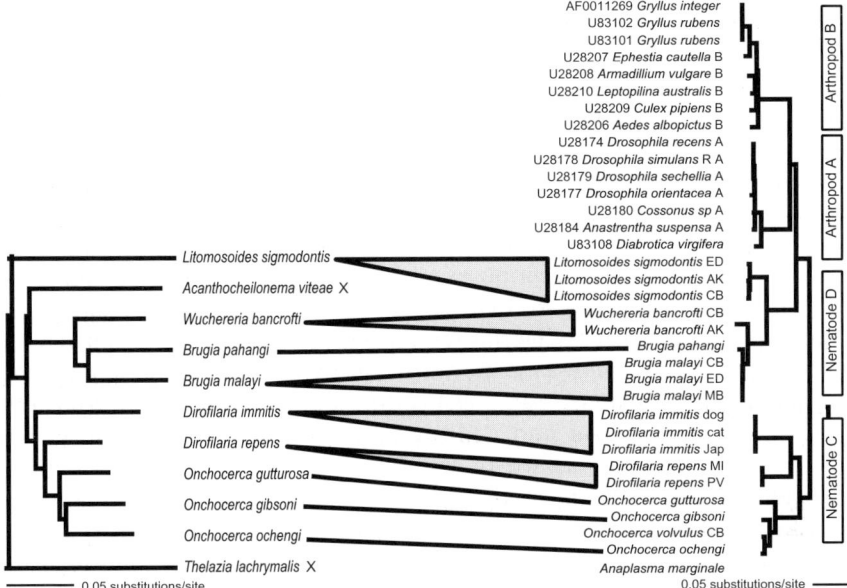

Figure 4 Nematode-bacterial symbiosis: Coevolution between filarial nematodes and their *Wolbachia* symbionts. On the left is a representation of a nematode phylogeny derived from the mitochondrial cytochrome oxidase I gene from species of filarial nematodes (data from Casiraghi *et al.*, 2001). *Thelazia lachrymalis* is a *Wolbachia*-free outgroup. As marked by the bold X, *Acanthocheilonema viteae* is also *Wolbachia*-free. On the right is a phylogeny of the *Wolbachia* symbionts of selected arthropods and filarial nematodes, derived from analysis of the ftsZ gene. The symbionts can be divided into four groups, A to D: two of these are associated with arthropods and two with nematodes. The orthologous surface protein gene from *Anaplasma marginale*, a pathogenic relative of the *Wolbachia*, has been used as an outgroup, but the length of the branch leading to *A. marginale* is such that the exact position of the root (between groups C and D, or between (C+D) and (A+B)) is still unresolved. Comparison of the two trees shows that the phylogeny of the arthropod symbionts does not match that of their hosts, while there is close correspondence between the filarial symbionts and hosts. Re-sequencing of the ftsZ gene from different isolates or laboratory lines of filarial nematode species revealed minimal divergence within a host species. *A. viteae* may have lost its symbiont: the onchocercine deer parasite *Onchocerca flexuosa* is also *Wolbachia*-free, but is robustly placed within the genus Onchocerca and thus has probably also lost the symbiont. The data used to derive the phylogenies is based on that published by Bandi *et al.* (Bandi *et al.*, 1998, 2001; Lo *et al.*, 2002).

be a mistake to search only in parasitic species for unique adaptations at the genetic level, and potentially misleading to compare parasitic species with free-living species that are not their closest (or close) sisters. Conversely, when members of a parasitic group share a peculiar genomic trait, it will be revealing to investigate whether the trait is causally or functionally related to parasitism in the light of the presence or absence of the trait in other nematode groups.

The field of nematode genomics is surprisingly rich considering its infancy. The complete genome sequence of the free-living rhabditid *C. elegans* was published in 1998 (Consortium, 1998, 1999), following a 10-year programme of industrial-scale research by biologists spurred by the utility of *C. elegans* as a powerful genetic model for development, neurobiology and cancer (amongst other things; Wood, 1988; Riddle *et al.*, 1997). The vision and dedication of this group of scientists has been recognised by the award of the 2002 Nobel Prize in Physiology or Medicine to Sydney Brenner, the instigator of the modern programme of *C. elegans* research (Brenner, 1974), Bob Horvitz, a *C. elegans* geneticist whose work underlies the discovery of the molecular mechanisms of programmed cell death (Ellis *et al.*, 1991), and John Sulston, the prime mover behind the definition of the cell lineage of *C. elegans* (Sulston and Horvitz, 1977; Sulston *et al.*, 1980, 1983), the physical mapping of its genome (Coulson *et al.*, 1986, 1988), and the complete genome sequencing project (Sulston *et al.*, 1992; Wilson *et al.*, 1994b; Consortium, 1998). The genome sequencing tools put in place for the *C. elegans* project have now also been utilised to sequence to near-completion the genome of *C. briggsae*, a closely related nematode (see http://genome.wustl.edu/projects/cbriggsae/). For the *C. elegans* research community, the *C. briggsae* genome sequence is extremely useful, as it allows the identification of shared coding and regulatory elements (Heschl and Baillie, 1990) which are difficult (or impossible) to reveal using bioinformatic tools on a single genome. For this work, the phylogenetic closeness of the two caenorhabditids is important. Based on analyses of a large number of protein-coding genes that have a clock-like evolutionary rate between *Drosophila melanogaster* and the two *Caenorhabditis* species, they are estimated to have last shared a common ancestor ~60 million years ago (Kent and Zahler, 2000; Coghlan and Wolfe, 2002). Motifs involved in gene regulation are short and often degenerate, and the passage of evolutionary time will obscure all but the most highly selected. It is hoped that the separation between *C. briggsae* and *C. elegans* is close enough that important sequences are still recognisable but distant enough that they will stand out as islands of conservation in a background sea of evolutionarily neutral noise.

For the rest of nematology, the focus on this small side branch of the bushy nematode tree is both enlightening and frustrating. It enlightens because it reveals the deeply shared biology of all nematodes, the genes and genetic pathways that underlie the nematode body plan, and the biology shared with other phyla and kingdoms. It frustrates because it hints at the kinds of insights that may be gained from genomics, but fails to deliver for the remainder of nematode diversity. To help redress this deficiency, several genomics-based programmes are underway (see Figure 5 and Table 2), examining the biology of diverse species from across the phylum (Blaxter, 1998, 2002; Burglin et al., 1998; Blaxter et al., 1999; Johnston et al., 1999; McCarter et al., 2000, 2001, 2002; Wixon et al., 2000; Parkinson et al., 2001). While most effort is currently focussed on gene discovery and functional identification, some advances have been made in understanding genomic organisation and regulation in species other than Caenorhabditis.

3.1. Genome evolution in nematodes

In comparison to other protostomes, nematodes in general have small genomes (Hammond and Bianco, 1992), though some, such as *Parascaris univalens*, have genome sizes similar to mammals (~2 gigabases). The model genome, that of *C. elegans*', is 100 megabases (Mb), and most species appear to have genomes of between 250 Mb and 60 Mb (see http://www.genomesize.com/nematodes.htm for a compilation of nematode genome size estimations). It should be noted that very few nematode genomes have been accurately sized (five cephalobes, nine rhabditids and nine others) and thus the current known range may not reflect true diversity. In comparison, the dipterans *D. melanogaster* and *Aedes aegypti* have sequenced genomes of 160 Mb and 280 Mb respectively, and ~300 Mb is regarded as standard for an arthropod genome.

Nematodes can have from one (*P. univalens*) to over 100 chromosomes, but for most the karyotype has n between 4 and 12 (Walton, 1959). In nematodes of clade V, most taxa have 5 or 6 autosomes and a single X chromosome (Blaxter, 2000a): sex determination acts through an X-chromosome dosage mechanism (Nicoll et al., 1997). In Spiruromorpha, *Brugia* spp. and other onchocercines have four autosomes and an XY sex determination pair (Sakaguchi et al., 1982, 1983). Tylenchs have very diverse karyotypes, particularly in groups that reproduce parthenogenetically, and high chromosome numbers are common, possibly due to polyploidy (Triantaphyllou, 1971). The chromosomes of ascarids are unique in Metazoa in that the karyotype of the somatic cells differs from that of

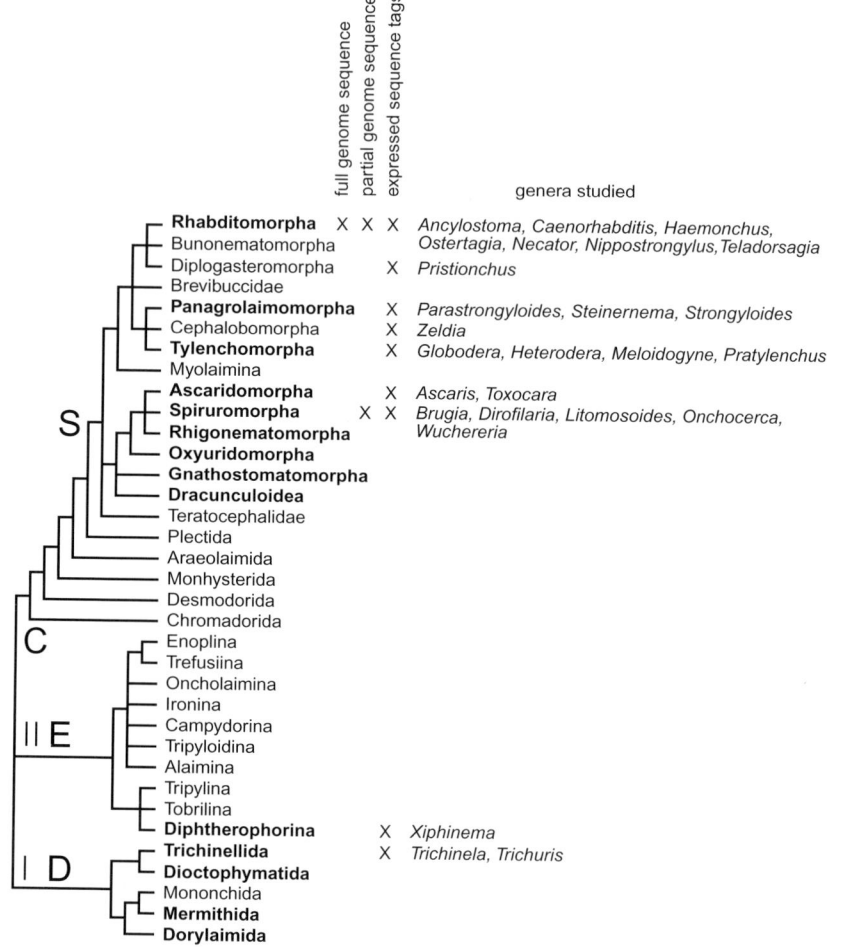

Figure 5 Genomics programmes on nematodes. The distribution of genomics effort in the Nematoda is illustrated in relation to the phylogenetic framework (as outlined in Figure 2). For each group, the presence of a complete genome sequence, a partial genome sequence or expressed sequence tag project is indicated. The genera studied (see Table 2 for species) are indicated to the right.

the germline (Tobler *et al.*, 1992; Muller and Tobler, 2000). The germline has one or a very few large chromosomes. During early development, in the founder cells of the somatic lineages, these chromosomes fragment into over 100 smaller pieces. In this process of chromosome or chromatin

Table 2 Expressed sequence tag sequences for nematodes*.

Major taxon	Species	Number of ESTs
Rhabditomorpha	*Ancylostoma caninum*	9331
	Ancylostoma ceylanicum	4476
	Caenorhabditis briggsae	2424
	Caenorhabditis elegans	189 632
	Haemonchus contortus	4906
	Necator americanus	961
	Nippostrongylus brasiliensis	978
	Ostertagia ostertagi	7008
	Teladorsagia circumcincta	315
Diplogasteromorpha	*Pristionchus pacificus*	8818
Panagrolaimomorpha	*Parastrongyloides trichosuri*	7963
	Steinernema feltiae	83
	Strongyloides stercoralis	11 392
	Strongyloides ratti	8992
Cephalobomorpha	*Zeldia punctata*	391
Tylenchomorpha	*Globodera pallida*	1832
	Globodera rostochiensis	5934
	Heterodera glycines	4327
	Heterodera schachtii	14
	Meloidogyne hapla	8815
	Meloidogyne incognita	12 715
	Meloidogyne javanica	5600
	Pratylenchus penetrans	2048
Ascaridomorpha	*Ascaris lumbricoides*	92
	Ascaris suum	27 871
	Toxocara canis	4206
Spiruromorpha	*Brugia malayi*	22 439
	Brugia pahangi	28
	Dirofilaria immitis	3908
	Litomosoides sigmodontis	198
	Onchocerca ochengi	60
	Onchocerca volvulus	14 922
	Wuchereria bancrofti	131
Trichinellida	*Trichinella spiralis*	10 372
	Trichuris muris	2125
Nematoda Total		385 307

*The GenBank dbEST database was searched on August 22, 2002. Additional ESTs are being deposited weekly.

diminution, DNA is lost from the somatic cells. The DNA excluded from the somatic cells includes non-coding, repetitive and retrotransposable element-rich segments, but also, importantly, includes some genes (Muller et al., 1982a, b; Aeby et al., 1986; Etter et al., 1991, 1994). For example, *Ascaris suum* (described in the literature as *Ascaris lumbricoides*, but as the material studied originated from pigs it is much more likely to be the very closely related *A. suum*) has two genes that encode the ribosomal protein S19. Only one is retained in the soma. The two genes are remarkably different in sequence, suggesting that ribosome function may differ between the germ line, with two distinct S19 proteins, and the soma, with only one (Etter et al., 1994). The breakpoints are not random, but reproducibly occur in particular segments of DNA. The molecular and cellular mechanism of breakage and repair of the chromosomes is yet unclear, as is the biological rationale for maintaining a large portion of the genome only in germline cells.

In general, nematode chromosomes are too small (~10 to 20 Mb of DNA) to have significant structure under the light microscope. Different-sized chromosomes can generally be distinguished in chromosome spreads of gonadal tissue, and some, such as the *Brugia malayi* Y, have distinctive morphologies (Sakaguchi et al., 1983), but most are just small dots. Chromosome banding techniques have been applied to *Caenorhabditis* species, and relatively robust patterns distinguished (Albertson and Thomson, 1976; Herman et al., 1976; Albertson, 1984, 1993; Albertson et al., 1995), but in general these techniques have not been used in nematodes.

3.1.1. *The Genome of* Caenorhabditis elegans

The *C. elegans* genome was sequenced using a physical clone map-based strategy (Coulson et al., 1986, 1988, 1996; Sulston et al., 1992), and was essentially completed in 1998 (Consortium, 1998, 1999). At that time it was the only metazoan genome sequenced, but since 1998, the genomes of *Homo sapiens*, *Mus musculus*, *Fugu rubripes*, *Ciona intestinalis*, *D. melanogaster* and *A. aegypti* have also been essentially finished, and near-complete genomes are available for several other animals. The genome sequence of *C. briggsae* has been produced to a 'first draft' quality, based on a seven-fold shotgun sequencing coverage (see http://genome.wustl.edu/projects/cbriggsae/). The *C. briggsae* genome is thus not complete, but comparisons with *C. elegans* and internal checks suggest that well over 95% of the genome is present in the draft product. Sequencing of the *C. elegans* genome was the largest project attempted at the time of its inception, and was made significantly easier by the existence of a robust and dense physical map, and the low repeat density of the genome. Even so, some regions of the genome proved refractory to cloning in standard bacterial and yeast systems, and the published paper describes over 95 Mb of sequence ascribed to chromosomal groups, and 2 Mb of 'unattached' sequence. The complete, contiguous sequence of each chromosome is now to hand after a further 4 years of effort, and is, as predicted, 100 Mb (see http://www.sanger.ac.uk/Projects/C_elegans/). The regions missing from the sequence at the time of publication were in general gene-poor and repeat rich and their addition to the genome dataset has not significantly changed the conclusions at first publication.

The *C. elegans* genome is 36% GC, and there are no isochores of significantly different base composition, or other regions of deviant base composition. *C. elegans* chromosomes are holocentric, and thus no localised, highly repetitive centromeres were found. The genome is organised as six chromosomes: five autosomes (chromosomes I to V) and one sex chromosome (designated X; see Table 3). The autosomes range in size from 12.8 Mb (chromosome (chr) III) to 20.8 Mb (chr V), and the X chromosome is 17.2 Mb. The autosomes differ from the sex chromosome in several respects. Firstly, the density of protein-coding genes is from one per 4.5 kb to one per 5.2 kb on the autosomes compared with one gene per 6.5 kb on the X. A correspondingly higher percentage of the autosome DNA is coding (26.3% compared with 19.8% of the X). The average coding length of genes (about 1.3 kb per gene) is the same on both autosomes and X. Conversely, tRNA genes are much more common on the X than on autosomes: 40% of the 903 tRNA genes are present on the 18% of total DNA that comprises the X. The protein-coding genes on the X are approximately equally likely to have a significantly similar homologue (detectable by database search) outside the phylum Nematoda compared to genes on autosomes (Figure 6) (see http://www.nematodes.org/SimiTri/SimiWorm.php for an up-to-date analysis of *C. elegans* protein similarity data; M.L. Blaxter and J. Parkinson, unpublished).

The autosomes and X chromosome also differ in distribution of repeat sequences. Repeat sequences can be classified into repeat genes, transposons (of several families, including DNA-based and RNA-based types), and other inverted, tandem and dispersed repeats with no clear function. On the autosomes, there is a pronounced tendency for repeats to cluster towards the ends of the chromosomes, in the left and right 'arms'. The autosome arms were originally defined genetically, as regions of chromosomal linkage groups that were gene-sparse, with high rates of recombination between loci. They contrasted with the chromosomal centres, or clusters, where many genetic loci were found to be very closely linked. The transition from centre

Table 3 The *Caenorhabditis elegans* genome: overall statistics.

Chromosome	Size (Mb)	Number of protein-coding genes*	Gene density (kb per gene)	tRNA genes	Percentage of DNA that is protein-coding	Percentage of predicted genes with significant database match outside nematodes
I	13.86	2803	4.94	74	26.87	48.55
II	14.72	3259	4.52	83	27.53	40.94
III	12.77	2508	5.09	94	26.29	49.67
IV	16.14	3094	5.22	84	23.29	44.77
V	20.82	4082	5.10	206	27.23	36.11
X	17.22	2631	6.54	362	19.80	43.30
Total	**95.53****	**18377**	**5.20**	**903**	**25.09**	**43.20**

Based on data from Consortium (1998).
* This set of totals is less than is currently predicted for the genome.
** 95.5 Mb of the genome sequence was allocated to chromosomes out of 97 Mb sequenced at publication in 1998. The now complete genome is just over 100 Mb.

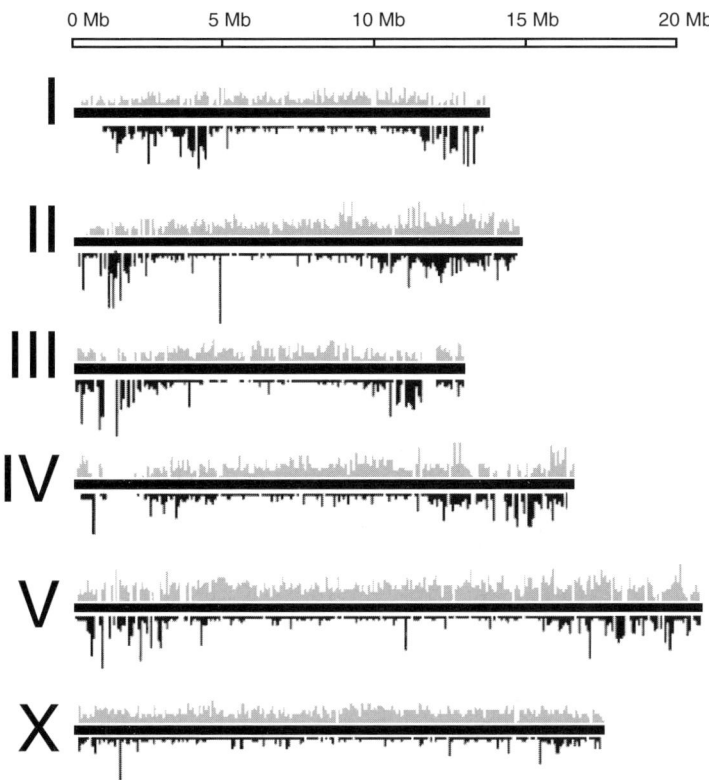

Figure 6 The *Caenorhabditis elegans* genome. This cartoon shows the six *C. elegans* chromosomes with graphs (in arbitrary units) of protein-coding gene density (light, above each chromosome) and tandemly repeated DNA density (below each chromosome, dark) illustrating the differences between the autosomes (I–V) and the X chromosome. Inverted repeats are more common on the autosomal arms, but are evenly distributed across the X chromosome. A scale in megabases (Mb) is given above. Derived from data in Consortium (1998).

to arm is rather sharp. Correlation of the genetic and physical, sequence map of *C. elegans* has been achieved by the cloning of the genes affected by mutations. This revealed that the relationship between genetic distance (measured in centiMorgans) and physical distance (measured in Mb) is S-shaped along each autosome (Barnes *et al.*, 1995). The chromosomal centres have a much smaller Mb/cM ratio than the arms. In contrast, to the autosomes, the X chromosome has a much more even distribution of repetitive elements, and this is matched by a more linear relationship between physical size and genetic length.

Table 4 Chromosome arms and chromosome centres in *Caenorhabditis elegans*.

Chromosome subdivision	Total size (Mb)	Number of protein-coding genes*	Gene density (kb per gene)	Percentage of DNA that is protein-coding	Percentage of predicted genes with significant database match outside nematodes
Centres	31.85	6985	4.56	30.36	49.52
Left arms	22.05	4851	4.55	24.07	37.08
Right arms	24.41	3910	6.24	22.88	39.46
Both arms**	46.46	8761	5.30	23.44	38.14
X Chromosome	17.22	2631	6.54	19.80	43.30

Based on data from Consortium (1998).
* This set of totals is less than is currently predicted for the genome.
** The designation of left versus right is based on historical accident. It is thus "interesting" that left and right arms differ (significantly) in gene density.

The differences between autosome arms and centres (Table 4) are expressed as differences in gene density, proportion of coding DNA and average length of each gene. Genes are less dense and are shorter on the arms. In addition, the genes on the arms tend to be less likely to have significant similarity to genes from other organisms. Only ~40% of genes on the arms have homologues, whereas ~50% of genes in the chromosome centres do. Genes on the X have a homologue 43% of the time. Genes on the arms are also more likely to be part of a gene family, and less likely to be expressed (as measured by presence in un-normalised expressed sequence tag datasets). The chromosome arms have been proposed to be the part of the genome where new genes are born, and where genome change, in terms of sequence change as well as rearrangement, is most frequent (Consortium, 1998; Hutter *et al.*, 2000). The centres are in contrast highly conserved and relatively stable. A similar pattern of genome organisation is also seen in other organisms, such as the linear chromosome of the bacterium *Streptomyces coelicolour*, where the arms encode exotic metabolic functions such as polyketide antibiotic synthesis (Bentley *et al.*, 2002), and in apicomplexan protozoans, where the chromosomal termini encode immune evasion-related genes.

3.1.2. *Patterns of Genome Evolution in* Caenorhabditis *Species*

C. elegans is a protandrous hermaphroditic species, with rare, functional males (Wood, 1988). The population genetics of wild *C. elegans* has not been studied in any detail, but it would appear from analysis of single nucleotide polymorphisms, transposon insertion patterns and mitochondrial DNA haplotypes, that there is significant population structure in the wild (Fitch and Thomas, 1997; Denver *et al.*, 2000; Koch *et al.*, 2000; Wicks *et al.*, 2001). There appear to have been many events of recombination between lines that today are separated by thousands of miles: the *C. elegans* strain most divergent from the reference genome strain N2 was isolated from Hawaii (strain CB4856). Mitochondrial haplotype analyses yield a similar pattern. The mutation rate in *C. elegans* has been estimated from both mitochondrial sequence data and genome comparisons: both measures indicate a very high rate (Keightley and Caballero, 1997; Davies *et al.*, 1999; Keightley *et al.*, 2000). Mitochondrial sequence analysis from independent mutation accumulation lines also suggest very high rates of ~14 mutations (9 substitutions) per site per million years, again an order of magnitude higher than observed for other animals. Comparison of *C. elegans* and *D. melanogaster* orthologous protein pairs to orthologues from *Saccharomyces cerevisiae* and *H. sapiens* shows that many *C. elegans* proteins have overall higher substitution rates, but only of the order of 1.2 to 2 fold higher (Mushegian *et al.*, 1998). This high rate is as yet unexplained mechanistically, but is the probable explanation of the long-known artefactual placing of *C. elegans* in molecular phylogenetic analyses of Metazoa using, for example, the small subunit ribosomal RNA gene (Philippe *et al.*, 1994).

The availability of the *C. briggsae* genome sequence, even in its draft form, allows extensive delineation of the patterns of genome evolution within the genus *Caenorhabditis*. The comparisons published thus far, based on matching protein-coding or other conserved DNA segments and examining their interrelationships, yield further evidence of highly elevated evolutionary rates (Kent and Zahler, 2000; Coghlan and Wolfe, 2002). While the language of comparison of genomes is derived from traditional genetics, the detail available from genome sequence requires close definition of terms. Linkage is the presence on the same physical DNA element (chromosome) of orthologous genes. Synteny is conservation of both linkage and gene order, while microsynteny is the presence of sets of conserved orthologues in close association on a chromosome. A comparison between *C. elegans* and ~12 Mb of the *C. briggsae* genome (Coghlan and Wolfe, 2002) revealed 756 instances of microsynteny, with from 2 to 109 genes matching between the two species involving DNA segments of 1.3 kb to 1040 kb. The longest region of conservation mapped to the *C. elegans* X chromosome, but the corresponding *C. briggsae* genomic contig did not appear to be X-associated. Using a stochastic model of chromosome evolution, it was estimated that about half of the rearrangements were transpositions (the movement of a gene or genes to a novel genomic environment) with inversions and translocations (reciprocal exchanges between chromatids) occurring in approximately equal numbers. Overall, the two species experienced an astonishing 42–102 chromosomal breakage events per million years, ~5000 times more frequent than that estimated for *H. sapiens* and ~10 times greater than that estimated for *Drosophila* species (Ranz *et al.*, 1997, 2001; Coghlan and Wolfe, 2002). Some dispersed repeat elements were preferentially associated with breakpoint junctions. Mechanistic explanations for this elevated rate focus on the short generation time of *C. elegans* and the small effective population size (because the nematodes are mostly self fertilising with occasional outbreeding).

3.1.3. *Patterns of Genomic Evolution Across the Nematoda*

Very little nematode genomic sequence is available other than that for *Caenorhabditis*. Other nematode genomes can have very different base compositions to the model nematode's 36% GC. For example, *B. malayi* has a GC content of only ~30% (McReynolds *et al.*, 1986; Maina *et al.*, 1987; Piessens *et al.*, 1987), while *Panagrellus silusiae* has 44% GC (Beauchamp *et al.*, 1979), and this in turn affects codon bias and the pattern of non-coding repeat DNAs observed. The genome of *C. elegans* has about 17% repetitive DNA, and similar relative amounts are seen in *B. malayi*. The pattern of repeats can differ significantly, however. Thus *B. malayi* has ~30 000 copies of a monomorphic 322 bp repeat, making up 10% of the genome (McReynolds *et al.*, 1986; Piessens *et al.*, 1987): no single *C. elegans* repeat is present at such genomic abundance.

From sequencing of specific genes, such as beta tubulin and globin, two features are evident. One is that mean intron size in other nematodes is often larger than in *C. elegans* (see Guiliano *et al.*, 2002). *C. elegans* introns are generally small, with a majority being less than 125 bp (Blumenthal and Steward, 1997). Other nematodes tend to lack very small introns (~37 to 60 bases) commonly found in *C. elegans*, and have mean intron sizes of ~300 bp. The splice site consensus for other nematodes appears to closely match that determined for *C. elegans* (Blumenthal and Steward, 1997). The second feature is that there have been many losses or gains of introns during nematode genome evolution. Between *C. elegans* and *C. briggsae*, over 250 introns were missing in one or other species in an 8 Mb pairwise comparison (Kent and Zahler, 2000). Between *C. elegans* and other species, the trend is for *C. elegans* to have fewer introns. For example, in the gene *pbr-1* (see below) *C. elegans* has 14 introns whereas *B. malayi* has 37 (Figure 7) (Guiliano *et al.*, 2002). In the globin genes, a dynamic pattern of intron gain and loss is observed, with the *C. elegans* myoglobin gene (*glb-1*) having one intron, while *Nippostrongylus brasiliensis glb-1* has three (Blaxter, 1993b, 1994a). In *A. suum* the myoglobin gene has no introns (Blaxter *et al.*, 1994b).

Larger-scale genome comparisons are now becoming possible between *B. malayi* and *C. elegans* through the work of the filarial genome project (Blaxter *et al.*, 1999, 2001; Williams *et al.*, 2000). Two components of the project are yielding significant DNA segments. The Institute for Genomic Research (Bethesda, USA) is following a whole-genome shotgun strategy to provide a draft genome coverage, and has released ~150 Mb of preliminary data derived from small-insert clones (see http://www.tigr.org/). The BaNG *B. malayi* project (see http://www.nematodes.org/) has been end-sequencing large-insert (bacterial artificial chromosome) clones, and also performing complete sequencing of selected genomic regions. The largest contiguous segment assembled thus far is 83 kb, or only 0.8% of the expected complete genome (Guiliano *et al.*, 2002), but comparison with *C. elegans* has been very revealing (Figure 8).

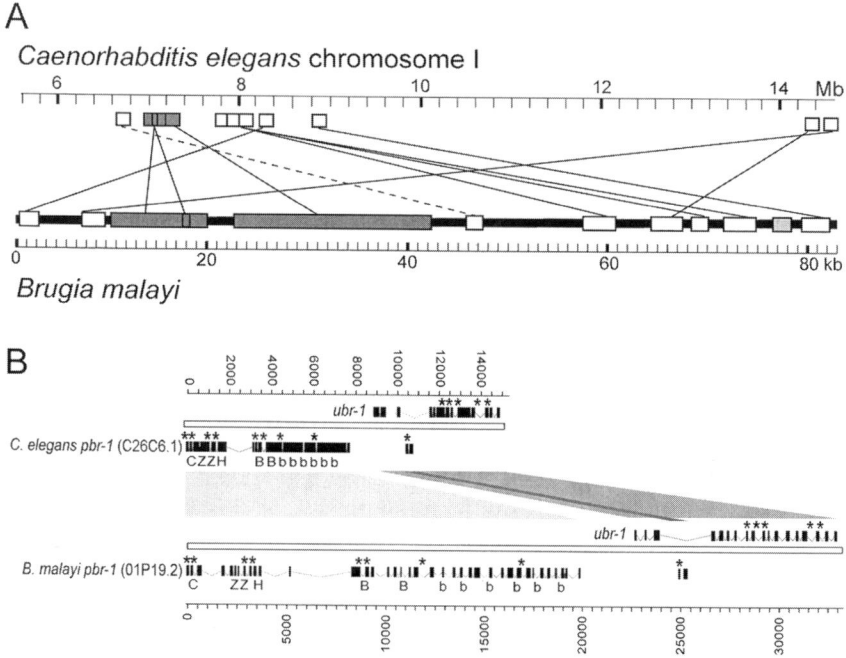

Figure 7 Linkage and microsynteny conservation between *Brugia malayi* and *Caenorhabditis elegans*. **A**. A diagram showing (above) *C. elegans* chromosome 1 (with a scale in megabases) and boxes (not to scale) indicating the relative positions of orthologues of *B. malayi* genes identified in an 83 kb contig (below) (Guiliano *et al.*, 2002). The *B. malayi* contig (scale in kb) is also annotated with boxes (to scale) linked to their *C. elegans* orthologues by lines. The dotted line indicates a pair of genes that may not be orthologous, as the *B. malayi* gene is more similar to a related gene on *C. elegans* chromosome III. The light grey box indicates a *B. malayi* gene with no *C. elegans* homologue. The dark grey boxes indicate a set of genes in a microsynteny cluster (see B below). **B**. A diagram of one of the microsynteny clusters identified (Guiliano *et al.*, 2002). Gene models, confirmed by cDNA sequencing, are indicated by boxes (exons) linked by brackets (introns). Both genomic segments are drawn at the same scale. *pbr* indicates "polybromodomain protein" and *ubr* "upstream of *pbr*". Bromodomains are associated with chromatin-binding complexes, and both the *C. elegans* and *B. malayi* predicted PBR-1 proteins have six bromo domains (B) or bromo-related domains (b). In addition, both have high mobility group domains (H), C2H2 type zinc finger domains (Z) and a C-terminal conserved domain (C). Similar proteins are also found in *D. melanogaster* and vertebrate genomes, but this pattern of domains is unique to the nematodes. The UBR-1 predicted proteins are 27% identical over their full lengths (847 amino acids in *B. malayi*) but have no significant similarity to any other proteins. In the longest intron of *ubr-1* in both nematodes is a small gene, *dap-1*, which is homologous to mammalian proteins associated with the cell death pathway. The asterisks above introns indicate that they are conserved (in terms of exact phase and position) between the *B. malayi* and *C. elegans* genes.

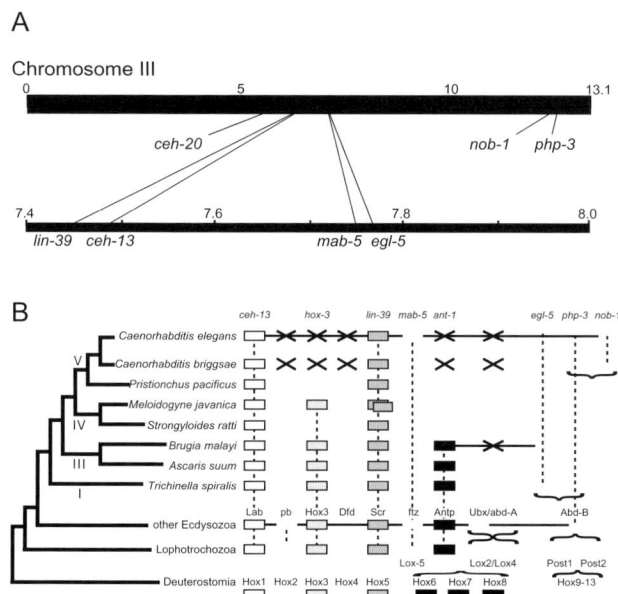

Figure 8 The evolution of nematode HOX genes. **A**. The *Caenorhabditis elegans* HOX gene "cluster". The six *C. elegans* HOX genes are scattered across 6 Mb of chromosome II (upper cartoon). Four are found as two pairs in the middle of the chromosome (lower cartoon), separated by 0.25 Mb of chromosome that contains 29 other, non-HOX genes. Scales in Mb. **B**. *C. elegans* has lost HOX genes relative to other nematodes (Aboobaker and Blaxter, 2003). The molecular phylogeny of the Nematoda (tree to the left, rooted by "other Ecdysozoa", Lophotrochozoa and Deuterostomia) was used to organise the pattern of presence of HOX genes in eight species of nematode from clades I, III, IV and V. HOX genes are indicated by boxes: different HOX orthology groups have different shading, and are linked by dashed vertical lines. The names of the HOX genes in the different superphyla are indicated. Several lineages have independent duplications of HOX genes. The duplications are indicated by curly brackets, and the HOX genes that result from these duplications have been shaded similarly (noting that these duplicated HOX genes have distinct functions). Some non-nematode taxa, particularly higher deuterostomes, have duplications of the whole HOX cluster. Two *lin-39* genes were found in *Meloidogyne javanica*. HOX genes known to be linked in the genomes of the different organisms are joined by bold horizontal lines. Genes definitively missing from a species/group are indicated by a bold X.

The 83 kb segment covers the genomic region flanking the *Bm-mif-1* gene, which encodes a homologue of the mammalian cytokine macrophage migration inhibitory factor. There are 12 protein-coding genes in the 83 kb segment, and eleven of these have a readily identified *C. elegans* orthologue (Figure 8A). All of the *C. elegans* orthologues are on chromosome I, suggesting that while many rearrangements have occurred since the last common ancestor of the two species there has been a tendency to retain linkage of these genes. Synteny conservation is less evident, as the *C. elegans* orthologues are spread over 6 Mb of chromosome I. However there are two segments that display microsynteny, suggesting that there may be strong selective constraint on separation of the genes involved (Figure 8B). A survey of sequences derived from both ends of large-insert *B. malayi* genomic clones also yielded an excess of matches that mapped to the same chromosome in *C. elegans* (Guiliano *et al.*, 2002).

The pattern observed around the *mif-1* locus may not be representative of the whole genome. Indeed, sequence surrounding the *Bm-alt-1* (abundant larval transcript) locus shows that *Bm-alt-1* lies in close proximity to the ribosomal RNA repeat of *B. malayi* (in *C. elegans* the rRNA repeat is on chr I) next to two protein-coding genes, *Bm-mog-1* (masculinisation of germline-1 homologue; *Ce-mog-1* is on chr II) and *Bm-prl-1* (pleiotropic regulator homologue; *Ce-prl-1* is on chr V) (Gomez-Escobar *et al.*, 2002). The whole-genome shotgun data will be very informative concerning the global pattern of chromosome change between the two species.

Many *C. elegans* genes are co-transcribed in operons, where a single upstream promoter directs transcription of an initially polycistronic messenger RNA that is resolved by *trans*-splicing with a specific set of spliced leader exons (SL2-like spliced leaders) (Zorio *et al.*, 1994; Blumenthal, 1998; Blumenthal *et al.*, 2002). The SL1 spliced leader is utilised in capping mRNAs that derive from many single-cistron transcription units, as well as being used to cap the 5' gene of most (but not all) operonic units. This pattern of gene organisation is currently unique to Nematoda and Platyhelminthes (Davis, 1995, 1996, 1997) in the Metazoa. In most *C. elegans* operons there is no obvious functional link between the co-expressed genes (Blumenthal *et al.*, 2002), though isolated examples of clearly related function have been identified (Page, 1999). Operons are present in other nematodes. In the free-living *Oscheius brevesophaga* an operon, linking two ribosomal protein genes, has been found that is also present in *C. elegans* (Evans *et al.*, 1997). In *O. brevesophaga* this operon is also resolved on transcription by *trans*-splicing to SL2-like leaders. SL2-like leaders have also been identified in the rhabditid parasite *Haemonchus contortus* (Redmond and Knox, 2001) and free-living *Pristionchus pacificus* (D. Guiliano and M. Blaxter, unpublished observations), but not elsewhere, suggesting that SL2-like leaders may be an innovation of the Rhabditina. In other nematodes, SL1 is present (Blaxter and Liu, 1996; Liu *et al.*, 1996), and operons appear to be resolved by *trans*-splicing to SL1. Several operonic structures have been identified in genomic sequencing from *Brugia malayi* that are shared with *C. elegans*, but other *C. elegans* operons are apparently absent (D. Guiliano and M. Blaxter, unpublished observations). A current working hypothesis is that genes that come to share a promoter sequence by being co-expressed in an operon will be unlikely to be separated by evolution, and thus that operons will have tended to accumulate through evolutionary time. This would suggest that there will be a core of shared operons, and unique sets of operons for each lineage.

3.1.4. An Example of Rapid Genomic Change: the Evolution of the Nematode HOX Gene Cluster

HOX genes have been implicated in anterior-posterior pattern formation in many animals (Akam *et al.*, 1994; Akam, 1998, 2000). They encode DNA-binding regulatory proteins that share a 60 amino acid domain (the homeodomain) and are present in most animals as a family of related genes that are all clustered in a small genomic segment. Comparison of HOX genes between phyla has defined a set of eight groups of orthologues present in most species. These orthologue groups have the property of controlling the expression of anterior-posterior patterning along the animal body axis during early embryogenesis collinearly with their arrangement on the chromosome (de Rosa *et al.*, 1999). Thus *hox-1/labial* orthologues pattern anterior structures, and are at one end of the HOX cluster, while the *AbdB/hox-9-13* orthologues are at the opposite end of the cluster and pattern posterior-most structures (see Figure 8). This linkage group has been conserved across metazoan evolution.

HOX genes have traditionally associated with segmental or metameric patterning, and thus their presence in unsegmented animals with minimal evidence of any metamerism, such as nematodes, was surprising. However, HOX genes appear to be one of the molecular structures that underpin the body plan of all animals, and thus their roles in 'lower' animals are thought also to be in anterior-posterior patterning, just not in the context of repeated structures. *C. elegans* has six recognisable HOX genes (Figure 8A) spread over 6 Mb of chr III (Wang *et al.*, 1993; Kenyon, 1994; Salser and Kenyon, 1994; Ruvkun and Hobert, 1998b; Van Auken *et al.*, 2000). The genes derive from just four of the eight distinct orthologue groups present in arthropods and deuterostomes. Their genomic context is also not as a 'cluster': they are found as three pairs of genes, with many genes of unrelated function separating them. The posterior pair, *nob-1* (no backend) and *php-1* (posterior HOX paralogue), appear to be the result of a relatively recent duplication in the lineage leading to *Caenorhabditis* and are members of the posterior *AbdB*-like group (Van Auken *et al.*, 2000). The central pair comprises *egl-5* (egg-laying defective, another

AbdB-like posterior gene) and *mab-5* (male abnormal, a member of the *ftz/Lox5* group). The anterior pair, *ceh-13* (*C. elegans* homeobox gene, a *labial/hox-1* orthologue) and *lin-39* (lineage defect, a *Scr/hox-5* orthologue), is inverted relative to their anterior-posterior patterning effects, suggesting a chromosomal inversion event. Of these genes, only *ceh-13* and *nob-1/php-3* have identified embryonic roles. *lin-39*, *mab-5* and *egl-5* have roles in later, post-embryonic development still associated with anterior-posterior pattern formation. Importantly, some of the post-embryonic cell lineages patterned by these HOX genes are serially repeated along the body axis of *C. elegans*.

Thus, *C. elegans* (and the draft genome of *C. briggsae*) has a reduced HOX gene complement that is not obviously clustered. This could be a relic of an ancestral genotype, with 'higher' animals evolving both more genes and a more tightly clustered HOX set (Kenyon, 1994), or the result of a relaxation of the constraints on the HOX cluster in nematodes, and evolution to lose clustering and genes. Elucidation of HOX gene content and linkage in other nematode species from across the diversity of the phylum has shown that the pattern in *C. elegans* is indeed the result of loss, a process that has happened piecemeal during the divergence of the Nematoda (Aboobaker and Blaxter, 2003). HOX genes that clearly are orthologues of groups missing from *C. elegans* are present in other nematodes, and in *B. malayi* one of these 'lost' genes has been shown to lie next to *egl-5* in the genome, as was predicted from its orthology assignment. There are still some orthologue groups not identified in nematodes, such as the *proboscipedia/hox-2* group, but as the analysis of other nematodes is based on polymerase-chain reaction screens, it may be that these groups are missing for technical rather than biological reasons. The collection of HOX orthologues from a phylogenetically diverse range of species also revealed that HOX genes in nematodes have experienced elevated evolutionary rates compared with arthropods, vertebrates and other animals. This change in tempo of evolution suggests that the nematode HOX genes may have had a major change in function, possibly related to their disengagement from an early embryogenesis-active, anterior-posterior pattern determination machine still functional in flies and mice, but redundant and discarded in nematodes (Aboobaker and Blaxter, 2003).

3.1.5. Nematode Mitochondria

As discussed above, estimates of the rate of molecular evolution in the mitochondrial genome of *C. elegans* suggested that both substitution and insertion/deletion events were occurring very rapidly. Sequencing of the complete or partial mitochondrial genomes of several nematode species has revealed further startling evidence of a high evolutionary rate. The first two nematode mitochondrial genomes to be sequenced, those of *C. elegans* and *A. suum*, were very similar in both sequence and gene order (Figure 9) (Okimoto *et al.*, 1992, 1994; Wolstenholme, 1992; Watanabe *et al.*, 1994). They differed from other animal mitochondrial genomes in lacking one of the mitochondrial ATPase subunits usually encoded there (gene ATP8), and in having tRNA genes with an unusual predicted tRNA secondary structure. Nematode mitochondrial genomes are 14–20 kb in size, circular (except in the mermithid *Romanomermis culcivorax*, see below), and encode two ribosomal RNA genes, 22 tRNAs and 12 (or 13 in the case of *T. spiralis* (Lavrov and Brown, 2001); see below) protein-coding genes. The gene order on the *C. elegans* and *A. suum* mitochondria is very different from that seen in other animals. The sequences of other rhabditid mitochondrial genomes (*Necator americanus* and *Ancylostoma duodenale* (Hu *et al.*, 2002)) are very similar to that from *C. elegans* in sequence and gene order. However, the genome sequences from *O. volvulus* (Keddie *et al.*, 1998, 1999) and *B. malayi* (J. Daub and M. Blaxter, unpublished observations) mitochondria share a novel gene order, despite being from the same clade (III) as *A. suum*. Many genome rearrangements would be required to transform the protein and tRNA gene order of the ascaridid genome into the onchocercine type. The genome sequences of additional spirurid mitochondria may be of utility in unravelling the phylogeny of this group.

The complete genome of the *T. spiralis* mitochondrion has a third gene order (Lavrov and Brown, 2001), but in this case the pattern is much more similar to that found in other metazoans, rather than being similar to other nematodes, and it includes the ATP8 gene missing from the secernentean mitochondria. In *R. culcivorax*, mapping of the apparently linear mitochondrial genome suggests that it is significantly rearranged compared with *T. spiralis* (Powers *et al.*, 1986, 1993; Beck and Hyman, 1988; Hyman *et al.*, 1988; Hyman and Slater, 1990; Hyman and Azevedo, 1996). Both these species are in clade I (Dorylaimia).

In the tylenchids, mitochondrial gene sequences from cyst and root knot nematodes are highly divergent from other nematodes. Indeed, within-species divergence in *Meloidogyne* is greater than the difference between *A. suum* and *C. elegans* suggesting that mutational rates in this lineage may be extraordinarily high (Pelonquin *et al.*, 1993; Powers *et al.*, 1993; Hugall *et al.*, 1994, 1997). The mitochondrial genome of *Globodera pallida* is unique in that it is fragmented into a population of sub-genomic circles of 7–10 kb, each encoding only a subset of the mitochondrial genes (Armstrong *et al.*, 2000). This multipartite genome might explain the observed high within-species divergence in mitochondrial genes, if paralogous copies were being

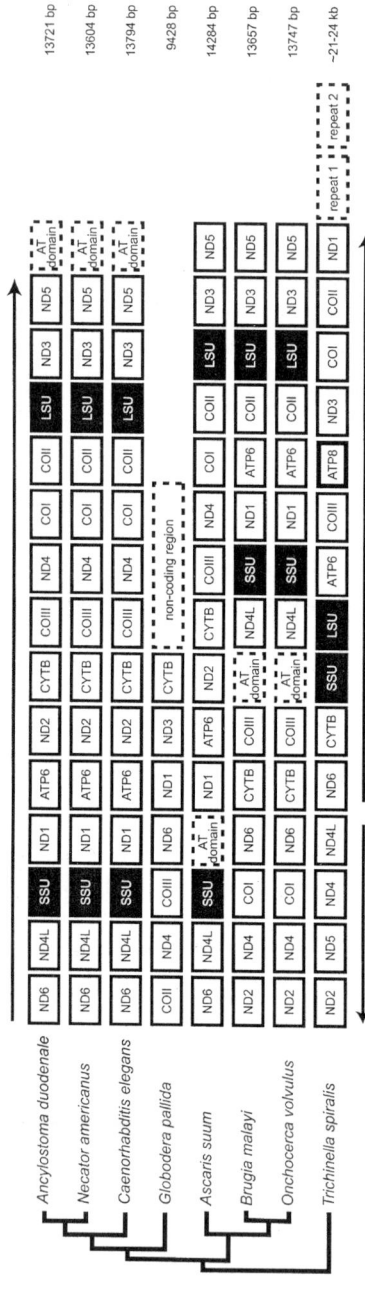

Figure 9 Nematode mitochondrial genomes. Diagram illustrating the changes in protein and ribosomal RNA gene order between nematode mitochondrial genomes. The tree to the left indicates the relationships of the species figured, based on Figure 2. Each protein-coding gene is indicated by a box (not to scale with gene length) and labelled with the gene name. Protein-coding genes have white box fills, RNA genes black fills. The AT domain is a non-coding segment that is probably the site of initiation of replication. In *Trichinella spiralis* there are two copies of a non-coding repeat element in the genome. The circular genomes have been arbitrarily linearised, between the AT domain and the ND6 gene for *C. elegans* and between genes to permit maximal alignment for the other genomes. The *Globodera pallida* genome is one of several mitochondrial circles present in this species. The lengths of the sequenced genomes are shown to the right. The tRNA genes have been omitted from the figure for clarity.

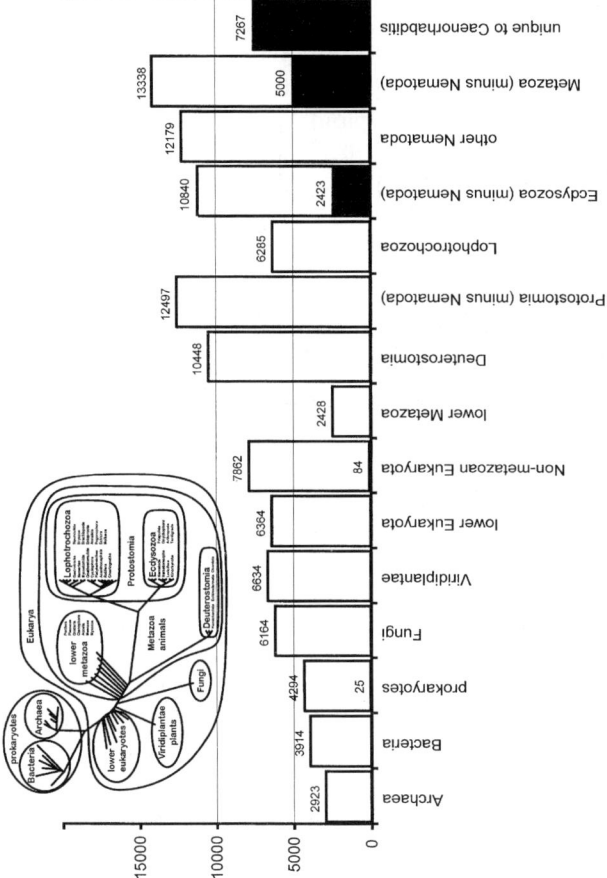

Figure 10 Comparison of the *C. elegans* proteome to other organisms. A set of taxon-limited databases were abstracted from GenBank/EMBL/DDBJ and used as targets for BLASTP (protein–protein) and TBLASTX (protein–nucleotide) searches with *C. elegans* WormPep proteins (release 89). Nucleotide databases were used for taxa, such as non-caenorhabditid nematodes, where there are no complete genomes and there is significantly more nucleotide data compared with protein data available. The BLAST results were processed using SimiTri (Parkinson and Blaxter, 2002). The solid bars indicate matches that are unique to the dataset, while the open bars indicate all matches to the dataset. The taxonomic spread of each dataset is indicated in the inset cartoon. For this figure a cut-off score of 50 was used: higher cut-offs reveal the same pattern, but with a greater skew to matches within Metazoa.

compared. The level of diversity in nematode mitochondrial genome structure and sequence is unparalleled in any other phylum.

3.1.6. Nematode Genome Evolution and Parasitism

None of the genomic patterns described above can be ascribed to the evolution of parasitism, as they are unique to particular lineages, some of which happen to be parasitic. Thus, it is hard to ascribe a parasitic function to the chromatin diminution observed in ascarids, and the rearrangements in onchocercine mitochondria do not appear to link to any particular adaptation to host exploitation. Other facets of genome organisation evolution such as the presence of polyploid lines in parthenogenetic tylenchid plant parasites (Triantaphyllou, 1983, 1985), or the utilisation of haplo-diploidy in sex determination in oxyurids (Adamson, 1989, 1994) may link to issues arising from life cycle dynamics in these parasites. At the current level of understanding, perhaps all that can be said is that nematode genomes have been discovered to be very dynamic in structure and organisation, and that the rate of change is reflected in the presence of apomorphic changes in crown lineages that may or may not be parasitic. It is clear that the evolution of parasitism in nematodes has not resulted in genome diminution, or simplification, as has been observed in bacterial and eukaryotic microbial genomes.

3.2. ORFeome/proteome evolution and the parasitic phenotype

Many textbooks have presented parasites as degenerate organisms that have lost the ability to exist without a host, that they are simplified derivatives of complex free-living ancestors (see for example von Brand, 1966 or Goldschmidt, 1937; see Brooks and McLennan, 1993, for a counter argument). This model can be extended to genomic considerations: it predicts that parasites will have lost genetic capacity to carry out functions that they now rely on hosts to perform. In bacterial pathogens (Fraser *et al.*, 1995, 1997; Andersson *et al.*, 1998) and some eukaryotic parasites this indeed appears to be the case. Mycoplasmas have evolved a minimal gene set and a tiny genome, and are obligate intracellular parasites that rely on host cells for most small-molecule synthetic biochemistry. *Encephalitozoon cuniculi*, a microsporidian intracellular parasite, also has a tiny genome that lacks many genes for basic, intermediate biochemistry (Katinka *et al.*, 2001).

I would argue that this view of parasitic phenotypes is rather narrow. While a parasite may rely on the host for supply of basic nutrients, and thus have diminished or no ability to carry out subsets of intermediary metabolism, it also has to ensure transmission between hosts, and deal effectively with host and environmental challenges. Thus, compared with *C. elegans* living in a bacterial heaven on an agar plate in a laboratory, a lymphatic filarial parasite such as *Brugia malayi* inhabits a very stressful niche. The parasite has to survive passage between two hosts, mosquito and vertebrate, with very different internal environments (in terms of osmolarity, pH and temperature for example), invade and establish in very different tissue types, invading through the mosquito gut to the flight muscles, then to the proboscis, and from the vertebrate skin to the lymphatics and thence, as larvae, to the blood. In addition it has to defend itself against a set of potentially lethal immune attacks, in the insect from the peritrophic membrane and gut digestion and the melanisation/engulfment effector arm of the immune system, and in the mammal from the innate, rapidly reacting immune system and the acquired, antigen-specific immune system. Lymphatic filariae have particular problems, in that they reside in one of the main thoroughfares of the active immune system, and thus are continuously exposed to possible attack. Thus we should expect parasites to have a mixture of specific losses or simplifications, as well as new complexities and amplifications: overall the balance may lie on the side of parasites being more, not less, complex.

3.2.1. How Many Genes do Nematodes Have?

Nematodes, with their outwardly simple body plan, were traditionally classified as 'lower metazoans'. With this prejudice came a conception that these lowly organisms must have few genes, fewer than the complex arthropods, such as *D. melanogaster*, with its millions of brain cells and embarrassment of bristly morphology, and at least an order of magnitude less than *H. sapiens*, with a billion brain cells and culture to match. Genome sequencing has levelled these perceptions. Initial, traditional genetics-based surveys of *C. elegans* suggested that there were perhaps 6000 genes that could be mutated to give a visible phenotype (Brenner, 1974). This estimate has been amply confirmed by subsequent screens, and by post-genome functional screens of all the predicted genes in the genome: there are estimated to be ~6500 genes identifiable in standard screens by phenotype (Johnsen and Baillie, 1991; Clark and Baillie, 1992; Stewart *et al.*, 1998; Fraser *et al.*, 2000; Zipperlen *et al.*, 2001). From the earliest days of the genome project, it became clear that the genome of *C. elegans* had at least double that number of protein-coding genes, and by the time of publication in 1998, the tally was 19 099 (Consortium, 1998). The current number is ~20 000: the uncertainty arises from the presence in the annotations of the genome of a set of genes whose existence is perhaps doubtful, and from the bioinformatic problems of recognising where one gene ends and another begins (Harrison et al., 2001; Hodgkin, 2001; Reboul et al., 2001; Mounsey et al., 2002). The *D. melanogaster* genome annotation yielded only about 14 000 genes (Adams et al., 2000), though, again, about a thousand more have been added since first publication (Gopal et al., 2001). Mammals have between 30 000 and 40 000 genes (Rubin et al., 2000; Lander et al., 2001; Venter et al., 2001; Waterston et al., 2002), while fungi and parasitic protozoa such as malaria and trypanosomatids have between 5000 and 12 000 (Myler et al., 2001; Gardner et al., 2002).

The numbers of predicted genes, and the complexity that they imply, does not sit well with traditional views of the relative complexity of the different animals. Humans, to us, are not 'just' 50% more complex than nematodes. One argument concerning raw gene counts has been that flies and humans have a much larger number of alternatively spliced genes, which means that, whatever the primary gene count, the predicted proteome (the total number of different proteins present in an organism) of these complex species is three- to four-fold higher than their gene count, while *C. elegans* lacks such extensive alternative splicing, and thus its proteome is about as complex as its gene count predicts. This view is not upheld by careful examination of expressed sequence tag datasets from these species, and while *C. elegans* may have an absolutely lower mean number of alternative splice forms per gene, this number (about 2) is not significantly lower than that seen for mammals or flies (about 2.5) (Brett et al., 2002). Another possibility is that the excess of genes in *C. elegans* is the result of mis- or over-prediction, but extensive cDNA cloning efforts suggest that at least 80% of the predicted genes are expressed under laboratory culture conditions (Reboul et al., 2001). A third is that the excess is mainly made up of pseudogenes, the result of gene duplication followed by mutational disablement. These non-functional genes may still be predicted by purely informatic methods from the genome sequence, but are not expressed and therefore not true genes (Mounsey et al., 2002). Promoter-trapping surveys of randomly selected genes also suggests that many putative promoter elements do not drive detectable transgene reporter expression under laboratory conditions, again suggesting that they may be non-functional (Young and Hope, 1993; Lynch et al., 1995; Hope et al., 1998). In this survey, the promoters that failed to drive expression had an over-representation of duplicated and *C. elegans*-unique genes, two classes that might be expected to harbour pseudogenes or mispredictions. Only ~12 000 of 17 000 genes represented on a cDNA-based DNA microarray were scored as positively expressed in any of a series of gene-expression hybridisation experiments (Kim, 2001; Kim *et al.*, 2001). From these observations it could be argued that the excess of putative genes is due to over-prediction: gene-finding algorithms are trained to find genes, not distinguish recently inactivated genes from active ones.

However, from other investigations it is clear that many mutations show a phenotype only in specific circumstances. Many genes appear to be redundant but this does not mean that they are non-functional. Two closely related genes may be able to replace each other in performing a function, and thus deletion of both is necessary to observe an effect (Kornfeld, 1997). Partial overlap, where the mild phenotype of one is exacerbated by the inactivation of the second, is also observed. In more extreme cases, whole pathways act redundantly. For example, in the specification of the hermaphrodite vulva, an inhibitory signal originating from the hypodermis is transmitted via two independent pathways: either pathway, involving several genes in a signalling cascade, is sufficient to give apparently wild-type function (Ferguson and Horvitz, 1989). Many genes have been identified by mutation only as enhancers or suppressors of other mutant phenotypes: in a wild-type background they have no phenotype. As genetic screens become more and more sophisticated, and ecologically realistic challenges are made of the mutagenised nematodes, more conditionally expressed genes have been identified. *C. elegans* is kept in a rather impoverished stimulatory environment in the laboratory, but a significant number of its genes are involved in environmental perception, and thus may have essential function in the wild (Troemel *et al.*, 1995; Bargmann, 1998; de Bono *et al.*, 2002).

The ~20 000 genes predicted can be split into classes according to function. In general, *C. elegans* has a broadly similar number of basic intermediary metabolism gene homologues, and regulatory gene homologues, as does *D. melanogaster*: both these protostomes have fewer regulatory genes than *H. sapiens*, reflecting the genome duplications in the lineage leading to humans (Rubin *et al.*, 2000; Muller *et al.*, 2002). However, *C. elegans* does have some strikingly large gene families that are less prominent in *D. melanogaster*. For example, *C. elegans* has over 250 nuclear hormone receptor genes (Bargmann, 1998; Sluder *et al.*, 1999; Sluder and Maina, 2001), many of which have now been implicated in a diversity of developmental control processes (Carmi *et al.*, 1998; Asahina *et al.*, 2000; Gissendanner and Sluder, 2000). Flies and humans have less than 50. Similarly, *C. elegans* has over 800 serpentine receptor genes (Troemel *et al.*, 1995; Robertson, 1998, 2000, 2001). These seven-transmembrane spanning proteins are thought to play roles in sensing of environmental chemicals, such as food and noxious signals. The abundance of these receptors is all the more surprising when it is recalled that *C. elegans* has only 302 neurons, and only ~30 of these are apparently sensory in nature (Wood, 1988). Each receptor gene is only expressed in one (or a very few) cells, and at low levels. This phenotype of restricted, low level expression might be the reason why many genes score negative in promoter screens and microarray hybridisations. Similarly, unless the correct behavioural screen is performed, knockouts of these genes will appear wild-type. Thus, the current best estimate of the number of open reading frames, or ORFs (hence the 'ORFeome') in the *C. elegans* genome is between 17 000 and 20 000, with the lower number from *in vivo* screens and the higher number from bioinformatic predictions.

(a) Conservation and divergence in the C. elegans *ORFeome.* Comparison of the *C. elegans* ORFeome with those of other phyla allows the 20 000 genes to be split into a number of categories (Mushegian *et al.*, 1998; Rubin *et al.*, 2000) (see Table 5). First are those shared by all the kingdoms of life (Bacteria, Eukarya and Archaea) that presumably perform functions basic to cellular metabolism. Second are those found in other eukaryotes, such as fungi, but not in non-eukaryotes. These genes, which include those for cytoskeletal proteins such as tubulin and actin, are presumably innovations that underlie eukaryote phenotypes. A third set is found in *C. elegans* and other bilaterian Metazoa only and thus probably identifies novelties required for animal phenotypes such as neural transmission and germ line- and gastrulation-based development. (There are very few sequences, and no complete genomes, available for non-bilaterian Metazoa, and thus it is at present impossible to make robust statements about genes relevant to metazoan origins *per se*.) The separation between the animal phyla is estimated to have occurred between 750 and 650 Mya, and thus ~1.4 billion years of sequence evolution separates human or fly genes from those of nematodes. As described above (Section 3.1.2; p. 129), *C. elegans* has an elevated rate of sequence change, and this is evident in whole ORFeome comparisons with *D. melanogaster*, *S. cerevisiae* and *H. sapiens* (Table 5). Because of this elevated rate, it is much harder to distinguish 'protostome-' and 'ecdysozoan-' specific genes: clear novelties can be identified but clade-specific features of genes belonging to conserved gene families may be obscured. The original publication of the *C. elegans* genome, in 1998, was before the completion of the *D. melanogaster* and *H. sapiens* genomes, before the vast majority of nematode EST sequencing, and when there were only a few bacterial and archaeal genomes available. Thus, early estimates of the number of nematode-unique, metazoa-shared

GENES, GENOMES AND PARASITISM IN THE NEMATODA

Table 5 Comparisons of the ORFeome of *Caenorhabditis elegans* to other model organisms.

Organism	Number of genes*	Number of distinct gene families**	Number of different domains**	Number of different domain architectures**	Percentage of unique genes/gene families**	Percentage of genes similar to homologues in other model organisms with an expectation value in BLAST of <e-10**			
						C. elegans	*D. melanogaster*	*H. sapiens*	*S. cerevisiae*
C. elegans	19 000	9500	1014	1000	50	–	39	36	20
D. melanogaster	14 000	8100	1035	1000	37	48	–	52	28
H. sapiens	~35 000	~14 000	1262	1800	30	40	45	–	13
S. cerevisiae	6000	4400	851	300	40	38	41	38	–

* The number of genes has been rounded to a low estimate. Different annotations predict different numbers for each organism. For example *S. cervisiae* is predicted to have between 5800 and 6300 genes.
** These predictions derive from Rubin *et al.* (2000) and Lander *et al.* (2002).

and universally conserved genes were limited by a lack of comparative data, and complicated by complex orthology–paralogy relationships discovered (Xie and Ding, 2000). We are performing ongoing recomparisons of the *C. elegans* proteome as represented by the database WormPep. We use BLAST to compare the entries in WormPep with a set of databases abstracted from the GenBank/EMBL/DDBJ nucleotide and protein databases. These comparisons are collated and viewed in a new tool, SimiTri (Parkinson and Blaxter, 2002), developed for the nematode EST programme (Parkinson et al., 2001) (Figure 10; see http://www.nematodes.org/SimiTri/SimiWorm.php).

About one-fifth of *C. elegans* proteins have significant similarity to prokaryote proteins, with ~3000 matching archaeal and ~4000 matching bacterial proteins. There are a few *C. elegans* nuclear genes that only have matches to prokaryote proteins (~25), but these matches are of low significance and may represent chance convergence. Nearly 40% of the proteome matches proteins of fungi, plants (Viridiplantae) and lower eukaryotes (the protistan phyla), but only ~80 proteins have unique matches to these taxa, including some that appear to have been acquired by horizontal transfer from fungi to *C. elegans*. Half of the proteome has similarity to deuterostome proteins (human, mouse, Japanese pufferfish (*Fugu*), and other taxa) but only 30% match potential proteins from the protostome superphylum Lophotrochozoa (annelids, molluscs, platyhelminths and allies). As *C. elegans* is a protostome, not a deuterostome, this apparent conflict merely highlights the lack of sequence from lophotrochozoan taxa. 60% match potential proteins from other nematodes (excluding the congeneric *C. briggsae*) and 65% have a match to some other sequenced gene: thus over 7000 *C. elegans* genes are still apparently unique to this nematode, despite the acquisition of gigabases of genome sequence since 1998.

(b) The C. elegans *ORFeome and other nematodes.* The EST datasets from other nematode species can be used to predict ORF sequences and thus partial proteomes for comparison to *C. elegans* (Parkinson et al., 2001, 2002). The number of distinct genes identified in each EST-sampled species depends on the number of ESTs generated, and the quality and number of cDNA libraries that are utilised. For *C. elegans*, over 190 000 ESTs derive from only ~55% of the genes (McCombie et al., 1992; Waterston et al., 1992; Kohara, 1997). These ESTs have been made from a limited number of libraries. In *B. malayi*, the 22 000 ESTs have been clustered and predicted to represent between 7500 and 8800 different genes (Williams et al., 2000; Blaxter et al., 2001). The uncertainty derives from problems assembling very short or low quality ESTs. There is no reason

to expect that *B. malayi* has significantly fewer genes than *C. elegans*, and thus these ESTs potentially identify ~45% of the *B. malayi* ORFeome. Confirmatory evidence that this is the case can be derived from EST matches to the sequenced segments of the *B. malayi* genome: about 40% of the predicted genes have EST matches (Gomez-Escobar *et al.*, 2002; Guiliano *et al.*, 2002).

As would be expected from the overall phylogeny, the predicted ORFeomes of strongylid parasites are more closely similar to *C. elegans* than are the proteomes of ascaridid, filarial and trichinellid species (Parkinson and Blaxter, 2002). Even so, there are still significant numbers of strongylid proteins that appear not to have a *C. elegans* homologue. For example, in *N. americanus* 42% of predicted genes from an adult nematode EST dataset were novel, lacking a *C. elegans* (or any other) homologue (Blaxter, 2000a; Daub *et al.*, 2000). A similar proportion of apparent novels was noted in analysis of the much larger *B. malayi* EST dataset: of 8800 predicted genes, over 6400 had no significant similarity to any *C. elegans* protein (Blaxter *et al.*, 2001).

Several methodological issues can confound such comparisons. Firstly, ESTs can be short and thus the maximal possible similarity may be below the cut-off for detection. Secondly, as most parasitic nematode ESTs are derived from the 5' end of the cDNA inserts, the sequence available may not extend far enough into the parasite open reading frame to provide a significant match. Thirdly, the evolutionary distance between *C. elegans* and the parasite may have allowed enough molecular change to accumulate in rapidly evolving genes to obscure homology. Analysis of *B. malayi* ESTs has confirmed that such issues may artificially elevate the numbers of apparent novels. Thus, when comparing all *B. malayi* ESTs to *C. elegans*, 43% of the genes with consensus sequences > 400 bases had matches, while only 2% of the genes with sequences < 200 bases had matches (where a match was scored at a BLAST probability of e^{-10}) (Blaxter *et al.*, 2001). In comparing the protein sequences derived from the 83 kb contig of *B. malayi* genomic DNA described above (see Section 3.1.3) it was noted that 11 of 12 predicted genes had homologues in *C. elegans* (Guiliano *et al.*, 2002). This elevated proportion could be a local fluke, but may also reflect an enhanced ability to detect homology when the complete gene sequence is available. The issue of elevated rates is also illustrated by genes in this section of the *B. malayi* genome, as some have only marginally significant pairwise matches to *C. elegans* genes that can robustly be called as orthologues due to their microsynteny and conserved intron boundaries.

Thus, simple similarity searches are not sufficient to identify homologues, and any estimate derived from the use of tools such as BLAST must be regarded as the lower bound. Despite these limitations, cross-species comparisons do reveal significant differences between datasets. In comparing the infective L3 gene expression profile of *B. malayi* to that of two other filarial nematodes, *Litomosoides sigmodontis* (Allen et al., 2000) and *O. volvulus* (Lizotte-Waniewski et al., 2000), it has been noted that while some genes are similarly highly expressed in each species, others are unique to one. This within-filarial comparison shows that, perhaps particularly in highly expressed genes encoding secreted products, even closely related taxa can differ significantly. This most likely reflects the effects of positive selection on functions related to the particular parasitic life cycle followed.

3.2.2. The 'Parasitome': Genes and Proteins Implicated in the Parasitic Phenotype

(a) Are parasites simpler? The comparison of genomic capacities between the partial parasite genomes available, and the free-living *C. elegans* suggests that parasitic nematodes are not radically simplified from their pre-parasitic ancestors. Indeed, comparing the predicted proteomes of *B. malayi* or *N. americanus* to *C. elegans* or other non-nematodes reveals genes present in the parasites and other metazoans, but absent from *C. elegans*. This shows that gene loss, like gene acquisition through duplication and diversification, is a constant part of the genomic evolutionary process in both parasitic and free-living organisms. Only in cases where there is strong selection on a parasite to replicate using minimal energy and rapidly, and where the host can be persuaded to give up complex metabolic resources to promote parasite growth, will there be genome reduction. Thus, intracellular pathogens such as mycoplasmas or *Rickettsia* (Andersson et al., 1998), or symbionts such as *Buchnera aphidicola* (Shigenobu et al., 2000; Tamas et al., 2002) or the mitochondrion (Lang et al., 1997), are able to lose genes. A metazoan parasite, residing for some or all of its life cycle outside a host cell, and often under extreme environmental conditions, probably cannot simplify. Indeed some parasites might be expected to be more complex than their pre-parasitic ancestors. *A. suum*, during its life cycle, experiences both aerobic and micro-aerobic or anaerobic conditions. Its glycolytic enzymes must therefore promote both anaerobic and aerobic glycolysis where relevant, and *A. suum* has duplicate genes for many glycolytic components with different kinetics for the forward, aerobic and reverse anaerobic reactions they carry out (Saruta et al., 1995; Duran et al., 1998; Amino et al., 2000).

In the absence of complete genome sequence it is impossible to be definitive about gene number in parasites. However, EST-based gene discovery in parasitic nematodes has been as effective as, and in many cases more effective than, similar analysis of *C. elegans* (see Section 3.2.1 (b) above; p. 146), suggesting that the gene count is unlikely to be significantly smaller. In genome sequencing we have noted that introns and intergenic regions tend to be longer in *B. malayi* than in *C. elegans* (Gomez-Escobar et al., 2002; Guiliano et al., 2002), suggesting that if the two genomes are of the same size (100 Mb) *B. malayi* may have fewer genes overall. Currently, it is not possible to distinguish between the possibility that *B. malayi* has fewer genes in a genome the same size as *C. elegans* and the possibility that the *B. malayi* genome is ~10% larger and has the same number of genes. Whole genome sequencing of *B. malayi* underway at The Institute for Genomic Research may offer answers to this.

(b) Genes with parasite-specific signatures. In the complexity of nematode genomes it is relatively easy to find genes specific to one parasitic species, or a parasitic group. It is harder to ascribe some function directly related to parasitism to these genes, as any comparison of free-living species will also yield a diverse set of genes unique to one or a subset of taxa. In a few cases, there may be direct functional evidence for a parasitic role, but for novel genes (those with no homologues elsewhere) it is often difficult to argue for specific roles. As interaction between nematode parasites and their hosts is almost exclusively extracellular and extraorganismal, one frequently observed feature of genes that may have acquired a role in host–parasite interaction is the acquisition of a secretory leader peptide. Thus, antioxidant enzymes found at the cuticle surface of filaria (see below) have secretory leader peptides, while their *C. elegans* orthologues do not. It may be fruitful to follow up all parasite genes that have a secretory leader peptide when their *C. elegans* homologues do not. Another feature that may mark genes important for parasitism is duplication. Genes may be duplicated so that one can perform a housekeeping role and the other a parasitic role, or may be multiply duplicated to form a gene family in a parasite where a single gene suffices for the free-living species. Duplication may permit expression at higher levels of critical activities.

The abundant larval transcript (*alt*) gene family of filarial nematodes is a case in point. The *alt-1* and *-2* genes were first identified as abundantly expressed transcripts of infective L3 of *B. malayi* (Gregory et al., 1997). Their protein products are secreted by the larvae on invasion of the definitive host. Subsequently, related genes with similar biology have been found to be highly expressed in infective L3 of several filaria, including

O. volvulus (Lizotte-Waniewski *et al.*, 2000), *Dirofilaria immitis* (Frank *et al.*, 1999) and *L. sigmodontis* (Allen *et al.*, 2000). These highly expressed *alt* genes have a secretory leader peptide followed by a small two-domain mature protein. The first domain is repetitive and rich in acidic amino acids, suggesting an extended hydrophilic structure, and the second has a conserved pattern of cysteine residues, suggesting a compact disulphide-bonded structure. The *B. malayi* and *O. volvulus* EST datasets include several sequences representing additional *alt* genes expressed at other life cycle stages, including one that lacks the acidic repetitive domain. *C. elegans* has a single *alt* gene that lacks an acidic repetitive domain, and whose cysteine-rich domain is most similar to the cysteine domain of the acidic domain lacking filarial genes (Gregory *et al.*, 2000; Gomez-Escobar *et al.*, 2002). Knockout of this gene in *C. elegans* has not yielded a phenotype (W.F. Gregory and M.L. Blaxter, unpublished), but a possible model is that the filariae have duplicated this gene, added an acidic domain, and this new combination has then been amplified to perform some yet unknown function in the host. This amplification may be specific to the filariae, as *A. suum* ESTs contain an acidic repeat domain lacking *alt* homologue, but none of the filarial type.

Nematode globin genes are another example of parasitism-specific gene duplication. *C. elegans* has a single globin gene, *glb-1*, encoding a 150-amino acid cytosolic protein expressed in anterior hypodermal cells (Kloek *et al.*, 1993a). Other nematodes also have this cytosolic or body wall globin (Blaxter, 1993b). Several novel globin isoforms are found in parasites. In *Mermis nigrescens*, the red pigment making up the 'eye' (a simple directional photoreceptor system) is a modified cytosolic globin gene, a duplicated daughter of the normal globin (Burr *et al.*, 1975, 2000; Burr and Harosi, 1987; Burr and Babinszki, 1990). *A. suum* also has two globin genes, a monomeric cytosolic globin (Blaxter *et al.*, 1994a) and a high molecular weight globin found in the pseudocoelomic fluid (von Brand, 1938; Davenport, 1949a, b; De Baere *et al.*, 1992; Sherman *et al.*, 1992b). This globin, renowned for its extraordinarily high affinity for oxygen (De Baere and Perutz, 1993; Gibson *et al.*, 1993; Kloek *et al.*, 1993b, c), is a translationally fused dimer of globin domains preceded by a signal peptide. Eight of these two-domain proteins associate to form the globin found in the fluid. The related ascaridid *Pseudoterranova decipiens* also has a two-domain globin (Dixon and Pohajdak, 1992), but *Toxocara canis* has a single-domain pseudocoelomic globin (P. Hunt and M. Blaxter, unpublished observations; see also (Kennedy *et al.*, 1989)). In strongylid nematodes such as *N. brasiliensis* there are also two globin genes (Davenport, 1949c; Blaxter *et al.*, 1994a): the cytosolic globin is joined by a second, single-domain secreted globin. In strongylids, the secreted globin is found in lacunae in the cuticle rather than in the pseudocoelom. The role of the *C. elegans glb-1* globin is unclear, as gene knockouts do not reveal a phenotype under normal conditions (P. Hunt and M.L. Blaxter, unpublished). In ascaridids, the pseudocoelomic globin may deliver oxygen to an essential aerobic process within the anaerobic or micro-aerobic environment. In strongylids, the cuticle globin, which has an affinity for oxygen lower than the cytosolic globin, appears to act as an oxygen scavenger, ensuring that the aerobic nematode is fully supplied even in the relatively oxygen-poor location of the gut (Blaxter, 1993b). 'Enzymatic' roles suggested for the globins of *Ascaris* (Sherman *et al.*, 1992a) are unlikely to be physiologically relevant as they are also effected by less oxygen-avid globins, and appear to derive from non-enzymatic side reactions resulting from superoxide generation. Interestingly, adults of the strongylid parasite *Syngamus trachea*, which reside in the trachea of bird hosts, also express a cuticle globin (Rose and Hwang, 1967; Rose and Kaplan, 1972), predicted to share the high affinity of gut parasitic species, at high levels: this may buffer the nematodes against the alternating atmospheric and exhaled air oxygen levels.

Some components of a parasite's environment may be less stressful than those of its pre-parasitic ancestors. For example, the lipid composition of a host is likely to be less variable than those of the diverse bacterial food sources a free-living bacteriovore exploits. Nematodes cannot synthesise sterols *de novo*, and require fatty acids. In *C. elegans* there are several different fatty acid and sterol-binding and carrier protein families that probably serve to collect and deliver lipids to required sites. Two of these lipid carrier gene types have a sequence pattern that suggests adaptation to varied targets. The fatty acid and retinol-binding gene family (*far*) has seven members in the *C. elegans* genome, each with very distinct sequence characteristics (Garofalo *et al.*, 2002; McDermott *et al.*, 2002). In contrast, it appears that *B. malayi* and *O. volvulus* adults express one major *far* gene (Tree *et al.*, 1995): this may reflect the relative simplicity of the filarial nematodes' lipid source. Similarly, the *C. elegans npa-1* gene encodes a polyprotein made up of 12 related but different small lipid-binding domains (McReynolds *et al.*, 1993; Kennedy *et al.*, 1995a, b). Each domain is separated by a subtilisin protease-like cleavage site, and the domains may therefore represent self-contained lipid-binding modules that collectively transport a diversity of substrates. A similar diverse *npa-1* gene has been sequenced from the strongylid *Dictyocaulus viviparus* (Kennedy *et al.*, 1995b). In filarial nematodes the *npa-1* gene also has many domains, but they are homogenous in sequence, suggesting convergence on a single optimal binding of a low complexity lipid source. The *A. suum npa-1* gene is comprised mainly of many identical domains but also has divergent domains (Kennedy *et al.*, 1995a).

(c) Evidence for variation in and selection on parasitism genes. Population-based research on viral, bacterial and protozoan parasites has been instrumental in identifying genes involved in host–parasite interaction. Genes under positive selection because of a core involvement in aspects of host immune evasion or recognition have been identified through particular signature patterns of rapid substitution and maintenance of large numbers of alleles in wild populations. Very little population genetic work has been carried out on nematode parasites, and the little that has been published has used genes or DNA segments that are not thought to play particularly significant roles in host–parasite interaction, such as microsattelite markers, intronic sequence, the ribosomal internal transcribed spacer, or mitochondrial genes (Powers *et al.*, 1986; Hyman and Slater, 1990; Blok and Phillips, 1995; Blouin *et al.*, 1995; Fisher and Viney, 1996, 1998; Hyman, 1996; Hyman and Azevedo, 1996; Hoekstra *et al.*, 1997; Roos *et al.*, 1998; Keddie *et al.*, 1999; Otsen *et al.*, 2000). It is thus premature to assess whether nematode parasites are more or less like other pathogens, though these neutral markers do show that parasite populations are very genetically diverse. It is of course essential that the variation in and dynamics of parasitic (and free-living) nematode populations are measured, and neutral, diverse markers such as microsatellites and other non-coding changes will be essential for these studies.

The major soluble surface glycoprotein of *B. malayi* and related filaria is a 29 kDa secreted glutathione peroxidase (GPX), postulated to have a role in immune evasion (see below) (Cookson *et al.*, 1992). Comparison of *gpx-1* sequences between species of *Brugia*, and with *Wuchereria bancrofti gpx-1* (Cookson *et al.*, 1993), revealed complete conservation of predicted peptide sequence between *B. malayi* and *Brugia pahangi*. At the nucleotide level, the *gpx-1* gene from these two species was 99% identical in exonic DNA (all changes were synonymous) and 90% identical in introns. Comparison with *W. bancrofti* revealed seven amino acid substitutions (the exons were 97.4% identical), and all these substitutions mapped either to the secretory leader peptide or to residues predicted to be on the external, solvent-accessible surface of the protein (Zvelebil *et al.*, 1993). This pattern of substitution could result from functional constraint and/or immune-mediated selection for antigenic difference. Whether there is within-species variation in this important protein is still unknown.

Examination of EST datasets reveals extensive within-species variation in some genes. *B. malayi* ESTs have all been generated from libraries derived from the inbred TRS Laboratory strain of parasites. Despite this, there is evidence for variation within this population in sequences for surface proteins such as the 15 kDa cystatin homologue (Gregory *et al.*, 1997). Allelic variation in noncoding regions has also been noted for other secreted protein genes, such as *alt-2* (see below) (Gomez-Escobar *et al.*, 2002). In other parasite EST datasets, particularly those from strongylid nematodes, the EST datasets reveal a number of variable genes. This pattern probably results from the use of poorly inbred or 'wild' populations of nematodes as the starting material for library construction. Thus, in *Haemonchus contortus* the most highly expressed L4 genes, *nim-1* and *-2*, have many allelic variants, up to seven per gene (C. Whitton and M. Blaxter, unpublished observations). Similarly, the cuticle globin gene-derived ESTs from *H. contortus* adults reveal several allelic variants. The functional significance, if any, of these variants is unknown, but studies on beta-tubulin, the target for benzimidazole drugs, has shown an extensive allelic variation in sequence that is selected upon by drug treatment (Kwa *et al.*, 1993a, b, 1994).

3.2.3. Arms Races Between Hosts and Parasites

The vertebrate immune system includes an innate and an adaptive arm. The innate arm is in all probability homologous to the innate immune systems of other animals. Innate immunity is induced or activated following recognition of particular chemical signatures of invading potential pathogens, such as conserved carbohydrate or lipid structures, and involves an immediate and non-specific response including the local or systemic production of small peptide antimicrobials, degradative enzymes and short-range, short-lived mediators such as nitrous oxide, oxy-radicals and halo-radicals. Adaptive immunity utilises a complex regulatory network of genes and gene products to firstly distinguish self from non-self, and then mount an appropriate and very specific response. The effector end of the response is mediated via the production of opsonising antibodies and the recruitment and activation of immune cells that kill or encapsulate pathogens. To survive in a host, therefore, a parasite not only has to be able to adapt its biochemistry to the available nutrients, but also counteract or evade the immune response.

Two general classes of defence are observed in nematode parasites (Blaxter *et al.*, 1992; Maizels *et al.*, 1993a, 2001a). The first is enzymatic, and is hypothesised to play a role in removing or counteracting the toxic products produced in response to infection. Thus, filarial nematodes express at their surfaces a set of antioxidant enzymes and anti-enzymes (Selkirk and Blaxter, 1990; Selkirk *et al.*, 1994). The major surface protein of *B. malayi* and other lymphatic filariae is a secreted glutathione peroxidase (GPX) that has lipid peroxidase activity (Tang *et al.*, 1995). It is unusual in both its activity spectrum and active site. It has a much more open substrate pocket that is more accepting of longer chain lipid peroxides than mammalian homologues, and has a cysteine at the active site in contrast to the seleno-cysteine found in other GPX. Also on the surface of *B. malayi* is a secreted Cu/Zn superoxide dismutase (SOD) (Tang et al., 1994; Ou et al., 1995a). The third soluble component detected at the *B. malayi* adult cuticle surface is a small secreted cystatin-type protease inhibitor (Lustigman et al., 1992). Together these three proteins may counteract local deposition of superoxide radicals (Ou et al., 1995b), may mop up lipid peroxides that would adversely affect the surface epicuticular lipid layer of the nematodes (Smith et al., 1996, 1998), and inhibit degradative proteases poised to attack the exposed cuticle or other structures. Anti-oxidant defences have also been characterised in other parasitic nematodes, such as *H. contortus*, where they are also hypothesised to play protective roles (Knox and Jones, 1992; Britton et al., 1994; Liddell and Knox, 1998; Newlands et al., 2001). EST analyses have identified as abundantly expressed genes several members of the thioredoxin peroxidase or peroxidoxin (TPX) family, particularly in mammalian-stage filaria (Klimowski et al., 1997; Schrum et al., 1998). In the free-living *C. elegans* the complete genome sequence reveals both intracellular and secreted SOD and GPX genes, and these have been implicated in protection against oxy-radical challenge and also in processes such as ageing and resistance of the dauer larva to environmental insult (Larsen, 1993). None of the extracellular forms appear to be expressed on the cuticle surface in *C. elegans* (Blaxter, 1993a), suggesting that the filariae have adapted an intrinsic defence system to counter extrinsic challenge. The cystatin inhibitors have been shown to play a role in cuticle moulting in *O. volvulus* (Lustigman, 1993) and also are present in *C. elegans*. Microfilariae (the L1 of filarial nematodes) of *B. malayi* also secrete an anti-enzyme in abundance, in this case a member of the serpin family of protease inhibitors (SPN-2) (Zang et al., 1999, 2000; Zang and Maizels, 2001). Interestingly, vector-stage infective L3 of *B. malayi* also synthesise an abundant serpin (SPN-1) (Yenbutr and Scott, 1995), of a different subfamily to SPN-2, that may play a role in parasite maintenance in the vector or in the early stages of infection.

Nematodes also have an innate immune system like other animals, and several anti-bacterial factor peptides (ABF) have been identified in *A. suum* (Kato and Komatso, 1996), and, by homology, in *N. americanus* and *C. elegans* (Daub et al., 2000). In *N. americanus* adults secretion of the ABF homologue may play some role in allowing the nematode to maintain itself despite the bacterial commensals in the gut. Other small, secreted peptides identified by EST sequencing in *N. americanus* may similarly have local roles in maintaining the feeding site (Blaxter, 2000b).

The second mode of defence expressed by nematode parasites is anti-immunity, through evasion or subversion (Maizels et al., 2001a, b). Bacterial and protozoan parasites have often evolved the ability to undergo antigenic variation, a trick that means they stay one step ahead of the adaptive immune system of their hosts for long enough to produce infective propagules. Nematodes are not known to undergo antigenic variation in the same manner, but it has been hypothesised that the limited diversity of possible protein antigens displayed at the surface of parasitic species, and the changing complement of these proteins on the surfaces of different parasitic stages of the same species, is an adaptation to avoid immune surveillance (Philipp *et al.*, 1980). However, as the free-living *C. elegans* also has a limited number of proteins exposed at its cuticle surface, and also changes the complement of proteins through development (Blaxter, 1993a), this adaptive link is not simply made. Some nematodes, such as *B. malayi* microfilariae, also adsorb host molecules to their surfaces, and these may act to mask nematode antigenic structures (Maizels *et al.*, 1984).

Nematode parasites may also actively intervene in the complex signalling of the mammalian immune system. *N. brasiliensis* adults secrete a number of acetylcholinesterase enzymes that are hypothesised to play a role in down-regulating the local inflammatory signalling pathways in the host gut (Blackburn and Selkirk, 1992; Grigg *et al.*, 1997; Hussein *et al.*, 1999b, 2000, 2002). *B. malayi* mammalian stages secrete a homologue of the mammalian cytokine macrophage migration inhibitory factor (MIF) (Pastrana *et al.*, 1998), first identified through EST sequencing (Williams *et al.*, 2000), and this factor may be a major player in the induction of immunotolerance to infection that is such a hallmark of filarial disease (Maizels and Lawrence, 1991; Allen and Maizels, 1996; Falcone *et al.*, 2001; Zang *et al.*, 2002). As *C. elegans* (and most other taxa examined, including protozoa and plants) also have MIF homologues (Marson, 2001), it is likely that the *B. malayi* MIF-1 implicated in immunosubversion has been recruited from a conserved role in within-organism homeostasis. Indeed, the *B. malayi* MIF-1 shares amino acid substitutions with mammalian MIFs that are absent from other nematode MIFs suggesting that it has evolved to be more mammal-like (Guiliano *et al.*, 2002). These examples suggest that secreted, highly expressed genes of animal parasitic stage nematodes are often implicated in anti-immune functions. In the nematode EST datasets are many highly expressed, potentially secreted molecules, that have no detectable similarity to proteins from other nematodes, or their hosts. These genes are prime candidates for investigation as potential immunomodulators and protective agents. For example, the hyper-abundant ALT proteins are prominent in filarial species but not in other nematodes: what is their role in immune evasion.

Plant parasitic nematodes have a different set of host defences to overcome (Bird and Wilson, 1994b; Bird and Koltai, 2000; Davis *et al.*,

2000) The standard non-self response of a plant under attack is to produce local necrosis that kills host cells as well as invading pathogens. These responses are under the control of a family of pathogen recognition genes that have been identified as resistance loci to many different sorts of pathogens (Ho et al., 1992; Kaloshian et al., 1995, 1998). These resistance genes are matched by virulence loci in the pathogens, the 'gene for gene' model (Staskawicz et al., 1995; Williamson and Hussey, 1996). Virulence genes have been mapped in the genetically tractable cyst nematode *Heterodera glycines* (Dong and Opperman, 1997) and resistance genes mapped and cloned from many plant species (Williamson et al., 1992; Cai et al., 1997; Devoto et al., 1999; Koltai et al., 2001).

The nematodes can either have a protective response to necrosis, by either secreting anti-enzymes that dampen the response or moving to a new feeding site when a response is mounted, or, as with the animal-parasitic taxa, interfering with or avoiding the host immune response. Indeed, one startling feature of many plant-parasitic nematode infections is the lack of pathological response from the plant: the nematode appears to be able to avoid tripping the triggers for necrosis despite its burrowing behaviours. In setting up its feeding site, a sedentary tylenchid plant-parasite must digest the plant cell walls and induce the selected cells to change metabolically and anatomically to support nematode growth (Jones, 1981). This appears to be achieved through secretion of enzymes and modulators through the feeding stylet from anterior dorsal and subventral glands (Hussey, 1989). Many of the components of the tylenchid plant parasitome have been identified molecularly through directed cloning (Bird and Wilson, 1994a; Ray et al., 1995; de Boer et al., 1996; Ding et al., 1998; Robertson et al., 1999; Gao et al., 2001a; Semblat et al., 2001; Wang et al., 2001) and EST strategies (McCarter et al., 2000, 2001, 2002), and display two surprising features. The first is the implication of plant parasitic nematode homologues of genes identified first in animal parasites as part of the host–parasite interface. Thus thioredoxin peroxidase (*tpx*) (Robertson et al., 2000) and lipid-binding (*far*) (Prior et al., 2001) genes are expressed in secretions from *Globodera rostochiensis*, suggesting that these plant parasites experience similar stresses and have similar ways of meeting nutritional requirements as their distant relatives parasitising animals. Fascinatingly, an allergen-like protein of unknown biological function homologous to proteins identified in filarial and strongylid secretions (Hawdon et al., 1996, 1999; Bin et al., 1999; Blaxter, 2000a; Daub et al., 2000; Blaxter et al., 2001; Murray et al., 2001) is secreted from *Heterodera glycines* and *Meloidogyne incognita* (Ding et al., 2000; Gao et al., 2001b). The second, more surprising finding is the presence in secretions of gene products related to plant cell wall degrading enzymes of bacteria. Glucanases and cellulases identified in *Meloidogyne* and

Heterodera are expressed from intron-containing nematode genes whose closest sequence relatives are found in saprobiotic microbes (Smant *et al.*, 1998; Wang *et al.*, 1999; Veronico *et al.*, 2001a, b). It would appear that the nematodes have acquired these genes from an environmental bacterium, and adapted them to their own use.

The setting-up and maintenance of the feeding site in cyst nematodes is a complex process (Wilson *et al.*, 1994a), and transgenic plant studies have revealed that certain plant genes are specifically responsive to nematode infection (Brenner *et al.*, 1998; Koltai *et al.*, 2001). One such gene, encoding a water channel protein normally expressed in the shoot, is ectopically turned on in root lesions by the nematode, and the plant gene promoter has a segment that is specifically responsive to the nematode (Opperman *et al.*, 1994). It is presumed that the nematode must secrete either some DNA-binding regulator, or a modifier of a host transcriptional regulator.

4. SUMMARY

From the above, I hope it is clear that the prospects for a molecular and evolutionary dissection of the adaptations of nematodes to plant and animal parasitism are very good. In the next few years there should be additional nematode genome sequences to mine for genes-of-interest, and comparative nematode genomics will be a major research topic. The 22 000 EST dataset from *B. malayi* defines ~8000 genes. The growing EST datasets from other parasitic nematodes should soon yield between 2000 and 10 000 genes per target species. The task for the future of parasitic nematode comparative genomics is to develop methods for functional testing of these genes that includes a phylogenetic perspective as well as an integrated view of parasite biology.

The embarrassment of riches in terms of numbers of genes identified must be matched by development of methods with which to test hypotheses of gene function in natural host–parasite systems. Due to complex life cycles and high fecundity, most nematode parasites are technically refractory to sensible genetic study, and thus 'classical' mutation-based analysis is unlikely to be applicable. However, emerging functional genomic and reverse genetic techniques have great promise. DNA-based genetic transformation of parasitic nematodes is in its infancy (Hashmi *et al.*, 1995; Higazi *et al.*, 2002) but with some development should permit testing of promoter function and residue-by-residue dissection of protein activity. As it can maintain foreign DNA as well as its own, the *C. elegans* system offers a tractable transgenesis test bed for conserved functions (Fire, 1986; Fire *et al.*, 1990; Mello *et al.*, 1991). Perhaps most exciting is the possibility of applying the reverse genetic technique of RNA interference (RNAi) to parasites (Montgomery and Fire, 1998; Timmons and Fire, 1998; Fire, 1999). In RNAi, introduced double-stranded RNA (dsRNA) induces the degradation of any corresponding mRNA through a sequence-specific RNAse activity. RNAi is effective in a wide range of eukaryotes, and in *C. elegans* has been used to screen over 80% of the genes in the genome for knockout phenotypes (Ashrafi *et al.*, 2003; Kamath *et al.*, 2003). RNAi has been applied to two parasitic nematode species. In *N. brasiliensis*, soaking adult animals in dsRNA corresponding to the secreted acetylcholinesterases induced a specific and lasting knock-down of mRNA and protein levels, and the treated nematodes were less able to maintain themselves when reimplanted into a rat host (Hussein *et al.*, 1999a, 2002). In *B. malayi*, RNAi was used to knock down expression of a drug target gene (beta tubulin) with lethal effect, and a microfilaria-specific gene (encoding a sheath component), resulting in failure of the later stages of embryogenesis and a reduction in fecundity of treated adult females (Aboobaker and Blaxter, 2003). RNAi should be applicable to all nematodes: dsRNA can be administered by soaking or by feeding bacteriovores with *Escherichia coli* expressing dsRNA from a recombinant plasmid.

Sequencing is cheap, and thus it should be possible, given the right collaborations between field scientists and molecular phylogeneticists, to trace the evolution of genes implicated in parasitism and compare their evolutionary trajectories to the background patterns of the rest of their genome. This phylogenetically biased genomics will be of especial interest to researchers wishing to derive universal molecular evolutionary signals of positive adaptation to parasitism, and may also allow the *de novo* identification of genes encoding important virulence factors. As nematodes have been coevolving with their hosts for many millions of years, we may also expect that the molecular tricks that nematodes use to evade host immunity will also reveal significant novelties in the logic of host immunity, benefiting the study and cure of other infections also. Thus anticoagulants, immunosuppressants and other regulators may be identified in nematodes and developed as novel pharmaceuticals (Cappello *et al.*, 1995).

As discussed above, phylogenetic analysis of nematode parasites in relation to their free-living relatives has already yielded some insights into the evolution of parasitism, and within-parasite group analyses have revealed patterns of change in the parasitic phenotype that are ripe for further analysis. Integration of a gene-centred or a genomic view with these phylogenies is important. For example, in the Rhabditina (Clade V) there are many examples of phoresis, necromeny, pathogenicity and true parasitism that currently appear to be independent events. More detailed investigation of the interrelationships of this complex group (Fitch, 1995) will hopefully yield dendrograms from which pattern and process can be discerned, mapping correlations between observed phenotypes of biology and behaviour and parasitic traits, and ultimately genetic changes underpinning them. Within the Strongyloidea, there are still major problems to be solved, such as the invasion strategy of the ancestral strongyloid (direct versus indirect) and the evolution of tissue migration. Within the nematodes of Spirurina (Clade III), the current molecular phylogeny suggests that taxa with complex, three-host, marine-based lifecycles arise from the base of the group: are these parasites relict indicators of the original spirurine or are they derived? In any case, the pattern of intermediate host usage and invasion/tissue migration strategies within the group remains a vital topic. Emerging gene and genomic data will need to be analysed in the light of an accepted species systematics: only in this framework will genetic and genomic novelties that define parasitic biology be identified.

ACKNOWLEDGEMENTS

I would like to thank my colleagues in the Nematode Genomics Group in Edinburgh (Jennifer Daub, Mark Dorris, Abebe Eyualem, Robin Floyd, David Guiliano, John Parkinson, Marian Thompson and Claire Whitton) and the Pathogens Sequencing Unit of the Wellcome Trust Sanger Institute (Neil Hall, Mike Quail and Bart Barrell) for support and data, Paul De Ley for his magisterial synthetic efforts in nematode phylogeny, and others who have supplied us with nematode material and libraries. The HOX evolution work was carried out in my laboratory by Aziz Aboobaker, the *Strongyloides* evolution by Mark Dorris and the synteny comparisons by David Guiliano. The *Brugia malayi* mitochondrial genome was sequenced by Jennifer Daub. I also would like to acknowledge the support of Jim McCarter, leader of the Washington University Nematode EST program, and the untiring work of the *C. elegans/C. briggsae* genome sequencing, annotation and database teams at the Sanger Institute (UK), the Washington University Genome Sequencing Center (USA), and in the WormBase Consortium (UK, USA). Research in my laboratory was funded by the UK Medical Research Council, The Wellcome Trust and the Linnean Society of London.

REFERENCES

Aboobaker, A.A. and Blaxter, M.L. (2003). Hox gene loss during dynamic evolution of the nematode cluster. *Current Biology* **13**, 37–40.
Adams, B.J., Burnell, A.M. and Powers, T.O. (1998). A phylogenetic analysis of the genus *Heterorhabditis* (Nemata: Rhabditidae) based on internal transcribed spacer 1 DNA sequence data. *Journal of Nematology* **30**, 22–39.
Adams, M.D., Celniker, S.E., Holt, R.A. *et al.* (2000). The genome sequence of *Drosophila melanogaster*. *Science* **287**, 2185–2195.
Adamson, M. (1994). Evolutionary patterns in life histories of Oxyurida. *International Journal for Parasitology* **24**, 1167–1177.
Adamson, M.L. (1989). Evolutionary biology of the Oxyurida (Nematoda): Biofacies of a haplodiploid taxon. *Advances in Parasitology* **28**, 175–226.
Adoutte, A., Balavoine, G., Lartillot, N. and de Rosa, R. (1999). Animal evolution. The end of the intermediate taxa? *Trends in Genetics* **15**, 104–108.
Adoutte, A., Balavoine, G., Lartillot, N., Lespinet, O., Prud'homme, B. and de Rosa, R. (2000). The new animal phylogeny: reliability and implications. *Proceedings of the National Academy of Science U. S. A.* **97**, 4453–4456.
Aeby, P., Spicher, A., de Chastonay, Y., Muller, F. and Tobler, H. (1986). Structure and genomic organization of proretrovirus-like elements partially eliminated from the somatic genome of *Ascaris lumbricoides*. *EMBO Journal* **5**, 3353–3360.
Aguinaldo, A.M.A., Turbeville, J.M., Linford, L.S., Rivera, M.C., Garey, J.R., Raff, R.A. and Lake, J.A. (1997). Evidence for a clade of nematodes, arthropods and other moulting animals. *Nature* **387**, 489–493.
Akam, M. (1998). Hox genes, homeosis and the evolution of segment identity: no need for hopeless monsters. *International Journal of Developmental Biology* **42**, 445–451.
Akam, M. (2000). Arthropods: developmental diversity within a (super) phylum. *Proceedings of the National Academy of Science U. S. A.* **97**, 4438–4441.
Akam, M., Averof, M., Castelli-Gair, J., Dawes, R., Falciani, F. and Ferrier, D. (1994). The evolving role of Hox genes in arthropods. *Development (Supplement)*, 209–215.
Albertson, D.G. (1984). Localization of the ribosomal genes in *Caenorhabditis elegans* chromosomes by in situ hybridization using biotin-labeled probes. *EMBO Journal* **3**, 1227–1234.
Albertson, D.G. (1993). Mapping chromosome rearrangement breakpoints to the physical map of *Caenorhabditis elegans* by fluorescent in situ hybridization. *Genetics* **134**, 211–219.
Albertson, D.G., Fishpool, R.M. and Birchall, P.S. (1995). Fluorescence in situ hybridization for the detection of DNA and RNA. *Methods in Cell Biology* **48**, 339–364.
Albertson, D.G. and Thomson, J.N. (1976). The pharynx of *Caenorhabditis elegans*. *Philosophical Transactions of the Royal Society of London. Series B Biological Sciences* **275**, 299–325.
Aleshin, V.V., Kedrova, O.S., Milyutina, I.A., Vladychenskaya, N.S. and Petrov, N.B. (1998a). Relationships among nematodes based on the analysis of 18S rRNA gene sequences: molecular evidence for a monophyly of chromadorian and secernentean nematodes. *Russian Journal of Nematology* **6**, 175–184.

Aleshin, V.V., Milyutina, I.A., Kedrova, O.S., Vladychenskaya, N.S. and Petrov, N.B. (1998b). Phylogeny of Nematoda and Cephalorhyncha derived from 18S rDNA. *Journal of Molecular Evolution* **47**, 597–605.

Allen, J.E., Daub, J., Guiliano, D., McDonnell, A., Lizotte-Waniewski, M., Taylor, D.W. and Blaxter, M. (2000). Analysis of genes expressed at the infective larval stage validates utility of *Litomosoides sigmodontis* as a murine model for filarial vaccine development. *Infection and Immunity* **68**, 5454–5458.

Allen, J.E. and Maizels, R.M. (1996). Immunology of human helminth infection. *International Archives of Allergy and Applied Immunology* **109**, 3–10.

Almeyda-Artigas, R.J., Bargues, M.D. and Mas-Coma, S. (2000). ITS-2 rDNA sequencing of *Gnathostoma* species (Nematoda) and elucidation of the species causing human gnathostomiasis in the Americas. *Journal of Parasitology* **86**, 537–544.

Amaral Zettler, L.A., Gomez, F., Zettler, E., Keenan, B.G., Amils, R. and Sogin, M.L. (2002). Microbiology: eukaryotic diversity in Spain's River of Fire. *Nature* **417**, 137.

Amino, H., Wang, H., Hirawake, H., Saruta, F., Mizuchi, D., Mineki, R., Shindo, N., Murayama, K., Takamiya, S., Aoki, T., Kojima, S. and Kita, K. (2000). Stage-specific isoforms of *Ascaris suum* complex. II: The fumarate reductase of the parasitic adult and the succinate dehydrogenase of free-living larvae share a common iron-sulfur subunit. *Molecular and Biochemical Parasitology* **106**, 63–76.

Anderson, R.C. (1992). *Nematode Parasites of Vertebrates*. Their development and transmission. C.A.B. International, Wallingford, UK.

Anderson, R.M. and May, R.M. (1982). Coevolution of hosts and parasites. *Parasitology* **85**, 411–426.

Andersson, S.G., Zomorodipour, A., Andersson, J.O., Sicheritz-Ponten, T., Alsmark, U.C., Podowski, R.M., Naslund, A.K., Eriksson, A.S., Winkler, H.H. and Kurland, C.G. (1998). The genome sequence of *Rickettsia prowazekii* and the origin of mitochondria. *Nature* **396**, 133–140.

Armstrong, M.R., Blok, V.C. and Phillips, M.S. (2000). A multipartite mitochondrial genome in the potato cyst nematode *Globodera pallida*. *Genetics* **154**, 181–192.

Asahina, M., Ishihara, T., Jindra, M., Kohara, Y., Katsura, I. and Hirose, S. (2000). The conserved nuclear receptor Ftz-F1 is required for embryogenesis, moulting and reproduction in *Caenorhabditis elegans*. *Genes and Cells* **5**, 711–723.

Ashford, R.W. and Barnish, G. (1989). *Strongyloides fuelleborni* and similar parasites in animals and man. In *Strongyloidiasis: a major roundworm infection of man*. (Grove, D.I., ed) pp. 271–287, London: Taylor and Francis.

Ashrafi, K., Chang, F.Y., Watts, J.L., Fraser, A.G., Kamath, R.S., Ahringer, J. and Ruvkun, G. (2003). Genome-wide RNAi analysis of *Caenorhabditis elegans* fat regulatory genes. *Nature* **421**, 268–272.

Ashton, F.T., Bhopale, V.M., Fine, A.E. and Schad, G.A. (1995). Sensory neuroanatomy of a skin-penetrating nematode parasite: *Strongyloides stercoralis*. I. Amphidial neurons. *Journal of Comparative Neurology* **357**, 281–295.

Ashton, F.T. and Schad, G.A. (1996). Amphids in *Strongyloides stercoralis* and other parasitic nematodes. *Parasitology Today* **12**, 187–194.

Baird, S.E., Fitch, D.H.A. and Emmons, S.W. (1994). *Caenorhabditis vulgaris* sp.n. (Nematoda: Rhabditidae): A necromenic associate of pill bugs and snails. *Nematologica* **40**, 1–11.

Bandi, C., Anderson, T.J.C., Genchi, C. and Blaxter, M.L. (1998). Phylogeny of *Wolbachia* bacteria in filarial nematodes. *Proceedings of the Royal Society of London. Series B Biological Sciences* **265**, 2407–2413.

Bandi, C., Trees, A.J. and Brattig, N.W. (2001). *Wolbachia* in filarial nematodes: evolutionary aspects and implications for the pathogenesis and treatment of filarial diseases. *Veterinary Parasitology* **98**, 215–238.

Bargmann, C.I. (1998). Neurobiology of the *Caenorhabditis elegans* genome. *Science* **282**, 2028–2033.

Barnes, T.M., Kohara, Y., Coulson, A. and Hekimi, S. (1995). Meiotic recombination, noncoding DNA and genomic organization in *Caenorhabditis elegans*. *Genetics* **141**, 159–179.

Baumann, P., Baumann, L., Clark, M.A. and Thao, M.L. (1998). *Buchnera aphidicola*: the endosymbiont of aphids. *ASM News* **64**, 203–209.

Beauchamp, R.S., Pasternak, J. and Straus, N.A. (1979). Characterization of the genome of the free-living nematode *Panagrellus silusiae*: absence of short period interspersion. *Biochemistry* **18**, 245–251.

Beck, J.L. and Hyman, B.C. (1988). Role of sequence amplification in the generation of nematode mitochondrial DNA polymorphism. *Current Genetics* **14**, 627–636.

Bentley, S.D., Chater, K.F., Cerdeno-Tarraga, A.M., Challis, G.L., Thomson, N.R., James, K.D., Harris, D.E., Quail, M.A., Kieser, H., Harper, D., Bateman, A., Brown, S., Chandra, G., Chen, C.W., Collins, M., Cronin, A., Fraser, A., Goble, A., Hidalgo, J., Hornsby, T., Howarth, S., Huang, C.H., Kieser, T., Larke, L., Murphy, L., Oliver, K., O'Neil, S., Rabbinowitsch, E., Rajandream, M.A., Rutherford, K., Rutter, S., Seeger, K., Saunders, D., Sharp, S., Squares, R., Squares, S., Taylor, K., Warren, T., Wietzorrek, A., Woodward, J., Barrell, B.G., Parkhill, J. and Hopwood, D.A. (2002). Complete genome sequence of the model actinomycete *Streptomyces coelicolor* A3(2). *Nature* **417**, 141–147.

Bin, Z., Hawdon, J., Qiang, S., Hainan, R., Huiqing, Q., Wei, H., Shu-hua, X., Tiehua, L., Xing, G., Zheng, F. and Hotez, P. (1999). *Ancylostoma* secreted protein 1 (ASP-1) homologues in human hookworms. *Molecular and Biochemical Parasitology* **98**, 143–149.

Bird, D.M. and Koltai, H. (2000). Plant parasitic nematodes: Habitats, hormones, and horizontally-acquired genes. *Journal of Plant Growth Regulation* **19**, 183–194.

Bird, D.M. and Wilson, M.A. (1994a). DNA sequence and expression analysis of root-knot nematode elicited giant cell transcripts. *Molecular Plant-Microbe Interactions* **7**, 419–424.

Bird, D.M. and Wilson, M.A. (1994b). Plant molecular and cellular responses to nematode infection. In: *NATO ARW: Advances in molecular plant nematology* (F. Lamberti, C. De Giorgi and D.M. Bird, eds), pp. 181–195. New York: Plenum Press.

Blackburn, C.C. and Selkirk, M.E. (1992). Characterisation of the secretory acetylcholinesterases from adult *Nippostrongylus brasiliensis*. *Molecular and Biochemical Parasitology* **53**, 79–88.

Blair, J.E., Ikeo, K., Gojobori, T. and Hedges, S.B. (2002). The evolutionary position of nematodes. *BMC Evolutionary Biology* **2**, 7.

Blaxter, M.L. (1993a). Cuticle surface proteins of wild type and mutant *Caenorhabditis elegans*. *Journal of Biological Chemistry* **268**, 6600–6609.

Blaxter, M.L. (1993b). Nemoglobins: Divergent nematode globins. *Parasitology Today* **9**, 353–360.

Blaxter, M.L. (1998). *Caenorhabditis elegans* is a nematode. *Science* **282**, 2041–2046.
Blaxter, M. (2000a). Genes and genomes of *Necator americanus* and related hookworms. *International Journal for Parasitology* **30**, 347–355.
Blaxter, M.L. (2000b). Genes and genomes of *Necator americanus* and related hookworms. *International Journal for Parasitology* **30**, 347–355.
Blaxter, M.L. (2002). Parasite Genomics. In: *Molecular Medical Parasitology* (R. Koumeniecki, J. Marr and T.W. Nilsen, eds), New York: Academic Press. pp. 3–28.
Blaxter, M. (2003). Molecular systematics: Counting angels with DNA. *Nature* **421**, 122–124.
Blaxter, M. and Liu, L. (1996). Nematode spliced leaders-ubiquity, evolution and utility. *International Journal for Parasitology* **26**, 1025–1033.
Blaxter, M.L., Page, A.P., Rudin, W. and Maizels, R.M. (1992). The nematode surface coat: Actively evading immunity. *Parasitology Today* **8**, 243–247.
Blaxter, M.L., Ingram, L. and Tweedie, S. (1994a). Sequence, expression and evolution of the globins of the nematode *Nippostrongylus brasiliensis*. *Molecular and Biochemical Parasitology* **68**, 1–14.
Blaxter, M.L., Vanfleteren, J.R., Xia, J. and Moens, L. (1994b). Structural characterization of an Ascaris myoglobin. *Journal of Biological Chemistry* **269**, 30181–30186.
Blaxter, M.L., De Ley, P., Garey, J.R., Liu, L.X., Scheldeman, P., Vierstraete, A., Vanfleteren, J.R., Mackey, L.Y., Dorris, M., Frisse, L.M., Vida, J.T. and Thomas, W.K. (1998). A molecular evolutionary framework for the phylum Nematoda. *Nature* **392**, 71–75.
Blaxter, M.L., Aslett, M., Daub, J., Guiliano, D. and The Filarial Genome Project (1999). Parasitic helminth genomics. *Parasitology* **118**, S39–S51.
Blaxter, M., Daub, J., Guiliano, D., Parkinson, J., Whitton, C. and The Filarial Genome Project (2001). The *Brugia malayi* genome project: expressed sequence tags and gene discovery. *Transactions of the Royal Society of Tropical Medicine and Hygiene* **96**, 1–11.
Bloemers, G.F., Hodda, M., Lambshead, P.J.D., Lawton, J.H. and Wanless, F.R. (1997). The effects of forest disturbance on diversity of tropical soil nematodes. *Oecologica* **111**, 575–582.
Blok, V.C. and Phillips, M.S. (1995). The use of repeat sequence primers for investigating genetic diversity between populations of potato cyst nematodes with different virulence. *Fundamental and Applied Nematology* **18**, 575–582.
Blouin, M.S., Yowell, C.A., Courtney, C.H. and Dame, J.B. (1995). Host movement and the genetic structure of populations of parasitic nematodes. *Genetics* **141**, 1007–1014.
Blumenthal, T. (1998). Gene clusters and polycistronic transcription in eukaryotes. *BioEssays* **20**, 480–487.
Blumenthal, T. and Steward, K. (1997). RNA processing and gene structure. In: *C. elegans II* (D. Riddle, T. Blumenthal, B. Meyer, and J. Priess, eds), pp. 117–145. Cold Spring Harbor, NY: Cold Spring Harbor Laboratory Press.
Blumenthal, T., Evans, D., Link, C.D., Guffanti, A., Lawson, D., Thierry-Mieg, J., Thierry-Mieg, D., Chiu, W.L., Duke, K., Kiraly, M. and Kim, S.K. (2002). A global analysis of *Caenorhabditis elegans* operons. *Nature* **417**, 851–854.
Bongers, T. (1990). The maturity index: an ecological measure of environmental disturbance based on nematode species composition. *Oecologica* **83**, 14–19.

Bovien, P. (1937). Some types of association between nematodes and insects. Bianco Luno A/S, KBHVN, Copenhagen.
Brenner, E.D., Lambert, K.N., Kaloshian, I. and Williamson, V.M. (1998). Characterization of LeMir, a root-knot nematode-induced gene in tomato with an encoded product secreted from the root. *Plant Physiology* **118**, 237–247.
Brenner, S. (1974). The genetics of *Caenorhabditis elegans*. *Genetics* **77**, 71–94.
Brett, D., Pospisil, H., Valcarcel, J., Reich, J. and Bork, P. (2002). Alternative splicing and genome complexity. *Nature Genetics* **30**, 29–30.
Britton, C., Knox, D.P. and Kennedy, M.W. (1994). Superoxide dismutase (SOD) activity of *Dictyocaulus viviparus* and its inhibition by antibody from infected and vaccinated bovine hosts. *Parasitology* **109**, 257–263.
Brooks, D.R. and McLennan, D.A. (1993). Parascript, parasites and the language of evolution. Smithsonian Institution press, Washington, USA.
Brusca, R.C. and Brusca, G.J. (1990). Invertebrates. Sinauer Associates, Sunderland, MA.
Bundy, D.A.P. (1997). This wormy world - then and now. *Parasitology Today* **13**, 407–408.
Burglin, T., Lobos, E. and Blaxter, M.L. (1998). *Caenorhabditis elegans* as a model for parasitic nematodes. *International Journal for Parasitology* **28**, 395–411.
Burr, A.H. and Harosi, F.I. (1987). Naturally crystalline hemoglobin of the nematode *Mermis nigrescens*. *Biophysical Journal* **47**, 527–536.
Burr, A.H., Hunt, P., Wagar, D.R., Dewilde, S., Blaxter, M.L., Vanfleteren, J.R. and Moens, L. (2000). A hemoglobin with an optical function. *Journal of Biological Chemistry* **275**, 4810–4815.
Burr, A.H., Schiefke, R. and Bollerup, G. (1975). Properties of a hemoglobin from the chromatrope of the nematode *Mermis nigrescens*. *Biochimica et Biophysica Acta* **405**, 404–411.
Burr, A.H.J. and Babinszki, C.P.F. (1990). Scanning motion, ocellar morphology and orientation mechanisms in the phototaxis of the nematode *Mermis nigrescens*. *Journal of Comparative Morphology A-Sensory, Neural and Behavioural Physiology* **167**, 257–268.
Burt, A. and Bell, G. (1987). Red Queen versus tangled bank model. *Nature* **330**, 118.
Cabaret, J. and Morand, S. (1990). Single and dual infections of the land snail *Helix aspersa* with *Muellerius capillaris* and *Alloionema appendiculatum* (Nematoda). *Journal of Parasitology* **76**, 579–580.
Cai, D., Kleine, M., Kifle, S., Harloff, H.J., Sandal, N.N., Marcker, K.A., Klein-Lankhorst, R.M., Salentijn, E.M., Lange, W., Stiekema, W.J., Wyss, U., Grundler, F.M. and Jung, C. (1997). Positional cloning of a gene for nematode resistance in sugar beet. *Science* **275**, 832–834.
Cappello, M., Vlasuk, G.P., Bergum, P.W., Huang, S. and Hotez, P.J. (1995). *Ancylostoma caninum* anticoagulant peptide: a hookworm-derived inhibitor of human coagulation factor Xa. *Proceedings of the National Academy of Science U. S. A.* **92**, 6152–6156.
Carmi, I., Kopczynski, J.B. and Meyer, B.J. (1998). The nuclear hormone receptor SEX-1 is an X-chromosome signal that determines nematode sex. *Nature* **396**, 168–173.
Casiraghi, M., Anderson, T.J., Bandi, C., Bazzocchi, C. and Genchi, C. (2001). A phylogenetic analysis of filarial nematodes: comparison with the phylogeny of *Wolbachia* endosymbionts. *Parasitology* **122 Pt 1**, 93–103.

Chen, J.Y., Oliveri, P., Li, C.W., Zhou, G.Q., Gao, F., Hagadorn, J.W., Peterson, K.J. and Davidson, E.H. (2000). Precambrian animal diversity: putative phosphatized embryos from the Doushantuo Formation of China. *Proceedings of the National Academy of Science U. S. A.* **97**, 4457–4462.

Chilton, N.B., Gasser, R.B. and Beveridge, I. (1995). Differences in ribosomal DNA sequence of morphologically indistinguishable species within the *Hypodontus macropi* complex (Nematoda: Strongyloidea). *International Journal for Parasitology* **25**, 647–651.

Chilton, N.B., Gasser, R.B. and Beveridge, I. (1997a). Phylogenetic relationships of Australian strongyloid nematodes inferred from ribosomal DNA sequence data. *International Journal for Parasitology* **27**, 1481–1494.

Chilton, N.L.B., Hoste, H., Hung, G.-C., Beveridge, I. and Gasser, R.B. (1997b). The 5.8S rDNA sequences of 18 species of bursate nematodes (order Strongylida): Comparison with Rhabditid and Tylenchid nematodes. *International Journal for Parasitology* **27**, 119–124.

Clark, D.V. and Baillie, D.L. (1992). Genetic analysis and complementation by germ-line transformation of lethal mutations in the *unc-22* IV region of *Caenorhabditis elegans*. *Molecular and General Genetics* **232**, 97–105.

Coghlan, A. and Wolfe, K.H. (2002). Fourfold faster rate of genome rearrangement in nematodes than in *Drosophila*. *Genome Research* **12**, 857–867.

Consortium (The *C. elegans* Genome Sequencing Consortium) (1998). Genome sequence of the nematode *C. elegans*: a platform for investigating biology. *Science* **282**, 2012–2018.

Consortium (The *C. elegans* Genome Sequencing Consortium) (1999). How the worm was won. *Trends in Genetics* **15**, 51–58.

Conway Morris, S. (1981). Parasites and the fossil record. *Parasitology* **82**, 489–509.

Conway Morris, S. (1993). The fossil record and the early evolution of the Metazoa. *Nature* **361**, 219–225.

Cookson, E., Blaxter, M.L. and Selkirk, M.E. (1992). Identification of the major soluble cuticular glycoprotein of lymphatic filarial nematode parasites as a secretory homologue of glutathione peroxidase. *Proceedings of the National Academy of Science U. S. A.* **89**, 5837–5841.

Cookson, E., Tang, L. and Selkirk, M.E. (1993). Conservation of primary structure of gp29, the major soluble cuticular glycoprotein, in three species of lymphatic filariae. *Molecular and Biochemical Parasitology* **58**, 155–160.

Coulson, A., Huynh, C., Kozono, Y. and Shownkeen, R. (1996). The physical map of the *Caenorhabditis elegans* genome. In: *Caenorhabditis elegans. Modern Biological Analysis of an Organism* (H.F. Epstein and D.C. Shakes, eds), pp. 533–550. San Diego, CA: Academic Press.

Coulson, A., Waterston, R., Kiff, J., Sulston, J. and Kohara, Y. (1988). Genome linking with yeast artificial chromosomes. *Nature* **335**, 184–186.

Coulson, A.R., Sulston, J.E., Brenner, S. and Karn, J. (1986). Towards a physical map of the genome of the nematode *C. elegans*. *Proceedings of the National Academy of Science U. S. A.* **83**, 7821–7825.

Daub, J., Loukas, A., Pritchard, D.I. and Blaxter, M. (2000). A survey of genes expressed in adults of the human hookworm, *Necator americanus*. *Parasitology* **120**, 171–184.

Davenport, H.E. (1949a). *Ascaris* haemoglobin as an indicator of the oxygen produced by isolated chloroplasts. *Proceedings of the Royal Society of London. Series B Biological Sciences* **136**, 281–290.

Davenport, H.E. (1949b). The haemoglobins of *Ascaris lumbricoides*. *Proceedings of the Royal Society of London. Series B Biological Sciences* **136**, 255–270.

Davenport, H.E. (1949c). The haemoglobins of *Nippostrongylus muris* (Yokagawa) and *Strongylus* spp. *Proceedings of the Royal Society of London. Series B Biological Sciences* **136**, 271–280.

Davies, E.K., Peters, A.D. and Keightley, P.D. (1999). High frequency of cryptic deleterious mutations in *Caenorhabditis elegans*. *Science* **285**, 1748–1751.

Davis, E.L., Hussey, R.S., Baum, T.J., Bakker, J., Schots, A., Rosso, M.N. and Abad, P. (2000). Nematode parasitism genes. *Annual Reviews of Phytopathology* **38**, 365–396.

Davis, R.E. (1995). *Trans*-splicing in flatworms. In: *Molecular Approaches to Parasitology*. Vol. 12 (J.C. Boothroyd and R. Komuniecki, eds), pp. 299–320. New York: Wiley-Liss.

Davis, R.E. (1996). Spliced leader RNA *trans*-splicing in metazoa. *Parasitology Today* **12**, 33–40.

Davis, R.E. (1997). Surprising diversity and distribution of spliced leader RNAs in flatworms. *Molecular and Biochemical Parasitology* **87**, 29–48.

De Baere, I., Liu, L., Moens, L., Van Beeumen, J., Gielens, C., Richelle, J., Trottmann, C., Finch, J., Gerstein, M. and Perutz, M. (1992). Polar zipper sequence in the high-affinity hemoglobin of *Ascaris suum*: Amino acid sequence and structural interpretation. *Proceedings of the National Academy of Science U. S. A.* **89**, 4638–4642.

De Baere, I. and Perutz, M. (1993). Formation of two hydrogen bonds from the globin to the heme-linked oxygen molecule in *Ascaris* hemoglobin. *Proceedings of the National Academy of Science U. S. A.* **91**, 1594–1597.

de Boer, J., Smant, G., Goverse, A., Davis, E.L., Overmars, H.A., Pomp, H.R., van Gent-Pelzer, M., Zilverentant, J.F., Stokkermans, J.P.W.G., Hussey, R.S., Gommers, F.J., Bakker, J. and Schots, A. (1996). Secretory granule proteins from the subventral esophegal glands of the potato cyst nematode identified by monoclonal antibodies to a protein fraction from second-stage juveniles. *Molecular Plant-Microbe Interactions* **9**, 39–46.

de Bono, M., Tobin, D.M., Davis, M.W., Avery, L. and Bargmann, C.I. (2002). Social feeding in *Caenorhabditis elegans* is induced by neurons that detect aversive stimuli. *Nature* **419**, 899–903.

De Ley, P. and Blaxter, M.L. (2002). Systematic position and phylogeny. In: *The Biology of Nematodes* (D. Lee, ed), Raeding: Harwood Academic Publishers.

De Ley, P., van de Velde, M.C., Mounport, D., Baujard, P. and Coomans, A. (1995). Ultrastructure of the stoma in Cephalobidae, Panagrolaimidae and Rhabditidae, with a proposal for a revised stoma terminology in Rhabditida. *Nematologica* **41**, 153–182.

de Meeus, T. and Renaud, F. (2002). Parasites within the new phylogeny of eukaryotes. *Trends in Parasitology* **18**, 247–251.

de Rosa, R., Grenier, J.K., Andreeva, T., Cook, C.E., Adoutte, A., Akam, M., Carroll, S.B. and Balavoine, G. (1999). Hox genes in brachiopods and priapulids and protostome evolution. *Nature* **399**, 772–776.

Denver, D.R., Morris, K., Lynch, M., Vassilieva, L.L. and Thomas, W.K. (2000). High direct estimate of the mutation rate in the mitochondrial genome of *Caenorhabditis elegans*. *Science* **289**, 2342–2344.

Devoto, A., Piffanelli, P., Nilsson, I., Wallin, E., Panstruga, R., von Heijne, G. and Schulze-Lefert, P. (1999). Topology, subcellular localization, and sequence diversity of the Mlo family in plants. *Journal of Biological Chemistry* **274**, 34993–35004.

Ding, X., Shields, J., Allen, R. and Hussey, R.S. (1998). A secretory cellulose-binding protein cDNA cloned from the root-knot nematode (*Meloidogyne incognita*). *Molecular Plant-Microbe Interactions* **11**, 952–959.

Ding, X., Shields, J., Allen, R. and Hussey, R.S. (2000). Molecular cloning and characterisation of a venom allergen AG5-like cDNA from *Meloidogyne incognita*. *International Journal for Parasitology* **30**, 77–81.

Dixon, B. and Pohajdak, B. (1992). Did the ancestral globin gene of plants and animals contain only two introns? *Trends in Biochemical Sciences* **17**, 486–488.

Dong, K. and Opperman, C.H. (1997). Genetic analysis of parasitism in the soybean cyst nematode *Heterodera glycines*. *Genetics* **146**, 1311–1318.

Dorris, M. and Blaxter, M.L. (2000). The small subunit ribosomal RNA sequence of *Strongyloides stercoralis*. *International Journal for Parasitology* **30**, 939–941.

Dorris, M., De Ley, P. and Blaxter, M.L. (1999). Molecular analysis of nematode diversity and the evolution of parasitism. *Parasitol Today* **15**, 188–193.

Dorris, M., Viney, M.E. and Blaxter, M.L. (2002). Molecular phylogenetic analysis of the genus *Strongyloides* and related nematodes. *International Journal for Parasitology* **32**, 1507.

Dropkin, V.H. (1989). Introduction to plant nematology. John Wiley and sons, Missouri.

Duran, E., Walker, D.J., Johnson, K.R., Komuniecki, P.R. and Komuniecki, R.W. (1998). Developmental and tissue-specific expression of 2-methyl branched-chain enoyl CoA reductase isoforms in the parasitic nematode, *Ascaris suum*. *Molecular and Biochemical Parasitology* **91**, 307–318.

Durette-Desset, M., Gasser, R.B. and Beveridge, I. (1994). The origins and evolutionary expansion of the Strongylida (Nematoda). *International Journal for Parasitology* **24**, 1139–1166.

Ekschmitt, K., Bakonyi, G., Bongers, M., Bongers, T., Boström, S., Dogan, H., Harrison, a., Nagy, P., O'donnell, A.G., Papatheodorou, E.M., Sohlenius, B., Stamou, G.P. and Wolters, V. (2001). Nematode community structure as indicator of soil functioning in European grassland soils. *European Journal of Soil Biology* **37**, 263–268.

Ellis, R.E., Yuan, J. and Horvitz, H.R. (1991). Mechanisms and functions of cell death. *Annual Reviews of Cell Biology* **7**, 663–698.

Etter, A., Aboutanos, M., Tobler, H. and Muller, F. (1991). Eliminated chromatin of *Ascaris* contains a gene that encodes a putative ribosomal protein. *Proceedings of the National Academy of Science U. S. A.* **88**, 1593–1596.

Etter, A., Bernard, V., Kenzelmann, M., Tobler, H. and Muller, F. (1994). Ribosomal heterogeneity from chromatin diminution in *Ascaris lumbricoides*. *Science* **265**, 954–956.

Evans, D., Zorio, D., MacMorris, M., Winter, C.E., Lea, K. and Blumenthal, T. (1997). Operons and SL2 trans-splicing exist in nematodes outside the genus *Caenorhabditis*. *Proceedings of the National Academy of Science U. S. A.* **94**, 9751–9756.

Falcone, F.H., Loke, P., Zang, X., MacDonald, A.S., Maizels, R.M. and Allen, J.E. (2001). A *Brugia malayi* homolog of macrophage migration inhibitory factor reveals an important link between macrophages and eosinophil recruitment during nematode infection. *Journal of Immunology* **167**, 5348–5354.

Felix, M.A., De Ley, P., Sommer, R.J., Frisse, L., Nadler, S.A., Thomas, W.K., Vanfleteren, J. and Sternberg, P.W. (2000). Evolution of vulva development in the Cephalobina (Nematoda). *Developmental Biology* **221**, 68–86.

Ferguson, E.L. and Horvitz, H.R. (1989). The multivulva phenotype of certain *Caenorhabditis elegans* mutants results from defects in two functionally redundant pathways. *Genetics* **123**, 109–121.

ffrench-Constant, R. and Bowen, D. (1999). *Photorhabdus* toxins: novel biological insecticides. *Current Opinion in Microbiology* **2**, 284–288.

ffrench-Constant, R.H., Waterfield, N., Burland, V., Perna, N.T., Daborn, P.J., Bowen, D. and Blattner, F.R. (2000). A genomic sample sequence of the entomopathogenic bacterium *Photorhabdus luminescens* W14: potential implications for virulence. *Appllied and Environmental Microbiology* **66**, 3310–3329.

Fine, A.E., Ashton, F.T., Bhopale, V.M. and Schad, G.A. (1997). Sensory neuroanatomy of a skin-penetrating nematode parasite *Strongyloides stercoralis*. II. Labial and cephalic neurons. *Journal of Comparative Neurology* **389**, 212–223.

Fire, A. (1986). Integrative transformation of *Caenorhabditis elegans*. *EMBO Journal* **5**, 2673–2680.

Fire, A. (1999). RNA-triggered gene silencing. *Trends in Genetics* **15**, 358–363.

Fire, A., Kondo, K. and Waterston, R. (1990). Vectors for low copy transformation of *C. elegans*. *Nucleic Acids Research* **18**, 4269–4270.

Fisher, M.C. and Viney, M.E. (1996). Microsatellites of the parasitic nematode *Strongyloides ratti*. *Molecular and Biochemical Parasitology* **80**, 221–224.

Fisher, M.C. and Viney, M.E. (1998). The population genetic structure of the facultatively sexual parasitic nematode *Strongyloides ratti* in wild rats. *Proceedings of the Royal Society of London. Series B Biological Sciences* **265**, 703–709.

Fitch, D.H.A. (1997). Evolution of male tail development in rhabditid nematodes related to *Caenorhabditis elegans*. *Systematic Biology* **46**, 145–179.

Fitch, D.H.A. (2000). Evolution of "Rhabditidae" and the Male Tail. *Nematology* **32**, 235–244.

Fitch, D.H.A., Bugaj-gaweda, B. and Emmons, S.W. (1995). 18S ribosomal gene phylogeny for some rhabditidae related to *Caenorhabditis elegans*. *Molecular Biology and Evolution* **12**, 346–358.

Fitch, D.H.A. and Thomas, W.K. (1997). Evolution. In: *C. elegans II* (D. Riddle, T. Blumenthal, B. Meyer and J. Priess, eds), pp. 815–850. Cold Spring Harbor, New York: Cold Spring Harbor Laboratory Press.

Floyd, R., Abebe, E., Papert, A. and Blaxter, M. (2002). Molecular barcodes for soil nematode identification. *Molecular Ecology* **11**, 839–850.

Frank, G.R., Wisnewski, N., Brandt, K.S., Carter, C.R., Jennings, N.S. and Selkirk, M.E. (1999). Molecular cloning of the 22–24 kDa excretory-secretory 22U protein of *Dirofilaria immitis* and other filarial nematode parasites: *Molecular and Biochemical Parasitology* **98**, 297–302.

Fraser, A.G., Kamath, R.S., Zipperlen, P., Martinez-Campos, M., Sohrmann, M. and Ahringer, J. (2000). Functional genomic analysis of *C. elegans* chromosome I by systematic RNA interference. *Nature* **408**, 325–330.

Fraser, C.M., Casjens, S., Huang, W.M., Sutton, G.G., Clayton, R., Lathigra, R., White, O., Ketchum, K.A., Dodson, R., Hickey, E.K., Gwinn, M., Dougherty, B., Tomb, J.F., Fleischmann, R.D., Richardson, D., Peterson, J., Kerlavage, A.R., Quackenbush, J., Salzberg, S., Hanson, M., van Vugt, R., Palmer, N., Adams, M.D., Gocayne, J., Weidman, J., Utterback, T., Watthey, L., McDonald, L.,

Artiach, P., Bowman, C., Garland, S., Fujii, C., Cotton, M.D., Horst, K., Roberts, K., Hatch, B., Smith, H.O. and Venter, J.C., et al. (1997). Genomic sequence of a Lyme disease spirochaete, *Borrelia burgdorferi*. *Nature* **390**, 580–586.
Gao, B., Allen, R., Maier, T., Davis, E.L., Baum, T.J. and Hussey, R.S. (2001a). Identification of putative parasitism genes expressed in the esophageal gland cells of the soybean cyst nematode *Heterodera glycines*. *Molecular Plant-Microbe Interactions* **14**, 1247–1254.
Gao, B., Allen, R., Maier, T., Davis, E.L., Baum, T.J. and Hussey, R.S. (2001b). Molecular characterisation and expression of two venom allergen-like protein genes in *Heterodera glycines*. *International Journal for Parasitology* **31**, 1617–1625.
Gardner, M.J., Hall, N., Fung, E., et al. (2002). Genome sequence of the human malaria parasite *Plasmodium falciparum*. *Nature* **419**, 498–511.
Garey, J.R. and Schmidt-Rhaesa, A. (1998). The essential role of "minor" phyla in molecular studies of animal evolution. *American Zoologist* **38**, 907–917.
Garofalo, A., Klager, S.L., Rowlinson, M.C., Nirmalan, N., Klion, A., Allen, J.E., Kennedy, M.W. and Bradley, J.E. (2002). The FAR proteins of filarial nematodes: secretion, glycosylation and lipid binding characteristics. *Molecular and Biochemical Parasitology* **122**, 161–170.
Gasser, R.B., Chilton, N.B., Hoste, H. and Beveridge, I. (1993). Rapid sequencing of rDNA from single worms and eggs of parasitic helminths. *Nucleic Acids Research* **21**, 2525–2526.
Gasser, R.B. and Newton, S.E. (2000). Genomic and genetic research on bursate nematodes: significance, implications and prospects. *International Journal for Parasitology* **30**, 509–534.
Gasser, R.B., Stewart, L.E. and Speare, R. (1996). Genetic markers in ribosomal DNA for hookworm identification. *Acta Tropica* **62**, 15–21.
Gaugler, R. and Kaya, H. (1990). The entomopathogenic nematodes in biological control. Boca Raton: CRC Press.
Geenen, P.L., Bresciani, J., Boes, J., Pedersen, A., Eriksen, L., Fagerholm, H.P. and Nansen, P. (1999). The morphogenesis of *Ascaris suum* to the infective third-stage larvae within the egg. *Journal of Parasitology* **85**, 616–622.
Gemmil, A.W., Viney, M.E. and Read, A.F. (1997). Host immune status determines sexuality in a parasitic nematode. *Evolution* **51**, 393–401.
Geraert, E., Mracek, Z. and Sudhaus, W. (1989). The facultative insect parasite *Pristionchus lheritieri* (syn. *P. pseudolheritieri*) (Nematoda: Diplogasterida). *Nematologica* **35**, 133–141.
Gibson, Q.H., Regan, R., Olson, J.S., Carver, T.E., Dixon, B., Pohajdak, B., Sharma, P.K. and Vinogradov, S.N. (1993). Kinetics of ligand binding to *Pseudoterranova decipiens* and *Ascaris suum* hemoglobins and to Leu-29 –>Tyr sperm whale myoglobin mutant. *Journal of Biological Chemistry* **268**, 16993–16998.
Giribet, G., Distel, D.L., Polz, M., Sterrer, W. and Wheeler, W.C. (2000). Triploblastic relationships with emphasis on the acoelomates and the position of Gnathostomulida, Cycliophora, Plathelminthes, and Chaetognatha: a combined approach of 18S rDNA sequences and morphology. *Systematic Biology* **49**, 539–562.
Giribet, G. and Wheeler, W.C. (1999). The position of arthropods in the animal kingdom: Ecdysozoa, islands, trees, and the "Parsimony ratchet". *Molecular Phylogenetics and Evolution* **13**, 619–623.

Gissendanner, C.R. and Sluder, A.E. (2000). nhr-25, the *Caenorhabditis elegans* ortholog of ftz-f1, is required for epidermal and somatic gonad development. *Developmental Biology* **221**, 259–272.

Goldschmidt, R. (1937). *Ascaris*. The biologist's story of life. New York: Prentice-Hall.

Goldstein, B. and Blaxter, M. (2002). Tardigrades. *Current Biology* **12**, R475.

Gomez-Escobar, N., Gregory, W.F., Britton, C., Murray, L., Daub, J., Blaxter, M.L. and Maizels, R.M. (2002). Abundant larval transcript (alt) -1 and -2 genes from *Brugia malayi* : duplication and diversity of genomic environments but conservation of 5' promoter sequences functional in *Caenorhabditis elegans*. *Molecular and Biochemical Parasitology* **125**, 59–71.

Gopal, S., Schroeder, M., Pieper, U., Sczyrba, A., Aytekin-Kurban, G., Bekiranov, S., Fajardo, J.E., Eswar, N., Sanchez, R., Sali, A. and Gaasterland, T. (2001). Homology-based annotation yields 1,042 new candidate genes in the *Drosophila melanogaster* genome. *Nature Genetics* **27**, 337–340.

Gregory, W.F., Atmadja, A.K., Allen, J.E. and Maizels, R.M. (2000). The abundant larval transcript-1 and -2 genes of *Brugia malayi* encode stage-specific candidate vaccine antigens for filariasis. *Infection and Immunity* **68**, 4174–4179.

Gregory, W.F., Blaxter, M.L. and Maizels, R.M. (1997). Differentially expressed, abundant, *trans*-spliced cDNAs from larval *Brugia malayi*. *Molecular and Biochemical Parasitology* **86**, 85–96.

Grigg, M.E., Tang, L., Hussein, A.S. and Selkirk, M.E. (1997). Purification and properties of monomeric (G1) forms of acetylcholinesterase secreted by *Nippostrongylus brasiliensis*. *Molecular and Biochemical Parasitology* **90**, 513–524.

Guiliano, D.B., Hall, N., Jones, S.J., Clark, L.N., Corton, C.H., Barrell, B.G. and Blaxter, M.L. (2002). Conservation of long-range synteny and microsynteny between the genomes of two distantly related nematodes. *Genome Biology* **3**, RESEARCH0057.

Haase, A., Stern, M., Wachtler, K. and Bicker, G. (2001). A tissue-specific marker of Ecdysozoa. *Development, Genes and Evolution* **211**, 428–433.

Haeckel, E. (1874). The Evolution of Man. New York.

Hammond, M.P. and Bianco, A.E. (1992). Genes and genomes of parasitic nematodes. *Parasitology Today* **8**, 299–305.

Harrison, P.M., Echols, N. and Gerstein, M.B. (2001). Digging for dead genes: an analysis of the characteristics of the pseudogene population in the *Caenorhabditis elegans* genome. *Nucleic Acids Research* **29**, 818–830.

Harvey, S.C., Gemmill, A.W., Read, A.F. and Viney, M.E. (2000). The control of morph development in the parasitic nematode *Strongyloides ratti*. *Proceedings of the Royal Society of London. Series B Biological Sciences* **267**, 2057–2063.

Harvey, S.C., Paterson, S. and Viney, M.E. (1999). Heterogeneity in the distribution of *Strongyloides ratti* infective stages among the faecal pellets of rats. *Parasitology* **119**, 227–235.

Harvey, S.C. and Viney, M.E. (2001). Sex determination in the parasitic nematode *Strongyloides ratti*. *Genetics* **158**, 1527–1533.

Hashmi, S., Ling, P., Hashmi, G., Reed, M., Gaugler, R. and Trimmer, W. (1995). Genetic transformation of nematodes using arrays of micromechanical piercing structures. *BioTechniques* **19**, 766–770.

Hawdon, J.M., Jones, B.F., Hoffman, D.R. and Hotez, P.J. (1996). Cloning and characterization of *Ancylostoma*-secreted protein. A novel protein associated with the transition to parasitism by infective hookworm larvae. *Journal of Biological Chemistry* **271**, 6672–6678.

Hawdon, J.M., Narasimhan, S. and Hotez, P.J. (1999). Ancylostoma secreted protein 2: cloning and characterization of a second member of a family of nematode secreted proteins from *Ancylostoma caninum*. *Molecular and Biochemical Parasitology* **99**, 149–165.

Hedges, S.B. (2002). The origin and evolution of model organisms. *Nature Reviews Genetics* **3**, 838–849.

Herman, R.K., Albertson, D.G. and Brenner, S. (1976). Chromosome rearrangements in *Caenorhabditis elegans*. *Genetics* **83**, 91–105.

Herniou, E.A., Pearce, A.C. and Littlewood, D.T.J. (1998). Vintage helminths yield valuable molecules. *Parasitology Today* **14**, 289–292.

Heschl, M.F.P. and Baillie, D.L. (1990). Functional elements and domains inferred from sequence comparisons of a heat shock gene in two nematodes. *Journal of Molecular Evolution* **31**, 3–9.

Higazi, T.B., Merriweather, A., Shu, L., Davis, R. and Unnasch, T.R. (2002). *Brugia malayi*: transient transfection by microinjection and particle bombardment. *Experimental Parasitology* **100**, 95–102.

Ho, J.Y., Weide, R., Ma, H.M., van Wordragen, M.F., Lambert, K.N., Koornneef, M., Zabel, P. and Williamson, V.M. (1992). The root-knot nematode resistance gene (Mi) in tomato: construction of a molecular linkage map and identification of dominant cDNA markers in resistant genotypes. *Plant Journal* **2**, 971–982.

Hodgkin, J. (2001). What does a worm want with 20,000 genes? *Genome Biology* **2**, COMMENT2008.

Hoekstra, R., Criado-Fornelio, A., Fakkeldij, J., Bergman, J. and Roos, M.H. (1997). Microsatellites of the parasitic nematode *Haemonchus contortus*: polymorphism and linkage with a direct repeat. *Molecular and Biochemical Parasitology* **89**, 97–107.

Hoerauf, A., Mand, S., Adjei, O., Fleischer, B. and Buttner, D.W. (2001). Depletion of *Wolbachia* endobacteria in *Onchocerca volvulus* by doxycycline and microfilaridermia after ivermectin treatment. *Lancet* **357**, 1415–1416.

Hoerauf, A., Nissen-Pahle, K., Schmetz, C., Henkle-Duhrsen, K., Blaxter, M.L., Buttner, D.W., Gallin, M.Y., Al-Qaoud, K.M., Lucius, R. and Fleischer, B. (1999). Tetracycline therapy targets intracellular bacteria in the filarial nematode *Litomosoides sigmodontis* and results in filarial infertility. *Journal of Clinical Investigation* **103**, 11–18.

Hoerauf, A., Volkmann, L., Hamelmann, C., Adjei, O., Autenrieth, I.B., Fleischer, B. and Buttner, D.W. (2000a). Endosymbiotic bacteria in worms as targets for a novel chemotherapy in filariasis. *Lancet* **355**, 1242–1243.

Hoerauf, A., Volkmann, L., Nissen-Paehle, K., Schmetz, C., Autenrieth, I., Buttner, D.W. and Fleischer, B. (2000b). Targeting of *Wolbachia* endobacteria in *Litomosoides sigmodontis*: comparison of tetracyclines with chloramphenicol, macrolides and ciprofloxacin. *Tropical Medicine and International Health* **5**, 275–279.

Hope, I.A., Arnold, J.M., McCarroll, D., Jun, G., Krupa, A.P. and Herbert, R. (1998). Promoter trapping identifies real genes in *C. elegans*. *Molecular and General Genetics* **260**, 300–308.

Hoste, H., Chilton, N.B., Gasser, R.B. and Beveridge, I. (1995). Differences in the second internal transcribed spacer (ribosomal DNA) between five species of *Trichostrongylus* (Nematoda: Trichostrongylidae). *International Journal for Parasitology* **25**, 75–80.

Hu, M., Chilton, N.B. and Gasser, R.B. (2002). The mitochondrial genomes of the human hookworms, *Ancylostoma duodenale* and *Necator americanus* (Nematoda: Secernentea). *International Journal for Parasitology* **32**, 145–158.

Hugall, A., Moritz, C., Stanton, J. and Wolstenholme, D.R. (1994). Low, but strongly structured mitochondrial DNA diversity in root knot nematodes (*Meloidogyne*). *Genetics* **136**, 903–912.

Hugall, A., Stanton, J. and Moritz, C. (1997). Evolution of the AT-rich mitochondrial DNA of the root-knot nematode, *Meloidogyne hapla*. *Molecular Biology and Evolution* **14**, 40–48.

Hugot, J.P. (1999). Primates and their pinworm parasites: the cameron hypothesis revisited. *Systematic Biology* **48**, 523–546.

Hugot, J.P., Gardner, S.L. and Morand, S. (1996). The Enterobiinae subfam. Nov. (Nematoda, Oxyurida) pinworm parasites of primates and rodents. *International Journal for Parasitology* **26**, 147–159.

Hussein, A.S., Chacon, M.R., Smith, A.M., Tosado-Acevedo, R. and Selkirk, M.E. (1999a). Cloning, expression, and properties of a nonneuronal secreted acetylcholinesterase from the parasitic nematode *Nippostrongylus brasiliensis*. *Journal of Biological Chemistry* **274**, 9312–9319.

Hussein, A.S., Grigg, M.E. and Selkirk, M.E. (1999b). *Nippostrongylus brasiliensis*: characterisation of a somatic amphiphilic acetylcholinesterase with properties distinct from the secreted enzymes. *Experimental Parasitology* **91**, 144–150.

Hussein, A.S., Kichenin, K. and Selkirk, M.E. (2002). Suppression of secreted acetylcholinesterase expression in *Nippostrongylus brasiliensis* by RNA interference. *Molecular and Biochemical Parasitology* **122**, 91–94.

Hussein, A.S., Smith, A.M., Chacon, M.R. and Selkirk, M.E. (2000). Determinants of substrate specificity of a second non-neuronal secreted acetylcholinesterase from the parasitic nematode *Nippostrongylus brasiliensis*. *European Journal of Biochemistry* **267**, 2276–2282.

Hussey, R.S. (1989). Disease-inducing secretions of plant-parasitic nematodes. *Annual Reviews of Phytopathology* **27**, 123–141.

Hutter, H., Vogel, B.E., Plenefisch, J.D., Norris, C.R., Proenca, R.B., Spieth, J., Guo, C., Mastwal, S., Zhu, X., Scheel, J. and Hedgecock, E.M. (2000). Conservation and novelty in the evolution of cell adhesion and extracellular matrix genes. *Science* **287**, 989–994.

Hyman, B.C. (1996). Molecular systematics and population biology of phytonematodes: some unifying principles. *Fundamental and Applied Nematology* **19**(4), 309–313.

Hyman, B.C. and Azevedo, J.L. (1996). Similar evolutionary patterning among repeated and single copy nematode mitochondrial genes. *Molecular Biology and Evolution* **13**, 221–232.

Hyman, B.C., Beck, J.L. and Weiss, K.C. (1988). Sequence amplification and gene rearrangement in parasitic nematode mitochondrial DNA. *Genetics* **120**, 707–712.

Hyman, B.C. and Slater, T.M. (1990). Recent appearance and molecular characterization of mitochondrial DNA deletions within a defined nematode pedigree. *Genetics* **124**, 845–853.

Johnsen, R.C. and Baillie, D.L. (1991). Genetic analysis of a major segment [LGV(left)] of the genome of *Caenorhabditis elegans*. *Genetics* **129**, 735–752.
Johnston, D.A., Blaxter, M.L., Degrave, W.M., Foster, J., Ivens, A.C. and Melville, S.E. (1999). Genomics and the biology of parasites. *BioEssays* **21**, 131–147.
Jones, M.G.K. (1981). The development and function of plant cells modified by endoparasitic nematodes. In *Plant Parasitic Nematodes*. Vol. 3 (B.M. Zuckerman and R.A. Rhoade, eds), pp. 255–279. New York: Academic Press.
Kaloshian, I., Lange, W.H. and Williamson, V.M. (1995). An aphid-resistance locus is tightly linked to the nematode-resistance gene, *Mi*, in tomato. *Proceedings of the National Academy of Science U. S. A.* **92**, 622–625.
Kaloshian, I., Yaghoobi, J., Liharska, T., Hontelez, J., Hanson, D., Hogan, P., Jesse, T., Wijbrandi, J., Simons, G., Vos, P., Zabel, P. and Williamson, V.M. (1998). Genetic and physical localization of the root-knot nematode resistance locus mi in tomato. *Molecular and General Genetics* **257**, 376–385.
Kamath, R.S., Fraser, A.G., Dong, Y., Poulin, G., Durbin, R., Gotta, M., Kanapin, A., Le Bot, N., Moreno, S., Sohrmann, M., Welchman, D.P., Zipperlen, P. and Ahringer, J. (2003). Systematic functional analysis of the *Caenorhabditis elegans* genome using RNAi. *Nature* **421**, 231–237.
Kampfer, S., Sturmbauer, C. and Ott, C.J. (1998). Phylogenetic analysis of rDNA sequences from adenophorean nematodes and implications for the Adenophorea-Secernetea controversy. *Invertebrate Biology* **117**, 29–36.
Katinka, M.D., Duprat, S., Cornillot, E., Metenier, G., Thomarat, F., Prensier, G., Barbe, V., Peyretaillade, E., Brottier, P., Wincker, P., Delbac, F., El Alaoui, H., Peyret, P., Saurin, W., Gouy, M., Weissenbach, J. and Vivares, C.P. (2001). Genome sequence and gene compaction of the eukaryote parasite *Encephalitozoon cuniculi*. *Nature* **414**, 450–453.
Kato, Y. and Komatso, S. (1996). ASABF, a novel, cysteine-rich antibacterial peptide isolated from the nematode *Ascaris suum*. Purification, primary structure and molecular cloning of cDNA. *Journal of Biological Chemistry* **271**, 30493–30498.
Keddie, E.M., Higazi, T., Boakye, D., Merriweather, A., Wooten, M.C. and Unnasch, T.R. (1999). *Onchocerca volvulus*: limited heterogeneity in the nuclear and mitochondrial genomes. *Experimental Parasitology* **93**, 198–206.
Keddie, E.M., Higazi, T. and Unnasch, T.R. (1998). The mitochondrial genome of *Onchocerca volvulus*: Sequence, structure and phylogenetic analysis. *Molecular and Biochemical Parasitology* **95**, 111–127.
Keeble, F. (1910). Plant-Animals. A study in symbiosis. The Cambridge University Press, Cambridge.
Keightley, P.D. and Caballero, A. (1997). Genomic mutation rates for lifetime reproductive output and lifespan in *Caenorhabditis elegans*. *Proceedings of the National Academy of Science U. S. A.* **94**, 3823–3827.
Keightley, P.D., Davies, E.K., Peters, A.D. and Shaw, R.G. (2000). Properties of ethylmethane sulfonate-induced mutations affecting life-history traits in *Caenorhabditis elegans* and inferences about bivariate distributions of mutation effects. *Genetics* **156**, 143–154.
Kelly, A., Little, M.D. and Voge, M. (1976). *Strongyloides fuelleborni*-like infections in man in Papua New Guinea. *American Journal of Tropical Medicine and Hygiene* **25**, 694–699.
Kennedy, M.W., Brass, A., McCruden, A.B., Price, N.C., Kelly, S.M. and Cooper, A. (1995a). The ABA-1 allergen of the parasitic nematode *Ascaris suum*: fatty acid

and retinoid binding function and structural characterization. *Biochemistry* **34**, 6700–6710.
Kennedy, M.W., Britton, C., Price, N.C., Kelly, S.M. and Cooper, A. (1995b). The DvA-1 polyprotein of the parasitic nematode *Dictyocaulus viviparus*. A small helix-rich lipid-binding protein. *Journal of Biological Chemistry* **270**, 19277–19281.
Kennedy, M.W., Qureshi, F., Fraser, E.M., Haswell-Elkins, M.R., Elkins, D.B. and Smith, H.V. (1989). Antigenic relationships between the surface-exposed; secreted and somatic materials of the nematode parasites *Ascaris lumbricoides*, *Ascaris suum* and *Toxocara canis*. *Clinical and Experimental Immunology* **75**, 493–500.
Kent, W.J. and Zahler, A.M. (2000). Conservation, regulation, synteny, and introns in a large-scale *C. briggsae-C. elegans* genomic alignment. *Genome Research* **10**, 1115–1125.
Kenyon, C. (1994). If birds can fly, why cant we? Homeotic genes and evolution. *Cell* **78**, 175–180.
Kim, S.K. (2001). *C. elegans*: mining the functional genomic landscape. *Nature Reviews Genetics* **2**, 681–689.
Kim, S.K., Lund, J., Kiraly, M., Duke, K., Jiang, M., Stuart, J.M., Eizinger, A., Wylie, B.N. and Davidson, G.S. (2001). A gene expression map for *Caenorhabditis elegans*. *Science* **293**, 2087–2092.
Kinchin, I.M. (1994). The Biology of Tardigrades. Portland Press, London.
Kiontke, K. (1996). The phoretic association of *Diplogaster coprophilia* Sudhaus and Rehfeld, 1990 (Diplogastridae) from cow dung with its carriers, in particular flies of the family Sepsidae. *Nematologica* **42**, 354–366.
Klimowski, L., Chandrashekar, R. and Tripp, C.A. (1997). Molecular cloning, expression and enzymatic activity of a thioredoxin peroxidase from *Dirofilaria immitis*. *Molecular and Biochemical Parasitology* **90**, 297–306.
Kloek, A.P., Sherman, D.R. and Goldberg, D.E. (1993a). Novel gene structure and evolutionary context of *Caenorhabditis elegans* globin. *Gene* **129**, 215–221.
Kloek, A.P., Yang, J., Matthews, F.S., Frieden, C. and Goldberg, D.E. (1993b). The tyrosine B10 hydroxyl is crucial for oxygen avidity of *Ascaris* hemoglobin. *Journal of Biological Chemistry* **268**, 17669–17671.
Kloek, A.P., Yiang, J., Matthews, F.S. and Goldberg, D.E. (1993c). Expression, characterisation and crystallization of oxygen-avid *Ascaris* hemoglobin domains. *Journal of Biological Chemistry* **268**, 17669–17671.
Knox, D.P. and Jones, D.G. (1992). A comparison of superoxide dismutase (SOD, EC:1.15.1.1) distribution in gastro-intestinal nematodes. *International Journal for Parasitology* **22**, 209–214.
Koch, R., van Luenen, H.G., van der Horst, M., Thijssen, K.L. and Plasterk, R.H. (2000). Single nucleotide polymorphisms in wild isolates of *Caenorhabditis elegans*. *Genome Research* **10**, 1690–1696.
Kohara, Y. (1997). Functional genomics of the nematode *C. elegans*. *Tanpakushitsu Kakusan Koso* **42**, 2907–2913.
Koltai, H., Dhandaydham, M., Opperman, C., Thomas, J. and Bird, D. (2001). Overlapping plant signal transduction pathways induced by a parasitic nematode and a rhizobial endosymbiont. *Molecular Plant-Microbe Interactions* **14**, 1168–1177.
Kornfeld, K. (1997). Vulval development in *Caenorhabditis elegans*. *Trends in Genetics* **13**, 55–61.

Kühne, R. (1996). Relations between free-living nematodes and dung-burying *Geotrupes* spp. (Coleoptera: Geotrupini). *Fundamental and Applied Nematology* **19**, 263–271.

Kwa, M.S., Kooyman, F.N., Boersema, J.H. and Roos, M.H. (1993a). Effect of selection for benzimidazole resistance in *Haemonchus contortus* on beta-tubulin isotype 1 and isotype 2 genes. *Biochemical and Biophysical Research Communications* **191**, 413–419.

Kwa, M.S.G., Veenstra, J.G. and Roos, M.H. (1993b). Molecular characterisation of β-tubulin genes present in benzimidazole-resistant populations of *Haemonchus contortus*. *Molecular and Biochemical Parasitology* **60**, 133–144.

Kwa, M.S.G., Veenstra, J.G. and Roos, M.H. (1994). Benzimidazole resistance in *Haemonchus contortus* is correlated with a conserved mutation at amino acid 200 in β-tubulin isotype 1. *Molecular and Biochemical Parasitology* **63**, 299–303.

Lambshead, J. (1993). Recent developments in marine benthic biodiversity research. *Oceanis* **19**, 5–24.

Lambshead, P.J., Brown, C.J., Ferrero, T.J., Hawkins, L.E., Smith, C.R. and Mitchell, N.J. (2003). Biodiversity of nematode assemblages from the region of the Clarion-Clipperton Fracture Zone, an area of commercial mining interest. *BMC Ecol* **3**, 1.

Lander, E.S., Linton, L.M., Birren, B., *et al.* (2001). Initial sequencing and analysis of the human genome. *Nature* **409**, 860–921.

Lang, B.F., Burger, G., O'Kelly, C.J., Cedergren, R., Golding, G.B., Lemieux, C., Sankoff, D., Turmel, M. and Gray, M.W. (1997). An ancestral mitochondrial DNA resembling a eubacterial genome in miniature. *Nature* **387**, 493–497.

Larsen, P.L. (1993). Aging and resistance to oxidative damage in *Caenorhabditis elegans*. *Proceedings of the National Academy of Science U. S. A.* **90**, 8905–8909.

Lavrov, D.V. and Brown, W.M. (2001). *Trichinella spiralis* mtDNA: a nematode mitochondrial genome that encodes a putative ATP8 and normally structured tRNAS and has a gene arrangement relatable to those of coelomate metazoans. *Genetics* **157**, 621–637.

Lawton, J.H., Bignell, D.E., Bloemers, G.F., Eggleton, P. and Hodda, M.E. (1996). Carbon flux and diversity of nematodes and termites in Cameroon forest soils. *Biodiversity and Conservation* **5**, 261–273.

Li, J., Ashton, F.T., Gamble, H.R. and Schad, G.A. (2000). Sensory neuroanatomy of a passively ingested nematode parasite, *Haemonchus contortus*: amphidial neurons of the first stage larva. *Journal of Comparative Neurology* **417**, 299–314.

Liddell, S. and Knox, D.P. (1998). Extracellular and cytoplasmic Cu/Zn superoxide dismutases from *Haemonchus contortus*. *Parasitology* **116**(Pt 4), 383–394.

Liu, L.X., Blaxter, M.L. and Shi, A. (1996). The 5S ribosomal RNA intergenic region of parasitic nematodes: variation in size and presence of SL1 RNA. *Molecular and Biochemical Parasitology* **83**, 235–239.

Lizotte-Waniewski, M., Tawe, W., Guiliano, D.B., Lu, W., Liu, J., Williams, S.A. and Lustigman, S. (2000). Identification of potential vaccine and drug target candidates by expressed sequence tag analysis and immunoscreening of *Onchocerca volvulus* larval cDNA libraries. *Infection and Immunity* **68**, 3491–3501.

Lo, N., Casiraghi, M., Salati, E., Bazzocchi, C. and Bandi, C. (2002). How many *Wolbachia* Supergroups Exist? *Molecular Biology and Evolution* **19**, 341–346.

Lopez-Garcia, P., Rodriguez-Valera, F., Pedros-Alio, C. and Moreira, D. (2001). Unexpected diversity of small eukaryotes in deep-sea Antarctic plankton. *Nature* **409**, 603–607.

Lustigman, S. (1993). Molting, enzymes and new targets for chemotherapy of *Onchocerca volvulus*. *Parasitology Today* **9**, 294–297.

Lustigman, S., Brotman, B., Huitman, T., Prince, A.M. and McKerrow, J.H. (1992). Molecular cloning and characterisation of onchocystatin, a cysteine protease inhibitor of *Onchocerca volvulus*. *Journal of Biological Chemistry* **267**, 17339–17346.

Lynch, A.S., Briggs, D. and Hope, I.A. (1995). Developmental expression pattern screen for genes predicted in the *C. elegans* genome sequencing project. *Nature Genetics* **11**, 309–313.

Maggenti, A. (1981). General Nematology. Springer-Verlag, New York.

Maina, C.V., Grandea III, A.G., Tuyen, L.T.K., Asikin, N., Williams, S.A. and McReynolds, L.A. (1987). *Dirofilaria immitis*: Genomic complexity and characterisation of a structural gene. In *Molecular Paradigms for Eradicating Helminthic Parasites* (A.J. MacInnis, ed), pp. 193–204. New York: Alan R. Liss Inc.

Maizels, R.M., Blaxter, M.L. and Scott, A.L. (2001a). Immunological genomics of *Brugia malayi*: filarial genes implicated in immune evasion and protective immunity. *Parasite Immunol* **23**, 327–344.

Maizels, R.M., Blaxter, M.L. and Selkirk, M.E. (1993a). Forms and functions of nematode surfaces. *Experimental Parasitology* **77**, 380–384.

Maizels, R.M., Bundy, D.A.P., Selkirk, M.E., Smith, D.F. and Anderson, R.M. (1993b). Immunological modulation and evasion by helminth parasites in human populations. *Nature* **365**, 797–805.

Maizels, R.M., Gomez-Escobar, N., Gregory, W.F., Murray, J. and Zang, X. (2001b). Immune evasion genes from filarial nematodes. *International Journal for Parasitology* **31**, 889–898.

Maizels, R.M. and Lawrence, R.A. (1991). Immunological tolerance: The key feature of human filariasis? *Parasitology Today* **7**, 271–276.

Maizels, R.M., Philipp, M., Dasgupta, A. and Partono, F. (1984). Human serum albumin is a major component of the surface of microfilariae of *Wuchereria bancrofti*. *Parasite Immunology* **6**, 185–190.

ManWarren, T., Gagliardo, L., Geyer, J., McVay, C., Pearce-Kelling, S. and Appleton, J. (1997). Invasion of intestinal epithelia *in vitro* by the parasitic nematode *Trichinella spiralis*. *Infection and Immunity* **65**, 4806–4812.

Marson, A.L., Tarr, D.E.K. and Scott, A.L. (2001). Macrophage migration inhibitory (*mif*) transcription is significantly elevated in *Caenorhabditis elegans* dauer larvae. *Gene* **278**, 53–62.

McCarter, J.P., Abad, J., Jones, J.T. and Bird, D.M. (2000). Rapid gene discovery in plant parasitic nematodes via expressed sequence tags. *Nematology* **2**, 719–731.

McCarter, J.P., Bird, D.M., Clifton, S.W. and Waterston, R.H. (2001). Nematode gene sequences, Update for December 2000. *Journal of Nematology* **32**, 331–333.

McCarter, J.P., Clifton, S.W., Bird, D.M. and Waterston, R.H. (2002). Nematode gene sequences, Update for June 2002. *Journal of Nematology* **34**, 71–74.

McCombie, W.R., Adams, M.D., Kelley, J.M., FitzGerald, M.G., Utterback, T.R., Khan, M., Dubnick, M., Kerlavage, A.R., Venter, J.C. and Fields, C. (1992). *Caenorhabditis elegans* expressed sequence tags identify gene families and potential disease gene homologues. *Nature Genetics* **1**, 124–131.

McDermott, L., Kennedy, M.W., McManus, D.P., Bradley, J.E., Cooper, A. and Storch, J. (2002). How helminth lipid-binding proteins offload their ligands to membranes: differential mechanisms of fatty acid transfer by the ABA-1 polyprotein allergen and Ov-FAR-1 proteins of nematodes and Sj-FABPc of schistosomes. *Biochemistry* **41**, 6706–6713.

McReynolds, L.A., DeSimone, S.M. and Williams, S.A. (1986). Cloning and comparison of repeated DNA sequences from the human filarial parasite *Brugia malayi* and the animal parasite *Brugia pahangi*. *Proceedings of the National Academy of Science U. S. A.* **83**, 797–801.

McReynolds, L.A., Kennedy, M.W. and Selkirk, M.E. (1993). The polyprotein allergens of nematodes. *Parasitology Today* **9**, 403–406.

Mello, C.C., Kramer, J.C., Stinchcomb, D. and Ambros, V. (1991). Efficient gene transfer in *C. elegans*: Extrachromosomal maintenance and integration of transforming sequences. *EMBO Journal* **10**, 3959–3970.

Montgomery, M.K. and Fire, A. (1998). Double-stranded RNA as a mediator in sequence-specific genetic silencing and co-suppression. *Trends in Genetics* **14**, 255–258.

Moon-van der Staay, S.Y., De Wachter, R. and Vaulot, D. (2001). Oceanic 18S rDNA sequences from picoplankton reveal unsuspected eukaryotic diversity. *Nature* **409**, 607–610.

Moreira, D. and Lopez-Garcia, P. (2002). The molecular ecology of microbial eukaryotes unveils a hidden world. *Trends in Microbiology* **10**, 31–38.

Mounsey, A., Bauer, P. and Hope, I.A. (2002). Evidence suggesting that a fifth of annotated *Caenorhabditis elegans* genes may be pseudogenes. *Genome Research* **12**, 770–775.

Muller, F. and Tobler, H. (2000). Chromatin diminution in the parasitic nematodes *Ascaris suum* and *Parascaris univalens*. *International Journal for Parasitology* **30**, 391–399.

Muller, F., Walker, P., Aeby, P., Neuhaus, H., Back, E. and Tobler, H. (1982a). Molecular cloning and sequence analysis of highly repetitive DNA sequences contained in the eliminated genome of *Ascaris lumbricoides*. *Progress in Clinical Biology Research* **85 Pt A**, 127–138.

Muller, F., Walker, P., Aeby, P., Neuhaus, H., Felder, H., Back, E. and Tobler, H. (1982b). Nucleotide sequence of satellite DNA contained in the eliminated genome of *Ascaris lumbricoides*. *Nucleic Acids Research* **10**, 7493–7510.

Muller, A., MacCallum, R.M. and Sternberg, M.J. (2002). Structural characterization of the human proteome. *Genome Research* **12**, 1625–1641.

Murray, J., Gregory, W.F., Gomez-Escobar, N., Atmadja, A.K. and Maizels, R.M. (2001). Expression and immune recognition of *Brugia malayi* VAL-1, a homologue of vespid venom allergens and Ancylostoma secreted proteins. *Molecular and Biochemical Parasitology* **118**, 89–96.

Mushegian, A.R., Garey, J.R., Martin, J. and Liu, L.X. (1998). Large-scale taxonomic profiling of eukaryotic model organisms: a comparison of orthologous proteins encoded by the human, fly, nematode, and yeast genomes. *Genome Research* **8**, 590–598.

Myler, P.J., Beverley, S.M., Cruz, A.K., Dobson, D.E., Ivens, A.C., McDonagh, P.D., Madhubala, R., Martinez-Calvillo, S., Ruiz, J.C., Saxena, A., Sisk, E., Sunkin, S.M., Worthey, E., Yan, S. and Stuart, K.D. (2001). The *Leishmania* genome project: new insights into gene organization and function. *Medical Microbiology and Immunology (Berlin)* **190**, 9–12.

Nadler, S.A. and Hudspeth, D.S. (2000). Phylogeny of the Ascaridoidea (Nematoda: Ascaridida) based on three genes and morphology: hypotheses of structural and sequence evolution. *Journal of Parasitology* **86**, 380–393.

Nealson, K.H. (1991). Luminescent bacteria symbiotic with entomopathogenic nematodes. In *Symbiosis as a Source of Evolutionary Innovation* (L. Margulis and R. Fester eds), pp. 205–218. Massachusetts: MIT Press, Cambridge.
Nelson, D.R. (1982). Developmental biology of the Tardigrada. In *Developmental Biology of Freshwater Invertebrates*. New York: Alan R. Liss Inc., pp. 363–398.
Newlands, G.F., Skuce, P.J., Knox, D.P. and Smith, W.D. (2001). Cloning and expression of cystatin, a potent cysteine protease inhibitor from the gut of *Haemonchus contortus. Parasitology* **122**, 371–378.
(1991). In *Manual of Agricultural Nematology* (Nickle, W.R. ed.), New York: Marcel Dekker, Inc.
Nicoll, M., Akerib, C.C. and Meyer, B.J. (1997). X-chromosome-counting mechanisms that determine nematode sex. *Nature* **388**, 200–204.
Nielsen, C. (1995). Animal evolution. Interrelationships of the living phyla. Oxford University Press, Oxford.
Nielsen, C. (2001). Animal evolution. Interrelationships of the living phyla. Oxford University Press, Oxford.
Okimoto, R., MacFarlane, J.L., Clary, D.O. and Wolstenholme, D.R. (1992). The mitochondrial genomes of two nematodes, *Caenorhabditis elegans* and *Ascaris suum. Genetics* **130**, 471–498.
Okimoto, R., Macfarlane, J.L. and Wolstenholme, D.R. (1994). The mitochondrial ribosomal RNA genes of the nematodes *Caenorhabditis elegans* and *Ascaris suum*: consensus secondary structure models and conserved nucleotide sets for phylogenetic analysis. *Journal of Molecular Evolution* **39**, 598–613.
Opperman, C.H., Taylor, C.G. and Conkling, M.A. (1994). Root-knot nematode-directed expression of a plant root-specific gene. *Science* **263**, 221–223.
Otsen, M., Plas, M.E., Groeneveld, J., Roos, M.H., Lenstra, J.A. and Hoekstra, R. (2000). Genetic markers for the parasitic nematode *Haemonchus contortus* based on intron sequences. *Experimental Parasitology* **95**, 226–229.
Ou, X., Tang, L., McCrossan, M., Henkle-Duhrsen, K. and Selkirk, M.E. (1995a). *Brugia malayi*: localisation and differential expression of extracellular and cytoplasmic CuZn superoxide dismutases in adults and microfilariae. *Experimental Parasitology* **80**, 515–529.
Ou, X., Thomas, R., Chacón, M.R., Tang, L. and Selkirk, M.E. (1995b). *Brugia malayi*: Differential susceptibility to and metabolism of hydrogen peroxide in adults and microfilariae. *Experimental Parasitology* **80**, 530–540.
Pace, N.R. (1997). A molecular view of microbial diversity and the biosphere. *Science* **276**, 734–740.
Page, A.P. (1999). A highly conserved nematode protein folding operon in *Caenorhabditis elegans* and *Caenorhabditis briggsae. Gene* **230**, 267–275.
Page, R.M. and Hafner, R.S. (1996). Molecular phylogenies and host-parasite cospeciation: gophers and lice as a model system. In *New Uses for New Phylogenies* (P.H. Harvey, A.J. Leigh-Brown, J. Maynard Smith and S. Nee, eds), pp. 255–270. Oxford: Oxford University Press.
Parkinson, J. and Blaxter, M.L. (2002). SimiTri - visualising similarity relationships for large groups of sequences. *Bioinformatics* **19**, 390–395.
Parkinson, J., Guiliano, D. and Blaxter, M. (2002). Making sense of EST sequences by CLOBBing them. *BMC Bioinformatics* **3**, 31.

Parkinson, J., Whitton, C., Guiliano, D., Daub, J. and Blaxter, M.L. (2001). 200,000 nematode ESTs on the net. *Trends in Parasitology* **17**, 394–396.

Pastrana, D.V., Raghavan, N., FitzGerald, P., Eisinger, S.W., Metz, C., Bucala, R., Schleimer, R.P., Bickel, C. and Scott, A.L. (1998). Filarial nematode parasites secrete a homologue of the human cytokine macrophage migration inhibitory factor. *Infection and Immunity* **66**, 5955–5963.

Pelonquin, J.J., Bird, D.M., Kaloshian, I. and Matthews, W.C. (1993). Isolates of *Meloidogyne hapla* with distinct mitochondrial genomes. *Journal of Nematology* **25**, 239–243.

Peterson, K.J. and Eernisse, D.J. (2001). Animal phylogeny and the ancestry of bilaterians: inferences from morphology and 18S rDNA sequences. *Evolution and Development* **3**, 170–205.

Philipp, M., Parkhouse, R.M.E. and Ogilvie, B.M. (1980). Changing proteins on the surface of a parasitic nematode. *Nature* **287**, 538–540.

Philippe, H., Chenuil, A. and Adoutte, A. (1994). Can the cambrian explosion be inferred through molecular phylogeny. *Development*, **Supplement**, 15–25.

Piessens, W.F., McReynolds, L.A. and Williams, S.A. (1987). Highly repeated DNA sequences as species-specific probes for *Brugia*. *Parasitology Today* **3**, 378–379.

Platonova, T.A. and Gal'tsova, V.V. (1976). Nematodes and their role in the meiobenthos. Nakua, Leningrad.

Powers, T.O., Harris, T.S. and Hyman, B.C. (1993). Mitochondrial DNA sequence divergence among *Meloidogyne incognita*, *Romanomermis culcivorax*, *Ascaris suum* and *Caenorhabditis elegans*. *Journal of Nematology* **25**, 564–572.

Powers, T.O., Platzer, E.G. and Hyman, B.C. (1986). Large mitochondrial genome and mitochondrial DNA size polymorphism in the mosquito parasite, *Romanomermis culicivorax*. *Current Genetics* **11**, 71–77.

Powers, T.O., Todd, T.C., Burnell, A.M., Murray, P.C.B., Fleming, C.C., Szalanski, A.L., Adams, B.A. and Harris, T.S. (1997). The rDNA internal transcribed spacer region as a taxonomic marker for nematodes. *Journal of Nematology* **29**, 441–450.

Prior, A., Jones, J.T., Blok, V.C., Beauchamp, J., McDermott, L., Cooper, A. and Kennedy, M.W. (2001). A surface-associated retinol- and fatty acid-binding protein (Gp-FAR-1) from the potato cyst nematode *Globodera pallida*: lipid binding activities, structural analysis and expression pattern. *Biochemical Journal* **356**, 387–394.

Ranz, J.M., Casals, F. and Ruiz, A. (2001). How malleable is the eukaryotic genome? Extreme rate of chromosomal rearrangement in the genus *Drosophila*. *Genome Research* **11**, 230–239.

Ranz, J.M., Segarra, C. and Ruiz, A. (1997). Chromosomal homology and molecular organization of Muller's elements D and E in the *Drosophila repleta* species group. *Genetics* **145**, 281–295.

Ray, C., Abbott, A.G. and Hussey, R.S. (1995). *Trans*-splicing of a *Meloidogyne incognita* mRNA encoding a putative oesophageal gland protein. *Molecular and Biochemical Parasitology* **68**, 93–101.

Read, A.F. and Skorping, A. (1995a). Causes and consequences of life history variation in parasitic nematodes. In *Ecology and Transmission Strategies of Entomopathogenic Nematodes (COST 819)* (C.T. Griffin, R.L. Gwynn and J.P. Masson, eds), pp. 58–68. Brussels: European Commission.

Read, A.F. and Skorping, A. (1995b). The evolution of tissue migration by parasitic nematode larvae. *Parasitology* **111**, 359–371.

Reboul, J., Vaglio, P., Tzellas, N., Thierry-Mieg, N., Moore, T., Jackson, C., Shin, I.T., Kohara, Y., Thierry-Mieg, D., Thierry-Mieg, J., Lee, H., Hitti, J., Doucette-Stamm, L., Hartley, J.L., Temple, G.F., Brasch, M.A., Vandenhaute, J., Lamesch, P.E., Hill, D.E. and Vidal, M. (2001). Open-reading-frame sequence tags (OSTs) support the existence of at least 17,300 genes in *C. elegans*. *Nature Genetics* **27**, 332–336.

Redmond, D.L. and Knox, D.P. (2001). *Haemonchus contortus* SL2 trans-spliced RNA leader sequence. *Molecular and Biochemical Parasitology* **117**, 107–110.

Richter, S. (1993). Phoretic association between the dauerjuveniles of *Rhabditis stammeri* (Rhabditidae) and life history stages of the burying beetle *Nicrophorus vespilloides* (Coleoptera: Silphidae). *Nematologica* **39**, 346–355.

Riddle, D., Blumenthal, T., Meyer, B. and Priess, J. (eds.) (1997) *C. elegans II*. Cold Spring Harbor, NY: Cold Spring Harbor Laboratory Press.

Riddle, D.L., Swanson, M.M. and Albert, P.S. (1981). Interacting genes in nematode dauer larva formation. *Nature* **290**, 668–671.

Robertson, H.M. (1998). Two large families of chemoreceptor genes in the nematodes *Caenorhabditis elegans* and *Caenorhabditis briggsae* reveal extensive gene duplication, diversification, movement, and intron loss. *Genome Research* **8**, 449–463.

Robertson, H.M. (2000). The large *srh* family of chemoreceptor genes in *Caenorhabditis* nematodes reveals processes of genome evolution involving large duplications and deletions and intron gains and losses. *Genome Research* **10**, 192–203.

Robertson, H.M. (2001). Updating the str and srj (stl) families of chemoreceptors in *Caenorhabditis* nematodes reveals frequent gene movement within and between chromosomes. *Chemical Senses* **26**, 151–159.

Robertson, L., Robertson, W.M. and Jones, J.T. (1999). Direct analysis of the secretions of the potato cyst nematode *Globodera rostochiensis*. *Parasitology* **119**(Pt 2), 167–176.

Robertson, L., Robertson, W.M., Sobczak, M., Helder, J., Tetaud, E., Ariyanayagam, M.R., Ferguson, M.A., Fairlamb, A. and Jones, J.T. (2000). Cloning, expression and functional characterisation of a peroxiredoxin from the potato cyst nematode *Globodera rostochiensis*. *Molecular and Biochemical Parasitology* **111**, 41–49.

Romaris, F. and Appleton, J.A. (2001). Invasion of epithelial cells by *Trichinella spiralis*: in vitro observations. *Parasite* **8**, S48–S50.

Romstad, A., Gasser, R.B., Nansen, P., Polderman, A.M. and Chilton, N.B. (1998). *Necator americanus* (Nematoda: Ancylostomatidae) from Africa and Malaysia have different ITS-2 rDNA sequences. *International Journal for Parasitology* **28**, 611–615.

Roos, M.H., Hoekstra, R., Plas, M.E., Otsen, M. and Lenstra, J.A. (1998). Polymorphic DNA markers in the genome of parasitic nematodes. *Journal of Helminthology* **72**, 291–294.

Rose, J.E. and Hwang, J.C. (1967). Hemoglobins of *Syngamus trachaea* and some hosts: Electrophoresis; ultracentrifugation and alkali denaturation. *Journal of Parasitology* **53**, 1061–1063.

Rose, J.E. and Kaplan, K.L. (1972). Purification, molecular weight, and oxygen equilibrium of hemoglobin from *Syngamus trachea*, the poultry gapeworm. *Journal of Parasitology* **58**, 903–906.

Rubin, G.M., Yandell, M.D., Wortman, J.R., et al. (2000). Comparative genomics of the eukaryotes. *Science* **287**, 2204–2215.
Ruvkun, G. and Hobert, O. (1998a). The genomic topography of *C. elegans* developmental control. *Science* **282**, 2033–2041.
Ruvkun, G. and Hobert, O. (1998b). The taxonomy of developmental control in *Caenorhabditis elegans*. *Science* **282**, 2033–2041.
Sakaguchi, Y., Tada, I., Ash, L.R. and Aoki, Y. (1982). Chromosomes of two species of filarial worms, *Brugia pahangi* and *Brugia malayi* (Filariidae: Nematoda). *Chromosome Information Service* **32**, 12–14.
Sakaguchi, Y., Tada, I., Ash, L.R. and Aoki, Y. (1983). Karyotypes of *Brugia pahangi* and *Brugia malayi* (Nematoda: Filarioidea). *Journal of Parasitology* **69**, 1090–1093.
Salser, S.J. and Kenyon, C. (1994). Patterning *C. elegans*: homeotic cluster genes, cell fates and cell migrations. *Trends in Genetics* **10**, 159–164.
Saruta, F., Kuramochi, T., Nakamura, K., Takamiya, S., Yu, Y., Aoki, T., Sekimizu, K., Kojima, S. and Kita, K. (1995). Stage-specific isoforms of complex II (succinate-ubiquinone oxidoreductase) in mitochondria from the parasitic nematode, *Ascaris suum*. *Journal of Biological Chemistry* **270**, 928–932.
Schrum, S., Bialonski, A., Marti, T. and Zipfel, P.F. (1998). Identification of a peroxidoxin protein (OvPXN-2) of the human parasitic nematode *Onchocerca volvulus* by sequential protein fractionation. *Molecular and Biochemical Parasitology* **94**, 131–135.
Selkirk, M.E. and Blaxter, M.L. (1990). Cuticular proteins of *Brugia* filarial parasites. *Acta Tropica* **47**, 373–380.
Selkirk, M.E., Tang, L., Ou, X., Cookson, E. and Chacon, M.R. (1994). Filarial anti-oxidant enzymes: Mediators of parasite persistence and potential targets for vaccination. *Parasite* **1 (1S)**, 19–20.
Semblat, J.P., Rosso, M.N., Hussey, R.S., Abad, P. and Castagnone-Sereno, P. (2001). Molecular cloning of a cDNA encoding an amphid-secreted putative avirulence protein from the root-knot nematode *Meloidogyne incognita*. *Molecular Plant-Microbe Interactions* **14**, 72–79.
Sherman, D.R., Guinn, B., Perdok, M.M. and Goldberg, D.E. (1992a). Components of sterol biosynthesis assembled on the oxygen-avid hemoglobin of *Ascaris*. *Science* **258**, 1930–1932.
Sherman, D.R., Kloek, A.P., Krishman, B.R., Guinn, B. and Goldberg, D.E. (1992b). *Ascaris* hemoglobin gene: plant-like structure reflects the ancestral globin gene. *Proceedings of the National Academy of Science U. S. A.* **89**, 11696–11700.
Shigenobu, S., Watanabe, H., Hattori, M., Sakaki, Y. and Ishikawa, H. (2000). Genome sequence of the endocellular bacterial symbiont of aphids *Buchnera* sp. APS. *Nature* **407**, 81–86.
Sidow, A. and Thomas, W.K. (1994). A molecular evolutionary framework for eukaryotic model organisms. *Current Biology* **4**, 596–603.
Sironi, M., Bandi, C., Sacchi, L., Di Sacco, B., Damiani, G. and Genchi, C. (1995). Molecular evidence for a close relative of the arthropod endosymbiont *Wolbachia* in a filarial worm. *Molecular and Biochemical Parasitology* **74**, 223–227.
Sluder, A.E. and Maina, C.V. (2001). Nuclear receptors in nematodes: themes and variations. *Trends in Genetics* **17**, 206–213.

Sluder, A.E., Mathews, S.W., Hough, D., Yin, V.P. and Maina, C.V. (1999). The nuclear receptor superfamily has undergone extensive proliferation and diversification in nematodes. *Genome Research* **9**, 103–120.

Smant, G., Stokkermans, J.P., Yan, Y., de Boer, J.M., Baum, T.J., Wang, X., Hussey, R.S., Gommers, F.J., Henrissat, B., Davis, E.L., Helder, J., Schots, A. and Bakker, J. (1998). Endogenous cellulases in animals: isolation of beta-1, 4-endoglucanase genes from two species of plant-parasitic cyst nematodes. *Proceedings of the National Academy of Science U. S. A.* **95**, 4906–4911.

Smith, V.P., Selkirk, M.E. and Gounaris, K. (1996). Identification and composition of lipid classes in surface and somatic preparations of adult *Brugia malayi*. *Molecular and Biochemical Parasitology* **78**, 105–116.

Smith, V.P., Selkirk, M.E. and Gounaris, K. (1998). *Brugia malayi*: resistance of cuticular lipids to oxidant-induced damage and detection of alpha-tocopherol in the neutral lipid fraction. *Experimental Parasitology* **88**, 103–110.

Smyth, J.D. and Smyth, M.M. (1980). Frogs as Host Parasite Systems I. An introduction to parasitology through the parasites of *Rana temporaria, R. esculenta and R. pipens*. London: Macmillan Press.

Sorci, G., Morand, S. and Hugot, J.-P. (1997). Host-parasite coevolution: comparative evidence for covariation of life history traits in primates and oxyurid parasites. *Proceedings of the Royal Society of London. Series B Biological Sciences* **264**, 285–289.

Spieler, M. and Schierenberg, E. (1995). On the development of the alternating free-living and parasitic generations of the nematode *Rhabdias bufonis*. *Invertebrate Reproduction and Development* **28**, 193–203.

Staskawicz, B.J., Ausubel, F.M., Baker, B.J., Ellis, J.G. and Jones, J.D.G. (1995). Molecular genetics of plant disease resistance. *Science* **268**, 661–667.

Stewart, H.I., O'Neil, N.J., Janke, D.L., Franz, N.W., Chamberlin, H.M., Howell, A.M., Gilchrist, E.J., Ha, T.T., Kuervers, L.M., Vatcher, G.P., Danielson, J.L. and Baillie, D.L. (1998). Lethal mutations defining 112 complementation groups in a 4.5 Mb sequenced region of *Caenorhabditis elegans* chromosome III. *Molecular and General Genetics* **260**, 280–288.

Strassen, O.z. (1896). Embryonalentwicklung der *Ascaris megalocephala*. *Archiv fur Entwicklungsmechanik der Organismus* **3**, 27–105, 131–190, pls 105–109.

Sudhaus, W. and Asakawa, M. (1991). First record of the larval parasitic nematode *Rhabditis orbitalis* from Japanese wood mice (*Apodemus* spp.). *Journal of Helminthology* **65**, 232–233.

Sulston, J., Du, Z., Thomas, K., Wilson, R., Hillier, L., Staden, R., Halloran, N., Green, P., Thierry-Mieg, J., Qiu, L., Dear, S., Coulson, A., Craxton, M., Durbin, R., Berks, M., Metzstein, M., Hawkins, T., Ainscough, R. and Waterston, R. (1992). The *C. elegans* genome sequencing project: A beginning. *Nature* **356**, 37–41.

Sulston, J. and Horvitz, H.R. (1977). Post-embryonic cell lineages of the nematode *Caenorhabditis elegans*. *Developmental Biology* **56**, 110–156.

Sulston, J., Horvitz, H.R. and Kimble, J. (1988). Cell lineage. In *The nematode Caenorhabditis elegans*. (Wood, W.B. ed), pp. 457–490, Cold Spring harbor, NY: Cold Spring Harbor Laboratory Press.

Sulston, J.E., Albertson, D.G. and Thomson, J.N. (1980). The *Caenorhabditis elegans* male: Postembryonic development of nongonadal structures. *Developmental Biology* **78**, 542–576.

Sulston, J.E., Schierenberg, E., White, J.G. and Thompson, J.N. (1983). The embryonic cell lineage of the nematode *Caenorhabditis elegans*. *Developmental Biology* **100**, 64–119.

Swofford, D.L., Olsen, G.J., Waddell, P.J. and Hillis, D.M. (1996). Phylogenetic Inference. In *Molecular Systematics* (D.M. Hillis, C. Moritz and B.K. Mable, eds), pp. 407–514. Sunderland, MA: Sinauer Associates.

Tamas, I., Klasson, L., Canback, B., Naslund, A.K., Eriksson, A.S., Wernegreen, J.J., Sandstrom, J.P., Moran, N.A. and Andersson, S.G. (2002). 50 million years of genomic stasis in endosymbiotic bacteria. *Science* **296**, 2376–2379.

Tang, L., Gounaris, K., Griffiths, C. and Selkirk, M.E. (1995). Heterologous expression and enzymatic properties of a selenium-independent glutathione peroxidase from the parasitic nematode *Brugia pahangi*. *Journal of Biological Chemistry* **270**, 18313–18318.

Tang, L., Ou, X., Henkle-Dhursen, K.J. and Selkirk, M.E. (1994). Extracellular and cytoplasmic CuZn superoxide dismutase from the *Brugia* lymphatic filarial parasites. *Infection and Immunity* **62**, 961–967.

Taylor, M.J., Bilo, K., Cross, H.F., Archer, J.P. and Underwood, A.P. (1999). 16S rDNA phylogeny and ultrastructural characterization of *Wolbachia* intracellular bacteria of the filarial nematodes *Brugia malayi, B. pahangi*, and *Wuchereria bancrofti*. *Experimental Parasitology* **91**, 356–361.

Telford, M.J. and Holland, P.W.H. (1997). Evolution of 28S ribosomal DNA in chaetognaths: Duplicate genes and molecular phylogeny. *Journal of Molecular Evolution* **44**, 135–144.

Thomas, W.K., Vida, J.T., Frisse, L.M., Mundo, M. and Baldwin, J. (1997). DNA sequences from formalin-fixed nematodes: integrating molecular and morphological approaches to taxonomy. *Journal of Nematology* **29**, 248–252.

Thomas, W.K. and Wilson, A.C. (1991). Mode and tempo of molecular evolution in the nematode *Caenorhabditis*: cytochrome oxidase II and calmodulin sequences. *Genetics* **128**, 269–279.

Timmons, L. and Fire, A. (1998). Specific interference by ingested dsRNA. *Nature* **395**, 854.

Tobler, H., Etter, A. and Muller, F. (1992). Chromatin diminution in nematode development. *Trends in Genetics* **8**, 427–432.

Tree, T.I.M., Gillespie, A.J., Shepley, K.J., Blaxter, M.L., Tuan, R.S. and Bradley, J.E. (1995). Characterisation of an immunodominant glycoprotein antigen of *Onchocerca volvulus* with homologues in other filarial nematodes and *Caenorhabditis elegans*. *Molecular and Biochemical Parasitology* **69**, 185–195.

Triantaphyllou, A.C. (1971). Genetics and Cytology. In *Plant Parasitic Nematodes*. Vol. 2 (B.M. Zuckerman, W.F. Mai and R.A. Rohde, eds), pp. 1–34. New York: Academic Press.

Triantaphyllou, A.C. (1983). Cytogenetic aspects of nematode evolution. In *Concepts in Nematode Systematics*. Vol. 22 (A.R. Stone, H.M. Platt and L.F. Kahlil, eds), pp. 55–71. London: Academic Press.

Triantaphyllou, A.C. (1985). Cytogenetics, cytotaxonomy and phylogeny of root-knot nematodes. In *An Advanced Treatise on Meloidgyne. Vol. I: Biology and Control* (J.N. Sasser and C.C. Carter, eds), pp. 113–126. Raleigh: North Carolina State University Graphics.

Troemel, E.R., Chou, J.H., Dwyer, N.D., Colbert, H.A. and Bargmann, C.I. (1995). Divergent seven transmembrane receptors are candidate chemosensory receptors in *C. elegans*. *Cell* **83**, 207–218.

Van Auken, K., Weaver, D.C., Edgar, L.G. and Wood, W.B. (2000). *Caenorhabditis elegans* embryonic axial patterning requires two recently discovered posterior-group Hox genes. *Proceedings of the National Academy of Science U. S. A.* **97**, 4499–4503.

Vandekerckhove, T.T., Coomans, A., Cornelis, K., Baert, P. and Gillis, M. (2002). Use of the Verrucomicrobia-specific probe EUB338-III and fluorescent in situ hybridization for detection of "Candidatus Xiphinematobacter" cells in nematode hosts. *Applied and Environmental Microbiology* **68**, 3121–3125.

Vandekerckhove, T.T., Willems, A., Gillis, M. and Coomans, A. (2000). Occurrence of novel verrucomicrobial species, endosymbiotic and associated with parthenogenesis in Xiphinema americanum-group species (Nematoda, Longidoridae). *International Journal of Systematic and Evolutionary Microbiology* **50 Pt 6**, 2197–2205.

Vanfleteren, J.R., Van de Peer, Y., Blaxter, M.L., Tweedie, S.A., Trotman, C., Lu, L., Van Hauwaert, M.L. and Moens, L. (1994). Molecular genealogy of some nematode taxa as based on cytochrome c and globin amino acid sequences. *Molecular Phylogenetics and Evolution* **3**, 92–101.

Venter, J.C., Adams, M.D., Myers, E.W., Li, P.W., Mural, R.J., Sutton, G.G., Smith, H.O., Yandell, M., Evans, C.A., Holt, R.A., Gocayne, J.D., Amanatides, P., Ballew, R.M., Huson, D.H., Wortman, J.R., Zhang, Q., Kodira, C.D., Zheng, X.H., Chen, L., Skupski, M., Subramanian, G., Thomas, P.D., Zhang, J., Gabor Miklos, G.L., Nelson, C., Broder, S., Clark, A.G., Nadeau, J., McKusick, V.A., Zinder, N., Levine, A.J., Roberts, R.J., Simon, M., Slayman, C., Hunkapiller, M., Bolanos, R., Delcher, A., Dew, I., Fasulo, D., Flanigan, M., Florea, L., Halpern, A., Hannenhalli, S., Kravitz, S., Levy, S., Mobarry, C., Reinert, K., Remington, K., Abu-Threideh, J., Beasley, E., Biddick, K., Bonazzi, V., Brandon, R., Cargill, M., Chandramouliswaran, I., Charlab, R., Chaturvedi, K., Deng, Z., Di Francesco, V., Dunn, P., Eilbeck, K., Evangelista, C., Gabrielian, A.E., Gan, W., Ge, W., Gong, F., Gu, Z., Guan, P., Heiman, T.J., Higgins, M.E., Ji, R.R., Ke, Z., Ketchum, K.A., Lai, Z., Lei, Y., Li, Z., Li, J., Liang, Y., Lin, X., Lu, F., Merkulov, G.V., Milshina, N., Moore, H.M., Naik, A.K., Narayan, V.A., Neelam, B., Nusskern, D., Rusch, D.B., Salzberg, S., Shao, W., Shue, B., Sun, J., Wang, Z., Wang, A., Wang, X., Wang, J., Wei, M., Wides, R., Xiao, C., Yan, C., *et al.* (2001). The sequence of the human genome. *Science* **291**, 1304–1351.

Veronico, P., Gray, L.J., Jones, J.T., Bazzicalupo, P., Arbucci, S., Cortese, M.R., Di Vito, M. and De Giorgi, C. (2001a). Nematode chitin synthases: gene structure, expression and function in *Caenorhabditis elegans* and the plant parasitic nematode *Meloidogyne artiellia*. *Molecular Genetics and Genomics* **266**, 28–34.

Veronico, P., Jones, J., Di Vito, M. and De Giorgi, C. (2001b). Horizontal transfer of a bacterial gene involved in polyglutamate biosynthesis to the plant-parasitic nematode *Meloidogyne artiellia*. *FEBS Letters* **508**, 470–474.

Viney, M.E. (1996). Developmental switching in the parasitic nematode *Strongyloides ratti*. *Proceedings of the Royal Society of London. Series B Biological Sciences* **263**, 201–208.

Viney, M.E. (1999). Exploiting the life cycle of *Strongyloides ratti*. *Parasitology Today* **15**, 231–235.

Viney, M.E., Matthews, B.E. and Walliker, D. (1992). On the biological and biochemical nature of cloned populations of *Strongyloides ratti*. *Journal of Helminthology* **66**, 45–52.

von Brand, T. (1938). The nature of the metabolic activities of intestinal helminths in their natural habitat: Aerobiosis or anaerobiosis? *Biodynamica* **2**, 1–13.
von Brand, T. (1966). Biochemistry of Parasites. New York: Academic Press.
Wägele, J.W., Erikson, T., Lockhart, P. and Misof, B. (1999). The Ecdysozoa: Artifact or monophylum? *Journal of Zoological Systematics and Evolutionary Research* **37**, 211–223.
Walton, A.C. (1959). Some parasites and their chromosomes. *Journal of Parasitology* **45**, 1–20.
Wang, B.B., Müller-Immergluck, M.M., Austin, J., Robinson, N.T., Chisholm, A. and Kenyon, C. (1993). A homeotic gene cluster patterns the anterioposterior body axis of *C. elegans*. *Cell* **74**, 29–42.
Wang, X., Allen, R., Ding, X., Goellner, M., Maier, T., de Boer, J.M., Baum, T.J., Hussey, R.S. and Davis, E.L. (2001). Signal peptide-selection of cDNA cloned directly from the esophageal gland cells of the soybean cyst nematode *Heterodera glycines*. *Molecular Plant-Microbe Interactions* **14**, 536–544.
Wang, X., Meyers, D., Yan, Y., Baum, T., Smant, G., Hussey, R. and Davis, E. (1999). In planta localization of a beta-1,4-endoglucanase secreted by *Heterodera glycines*. *Molecular Plant-Microbe Interactions* **12**, 64–67.
Watanabe, Y., Tsurui, H., Ueda, T., Furushima, R., Takamiya, S., Kita, K., Nishikawa, K. and Watanabe, K. (1994). Primary and higher order structures of nematode (*Ascaris suum*) mitochondrial tRNAs lacking either the T or D stem. *Journal of Biological Chemistry* **269**, 22902–22906.
Waterston, R., Martin, C., Craxton, M., Huynh, C., Coulson, A., Hillier, L., Durbin, R., Green, P., Shownkeen, R., Halloran, N., Metzstein, M., Hawkins, T., Wilson, R., Berks, M., Du, Z., Thomas, K., Thierry-Mieg, J. and Sulston, J. (1992). A survey of expressed genes in *Caenorhabditis elegans*. *Nature Genetics* **1**, 114–123.
Waterston, R.H., Lindblad-Toh, K., Birney, E., Rogers, J., Abril, J.F., Agarwal, P., Agarwala, R., Ainscough, R., Alexandersson, M., An, P., Antonarakis, S.E., Attwood, J., Baertsch, R., Bailey, J., Barlow, K., Beck, S., Berry, E., Birren, B., Bloom, T., Bork, P., Botcherby, M., Bray, N., Brent, M.R., Brown, D.G., Brown, S.D., Bult, C., Burton, J., Butler, J., Campbell, R.D., Carninci, P., Cawley, S., Chiaromonte, F., Chinwalla, A.T., Church, D.M., Clamp, M., Clee, C., Collins, F.S., Cook, L.L., Copley, R.R., Coulson, A., Couronne, O., Cuff, J., Curwen, V., Cutts, T., Daly, M., David, R., Davies, J., Delehaunty, K.D., Deri, J., Dermitzakis, E.T., Dewey, C., Dickens, N.J., Diekhans, M., Dodge, S., Dubchak, I., Dunn, D.M., Eddy, S.R., Elnitski, L., Emes, R.D., Eswara, P., Eyras, E., Felsenfeld, A., Fewell, G.A., Flicek, P., Foley, K., Frankel, W.N., Fulton, L.A., Fulton, R.S., Furey, T.S., Gage, D., Gibbs, R.A., Glusman, G., Gnerre, S., Goldman, N., Goodstadt, L., Grafham, D., Graves, T.A., Green, E.D., Gregory, S., Guigo, R., Guyer, M., Hardison, R.C., Haussler, D., Hayashizaki, Y., Hillier, L.W., Hinrichs, A., Hlavina, W., Holzer, T., Hsu, F., Hua, A., Hubbard, T., Hunt, A., Jackson, I., Jaffe, D.B., Johnson, L.S., Jones, M., Jones, T.A., Joy, A., Kamal, M., Karlsson, E.K., *et al.* (2002). Initial sequencing and comparative analysis of the mouse genome. *Nature* **420**, 520–562.
Werren, J.H. (1997). Biology of *Wolbachia*. *Annual Reviews of Entomology* **42**, 587–609.
Wicks, S.R., Yeh, R.T., Gish, W.R., Waterston, R.H. and Plasterk, R.H. (2001). Rapid gene mapping in *Caenorhabditis elegans* using a high density polymorphism map. *Nature Genetics* **28**, 160–164.

Williams, S.A., Lizotte-Waniewski, M.R., Foster, J., Guiliano, D., Daub, J., Scott, A.L., Slatko, B. and Blaxter, M.L. (2000). The filarial genome project: analysis of the nuclear, mitochondrial and endosymbiont genomes of *Brugia malayi*. *International Journal for Parasitology* **30**, 411–419.

Williamson, V.M., Ho, J.-Y. and Ma, H.M. (1992). Molecular transfer of nematode resistance genes. *Journal of Nematology* **24**, 234–241.

Williamson, V.M. and Hussey, R.S. (1996). Nematode pathogenesis and resistance in plants. *Plant Cell* **8**, 1735–1745.

Wilson, M.A., Bird, D.M. and van der Knapp, E. (1994a). A comprehensive subtractive cDNA cloning approach to identify nematode-induced transcripts in tomato. *Molecular Plant Pathology* **84**, 299–303.

Wilson, R., Ainscough, R., Anderson, K., Baynes, C., Berks, M., Bonfield, J., Burton, J., Connell, M., Copsey, T., Cooper, J., Coulson, A., Craxton, M., Dear, S., Du, Z., Durbin, R., Favello, A., Fraser, A., Fulton, L., Gardner, A., Green, P., Hawkins, T., Hillier, L., Jier, M., Johnston, L., Jones, M., Kershaw, J., Kirsten, J., Laisster, N., Latrielle, P., Lightning, J., Lloyd, C., Mortimore, B., O'Callaghan, M., Parsons, J., Percy, C., Rifken, L., Roopra, A., Saunders, D., Shownkeen, R., Sims, M., Smaldon, N., Smith, A., Smith, M., Sonnhammer, E., Staden, R., Sulston, J., Thierry-Mieg, J., Thomas, K., Vaudin, M., Vaughan, K., Waterston, R., Watson, A., Weinstock, L., Wilkinson-Sproat, J. and Wohldman, P. (1994b). 2.2 Mb of contiguous nucleotide sequence from chromosome III of *C. elegans*. *Nature* **368**, 32–38.

Winnepenninckx, B., Backeljau, T., Mackey, L.Y., Brooks, J.M., DeWachter, R., Kumar, S. and Garey, J.R. (1995). 18S rRNA data indicate that Aschelminthes are polyphyletic in origin and consist of at least three distinct clades. *Molecular Biology and Evolution* **12**, 1132–1137.

Wixon, J., Blaxter, M., Hope, I., Barstead, R. and Kim, S. (2000). *Caenorhabditis elegans*. *Yeast* **17**, 37–42.

Wolstenholme, D.R. (1992). Genetic novelties in mitochondrial genomes of multicellular animals. *Current Opinion in Genetics and Development* **2**, 918–925.

Wood, W.B. (ed.) (1988) *The Nematode Caenorhabditis elegans*. New York: Cold Spring Harbor Laboratory.

Xie, H., Bain, O. and Williams, S.A. (1994a). Molecular phylogenetic studies on *Brugia* filariae using Hha-1 repeat sequences. *Parasite* **1**, 255–260.

Xie, H., Bain, O. and Williams, S.A. (1994b). Molecular phylogenetic studies on filarial parasites based on 5S ribosomal spacer sequences. *Parasite* **1**, 141–151.

Xie, T. and Ding, D. (2000). Investigating 42 candidate orthologous protein groups by molecular evolutionary analysis on genome scale. *Gene* **261**, 305–310.

Yamada, M., Matsuda, S., Nakazawa, M. and Arizono, N. (1991). Species-specific differences in heterogonic development of serially transferred free-living generations of *Strongyloides planiceps* and *Strongyloides stercoralis*. *Journal of Parasitology* **77**, 592–594.

Yenbutr, P. and Scott, A.L. (1995). Molecular cloning of a serine proteinase inhibitor from *Brugia malayi*. *Infection and Immunity* **63**, 1745–1753.

Young, J.M. and Hope, I.A. (1993). Molecular markers of differentiation in *Caenorhabditis elegans* obtained by promoter trapping. *Developmental Dynamics* **196**, 124–132.

Zang, X., Atmadja, A.K., Gray, P., Allen, J.E., Gray, C.A., Lawrence, R.A., Yazdanbakhsh, M. and Maizels, R.M. (2000). The serpin secreted by *Brugia malayi* microfilariae, Bm-SPN-2, elicits strong, but short-lived, immune responses in mice and humans. *Journal of Immunology* **165**, 5161–5169.

Zang, X. and Maizels, R.M. (2001). Serine proteinase inhibitors from nematodes and the arms race between host and pathogen. *Trends in Biochemical Science* **26**, 191–197.

Zang, X., Taylor, P., Wang, J.M., Meyer, D.J., Scott, A.L., Walkinshaw, M.D. and Maizels, R.M. (2002). Homologues of human macrophage migration inhibitory factor from a parasitic nematode. Gene cloning, protein activity, and crystal structure. *Journal of Biological Chemistry* **277**, 44261–44267.

Zang, X., Yazdanbakhsh, M., Jiang, H., Kanost, M.R. and Maizels, R.M. (1999). A novel serpin expressed by blood-borne microfilariae of the parasitic nematode *Brugia malayi* inhibits human neutrophil serine proteinases. *Blood* **94**, 1418–1428.

Zipperlen, P., Fraser, A.G., Kamath, R.S., Martinez-Campos, M. and Ahringer, J. (2001). Roles for 147 embryonic lethal genes on *C. elegans* chromosome I identified by RNA interference and video microscopy. *EMBO Journal* **20**, 3984–3992.

Zorio, D.A., Cheng, N.N., Blumenthal, T. and Spieth, J. (1994). Operons as a common form of chromosomal organization in *C. elegans*. *Nature* **372**, 270–272.

Zvelebil, M.J.J.M., Tang, L., Cookson, E., Selkirk, M.E. and Thornton, J.M. (1993). Molecular modelling and epitope prediction of gp29 from lymphatic filariae. *Molecular and Biochemical Parasitology* **58**, 145–154.

Life Cycle Evolution in the Digenea: a New Perspective from Phylogeny

Thomas H. Cribb[1], Rodney A. Bray[2], Peter D. Olson[2], D. Timothy J. Littlewood[2]

[1]*Department of Microbiology & Parasitology and Centre for Marine Studies, The University of Queensland, Brisbane, 4072 Australia*
[2]*Parasitic Worms Division, Department of Zoology, The Natural History Museum, Cromwell Road, London SW7 5BD, UK*

Abstract	198
1. Introduction	198
2. Methods	199
2.1. The Tree	200
2.2. The Life Cycle Database	200
2.3. Mapping Life Cycle Characters	202
3. Background to the Digenea	203
3.1. Relationships within the Neodermata	203
3.2. The Digenea in Outline	204
3.3. Aspidogastrea – The Sister Group to the Digenea	206
3.4. A New Phylogeny of the Trematoda	207
4. Mapping and Interpreting Life Cycle Traits	209
4.1. First Intermediate Hosts	209
4.2. Second Intermediate Hosts	214
4.3. Definitive Hosts	217
4.4. Infection Processes – Miracidial Behaviour	220
4.5. Rediae and Sporocysts	220
4.6. Cercarial Tails	223
4.7. Infection Processes – Cercarial Behaviour	224
4.8. A Hypothesis of Evolution within the Plagiorchiida	227
4.9. Hypotheses of Evolution within the Diplostomida	234
5. Problems	240
5.1. Shortcomings of the Parsimony Approach	240
5.2. Conflict between Hypotheses	241
5.3. Ten Questions	243

Appendix 244
    Order Diplostomida 244
    Order Piagiorchiida 245
Acknowledgements 249
References 249

ABSTRACT

We use a new molecular phylogeny, developed from small and large subunit ribosomal RNA genes, to explore evolution of the digenean life cycle. Our approach is to map character states on the phylogeny and then use parsimony to infer how the character evolved. We conclude that, plesiomorphically, digenean miracidia hatched from eggs and penetrated gastropod first intermediate hosts externally. Fork-tailed cercariae were produced in rediae and emerged from the snail to be eaten directly by the teleost definitive host. These plesiomorphic characters are seen in extant Bivesiculidae. We infer that external encystment and the use of second intermediate hosts are derived from this behaviour and that second intermediate hosts have been adopted repeatedly. Tetrapod definitive hosts have also been adopted repeatedly. The new phylogeny proposes a basal dichotomy between 'Diplostomida' (Diplostomoidea, Schistosomatoidea and Brachylaimoidea) and 'Plagiorchiida' (all other digeneans). There is no evidence for coevolution between these clades and groups of gastropods. The most primitive life cycles are seen in basal Plagiorchiida. Basal Diplostomida have three-host life cycles and are associated with tetrapods. The blood flukes (Schistosomatoidea) are inferred to have derived their two-host life cycles by abbreviating three-host cycles. Diplostomida have no adult stages in fishes except by life cycle abbreviation. We present and test a radical hypothesis that the blood-fluke cycle is plesiomorphic within the Diplostomida.

1. INTRODUCTION

The Digenea is one of two subclasses of the Trematoda that, together with the classes Cestoda and Monogenea, form the Neodermata within the phylum Platyhelminthes. These three classes of flatworms are entirely parasitic. Of the three, the Digenea, or flukes, have the most complex life cycles by far. They may have from one to four hosts, three distinct generations, many morphologically distinct forms, and they infect their hosts in many different ways. This complexity has been of interest to helminthologists from the moment their life cycles began to be elucidated. Two main classes of questions have been explored: How did the complex digenean life cycle as a whole arise? and How has the variation within the group evolved? By no means has a consensus emerged that answers either question satisfactorily. Indeed, even where consensus appears to have been reached on a particular point, the explanation is not necessarily well founded.

Until convincing phylogenetic hypotheses began to be published in the 1980s for the Platyhelminthes as a whole and the Digenea in particular, attempts to understand the evolution of life cycles within the phylum were seriously hampered. Hypotheses could only be developed as plausible narratives that accounted for what was known of the biology of the animals. The lack of methodological rigour inherent in the narrative approach meant that there were no real means of testing these hypotheses objectively. Despite this, earlier authors made remarkable progress; the ideas of Pearson (1972; 1992), Rohde (1972), Cable (1974; 1982), Gibson (1987) and many others summarised by these authors remain highly influential and often intuitively persuasive. If an advance in understanding is taking place (certainly no one would contend that the debates are concluded), then it is by the application of phylogenetic systematics and molecular data sets that allow the old ideas, and many new ones, to be tested with greater objectivity than was possible previously. Our goal here is to understand the origins and evolution of the digenean life cycle by examining its variation in the light of a phylogeny derived wholly independently of the life cycle characters themselves.

2. METHODS

Our inferences are based entirely on the phylogeny of the Digenea as inferred from an independent, molecular data set (Olson *et al.*, 2003); not on a tree derived from the life cycle characters we examine herein, or any other combination of morphological characters. We make no assumptions *a priori* about character polarity, for example, but instead allow the positions of the taxa exhibiting the characters themselves to determine which states are plesiomorphic and which are derived. It is implicit then that we accept the hypothesis presented in Olson *et al.* (2003), and the classification that follows from it, as the basis for all inferences made herein. Although the character state patterns we discuss are sometimes complicated or equivocal and thus require considerable discussion, there is no special pleading for a particular pattern by reference to other factors that may argue for one course of evolution over another; in all cases, the molecular-based phylogeny is the final arbitrator.

2.1. The Tree

The phylogeny of the Digenea (Olson *et al.*, 2003) was based on a Bayesian analysis (Huelsenbeck *et al.*, 2001) of 170 taxa characterized for both the large (variable domains D1–D3) and small (complete) subunit nuclear ribosomal genes giving a total of 2648 alignable positions. This constitutes by far the largest molecular-based hypothesis for the group to date and includes considerable representation of their diversity among marine and freshwater teleosts, elasmobranchs, and tetrapod hosts. In addition, seven species representing seven different aspidogastrean genera were used to root the tree. Although multiple hypotheses based on different analyses are presented in their paper (Olson *et al.*, 2003), the Bayesian analysis including all taxa and all characters is reproduced here (Figure 1). The same figure was used as the basis for a discussion and revision of digenean classification. Their classification amends and expands that proposed in volume one of *Keys to the Trematoda* (Gibson *et al.*, 2002) and forthcoming volumes, and includes a number of new higher-taxon names, altered memberships, and other systematic changes that the non-specialist is likely to be unfamiliar with. For example, we do not recognize the traditional orders Echinostomida and Strigeida. Instead, a basal dichotomy is proposed, separating the Diplostomida (comprised of only, but not includiing all, 'strigeids') and the Plagiorchiida (in which 'echinostomids', 'plagiorchiids', and the remaining 'strigeids' are intermixed). Smaller clades are recognized as suborders, superfamilies, and so on. *Keys to the Trematoda* (Gibson *et al.*, 2002) provide a revised classification only to the level of superfamily and, generally speaking, the superfamilial circumscriptions based on morphology (Gibson *et al.*, 2002) and those based on molecules (Olson *et al.*, 2003) are in agreement. Section 3.4 includes a more detailed discussion on the tree and the classification.

2.2. The Life Cycle Database

Life cycle trait character states are derived from a database of life cycles for the Digenea that we have compiled. This database currently comprises published information on approximately 1350 species (including many

LIFE CYCLE EVOLUTION IN THE DIGENEA 201

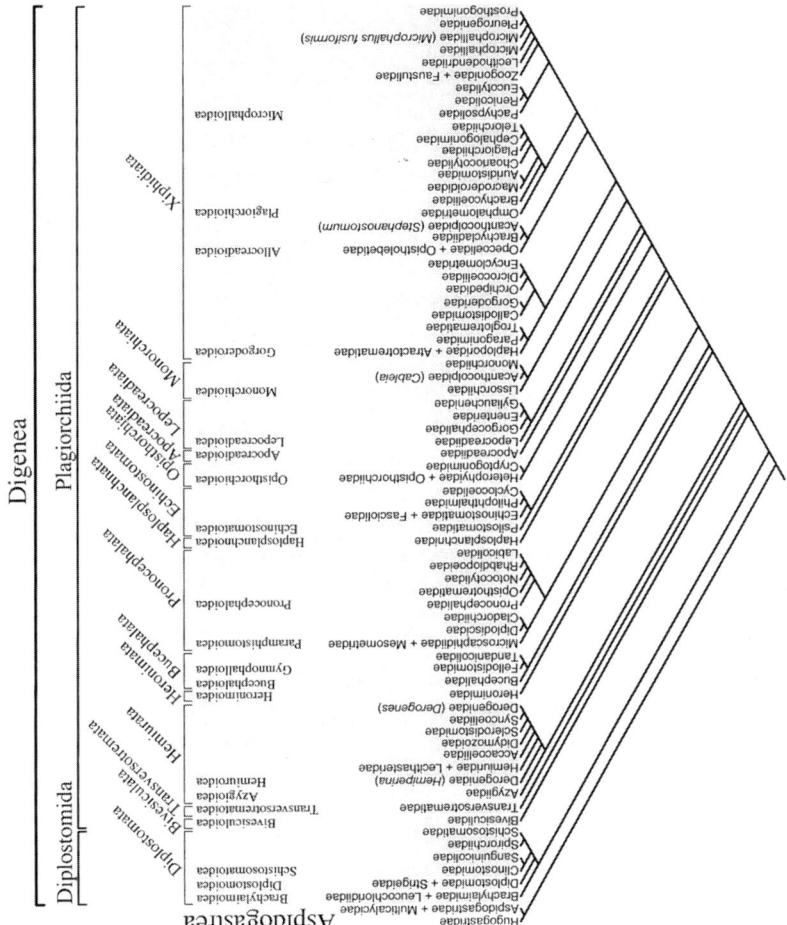

Figure 1 Phylogeny of the Digenea according to Olson *et al.* (2003), based on a Bayesian analysis of nuclear ribosomal genes and indicating their recent assessment of the group's classification.

cercariae which have not been associated with sexual adults) and compiles information on the identity of hosts and the behaviour and morphology of the life cycle stages. There is such great variability within life cycles of the Digenea that we have been forced in some cases to make assumptions when comparing states for higher taxa (e.g. families and superfamilies). For example, life cycles of some opecoelids, where an arthropod is the definitive host, are disregarded as we conclude that they have evolved by abbreviation of the life cycle (Poulin and Cribb, 2002). In most cases such exceptions are mentioned in the text but, when this is not so, we have striven to assume nothing that has been open to serious challenge. The literature is now very large and we are likely to have missed several significant studies. We have, however, endeavoured to compile a complete listing.

2.3. Mapping Life Cycle Characters

We infer the evolution of individual life cycle traits by examining their distribution on the phylogenetic tree. For character mapping we follow the methods outlined by Maddison and Maddison (2000) in the MacClade manual, and we encourage readers unfamiliar with this topic to start with their review. We paraphrase the basic principles here. Put simply, we pursue the following general question: given the topology of the phylogenetic tree, the states observed in the terminal taxa, and the assumptions (if any) regarding character evolution, what assignments of states to the internal nodes of the tree require the fewest evolutionary steps? When all extant members of one clade share character state 1, then it is simplest to infer that all the ancestral members of the clade also had state 1. Of course, it may be more complex than this, and although many algorithms for character mapping have been developed incorporating various nuances in the way characters are allowed to evolve, we have used a manual approach throughout, deducing plesiomorphic (ancestral) character states whenever possible. We have few preconceived ideas of how life cycle characters may evolve from one state to another, and such characters are undoubtedly difficult to homologize. For instance, we do not know if losing a host is as easy as acquiring one and whether the definitive host in a two-host life cycle is homologous with the definitive host in a three-host life-cycle. Thus, we have not weighted or dictated the direction of character state transformations. The acquisitions of a host, the appearance of an attachment organ or the development of a behavioural strategy are all life cycle characters worthy of mapping, but clearly they are not of equal rank. Indeed, the problem in employing phylogenies in understanding behaviours is a discipline unto itself (Martins, 1996). We do not have an explicit matrix of characters to hand that lends itself to a character mapping program such as MacClade. Instead, we make use of a growing descriptive database.

3. BACKGROUND TO THE DIGENEA

Understanding evolution of life cycle traits in the Digenea requires an understanding of the relationships in the Neodermata as a whole, the nature of the Digenea itself (especially its life cycle), the nature of the sister-group to the Digenea (the Aspidogastrea), and relationships within the Digenea. We briefly review these subjects below.

3.1. Relationships within the Neodermata

It is now well accepted that the Class Trematoda forms part of a clade of completely parasitic platyhelminths, the Neodermata, together with the Cestoda and the Monogenea. The Neodermata is separated from a myriad of turbellarian taxa by combinations of morphological and molecular characters (Littlewood *et al.*, 1999). The Neodermata thus encompasses all major platyhelminth taxa that parasitize vertebrates.

The higher classification and basic phylogenetic relationships of the Neodermata are considered broadly resolved by most authors. The system is simple. The Class Trematoda forms the sister taxon to the Cercomeromorpha which is composed of the Monogenea, Gyrocotylidea, Amphilinidea and Eucestoda (Figure 2). Lockyer *et al.* (2003) have new analyses that challenge the cercomeromorph theory and suggest that the Cestoda (= Gyrocotylidea + Amphilinidea + Eucestoda) may be the sister taxon to the Trematoda. Surprisingly, this controversial hypothesis has little effect on the analysis presented below and it is thus not considered further here.

The topology of the relationships of the Neodermata provides a simple example of the mapping and interpretation of life cycle traits. The parsimonious hypothesis for the adoption of these hosts is shown in Figure 2. It can be inferred that the Neodermata as a whole adopted vertebrate parasitism, that the Trematoda subsequently adopted parasitism of molluscs as intermediate hosts, and that the Eucestoda + Amphilinidea adopted parasitism of crustacean intermediate hosts. This inference was first made by Littlewood *et al.* (1999).

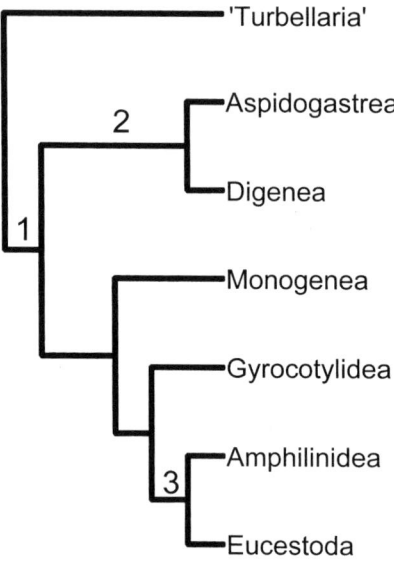

Figure 2 Phylogeny of the Neodermata and a parsimonious hypothesis for the adoption of definitive host groups.

3.2. The Digenea in Outline

The Digenea are one of the most successful groups of parasitic flatworms. Digeneans occur in significant numbers in all classes of vertebrates, although they are surprisingly rare in chondrichthyans, in which cestodes and monogeneans are the dominant platyhelminths (see Bray and Cribb, 2003). As adults they are primarily parasites of the intestine, but representatives are also found in fish under the scales, on the gills, in the swim bladder, in the body cavity, the urinary bladder, the gall bladder, in the flesh, connective tissue, ovary and in the circulatory system. In tetrapods, they may also be found in extra-intestinal sites including the circulatory system, lungs, air sacs, oesophagus, urinary bladder, liver, eye and ovary. Morphological variation is found in the form and position of the suckers, the digestive tract and the reproductive system. Despite this variation, the sexual adults of phylogenetically distinct taxa may resemble each other closely (e.g. Allocreadiidae and Opecoelidae, Heterophyidae and Microphallidae), whereas their life cycle stages are obviously different.

The digenean life cycle usually involves both free-living and parasitic stages and always incorporates both asexual and sexual multiplication. Most cycles are at least partially aquatic and alternate between a mollusc and a vertebrate. Sexual adults occur in any of the sites mentioned above where they produce eggs that pass to the external environment. These eggs typically hatch to release a motile, short-lived, non-feeding, ciliated larva, the miracidium. The miracidium swims and penetrates a molluscan intermediate host in which it sheds its ciliated outer cells and develops into a mother sporocyst. The mother sporocyst is a simple sac that lacks any trace of feeding structures or gonads. It produces a second intra-molluscan generation asexually. This generation comprises either multiple daughter sporocysts or rediae. Daughter sporocysts resemble the mother sporocyst, whereas rediae have a mouth, a pharynx and a short saccular gut. In the case of both second intra-molluscan generation types, i.e., daughter sporocysts and rediae, there is another round of asexual reproduction. In this case, the progeny are usually cercariae, the larvae of the generation that will ultimately become the sexual adult. The cercaria is usually a tailed form that emerges actively from the mollusc. After emergence, the cercaria behaves in one of several distinct ways that ultimately leads to active or passive infection of the vertebrate definitive host. Often the vertebrate is infected by ingestion of a metacercaria associated with a second intermediate host.

Significant departures from the developmental pattern described above are numerous. In many digenean taxa (e.g. the superfamilies Brachylaimoidea, Hemiuroidea, Opisthorchioidea, Pronocephaloidea), the molluscan host eats the egg and is then penetrated internally. In the Pronocephaloidea, this process is specialized to the point of a process of mechanical injection (Murrills *et al.*, 1985). The interaction between the intra-molluscan generations may be complex. In some families (e.g. Heronimidae and Bucephalidae), it is possible that there is only a mother sporocyst. Beyond this, the distinction between daughter sporocysts and rediae is not as clear as often thought. In some life cycles, rediae have been shown to lose their digestive systems during ontogeny so as to become sporocysts secondarily. Some authors have considered daughter sporocysts as paedogenetic rediae (Matthews, B.F. 1980). In addition, the progression between the intra-molluscan generations described above may not be as simple as outlined. There are many reports of rediae or sporocysts producing more rediae or sporocysts as well as cercariae. This plasticity in the life cycle was demonstrated most dramatically by Dönges (1971), who showed that cercaria-producing rediae of echinostomatids were always capable of reverting to the production of rediae if they were transplanted to an uninfected gastropod. Finally, life cycles of some cyathocotylids have been described in which sporocysts have been found to produce miracidia that emerge from the snail (Sewell, 1922; Barker and Cribb, 1993), presumably to infect other snails. The evolution of such complexity and variation has been fertile ground for more than a century of conjecture and debate.

3.3. Aspidogastrea – The Sister Group to the Digenea

The Trematoda comprises two subclasses – the Aspidogastrea and the Digenea. The Aspidogastrea comprises just four families and perhaps 80 species (Rohde, 2001) in contrast to well over 100 families and certainly over 10,000 species of digeneans (Gibson et al., 2002). Many authors, including Rohde (2001), have considered the Aspidogastrea to be 'archaic' relatives of the Digenea because of the relative simplicity of their life cycles compared to those of digeneans. As they are the sister group to the Digenea, it is necessary to have a basic understanding of their life cycles.

Of the four families (Aspidogastridae, Multicalycidae, Rugogastridae and Stichocotylidae), complete life cycles are known only for the Aspidogastridae. This, by far the largest family, occurs as sexual adults in molluscs, teleosts and turtles. Sexual adults typically live in the intestine of vertebrates and pass eggs in the faeces. These hatch to liberate a ciliated cotylocidium that swims to a molluscan host to which it attaches. The cotylocidium sheds its cilia and develops directly to an infective stage in which the characteristic loculated ventral sucker has formed something resembling its final condition. Parasitism of the mollusc is usually by attachment to its external surfaces, for example within the mantle cavity (Ferguson et al., 1999). In a number of cases, however (see summary by Rohde, 1972), the parasite enters pores of the molluscs and may be found in the pericardial cavity, kidney cavity, and even inside the gill filaments. The vertebrate is always infected by ingestion of the mollusc.

There are two significant variations on this pattern. The first was described for *Lobatostoma manteri* (Rohde, 1973, 1975; Rohde and Sandland, 1973). In this cycle, the egg does not hatch until the gastropod host eats it. It then hatches within the mollusc gut and migrates into the cavity of the digestive gland of the gastropod, where it develops to a juvenile; the life cycle is again completed by the ingestion of the mollusc. The juvenile is in the gut, not within the tissues as apparently always occurs in the infections of digeneans within molluscs. The second significant variation is that several aspidogastrids develop to sexual maturity on their molluscan hosts. In some life cycles, the vertebrate host is thus facultative and, in a few species, there is no known vertebrate host. The life cycles of the other families are not well known but there is no indication of significant differences except that the sole stichocotylid, *Stichocotyle nephropis*, occurs as juveniles encapsulated in the crustacean *Nephrops* which presumably acts as a second intermediate host since the final hosts are rays (*Raja clavata*).

Key features of the Aspidogastrea relative to the Digenea are that they are external rather than visceral parasites of molluscs, lack asexual reproduction entirely, and have no stage comparable to the cercaria. Thus, although certainly the sister-taxon to the Digenea, the aspidogastrean life cycle is so different from that of digeneans that it may offer little insight into the evolution of the digenean life cycle.

3.4. A New Phylogeny of the Trematoda

The phylogeny of the Trematoda used here (Figure 1) is that of (Olson *et al.*, 2003), which includes a revised classification of the group (their Figure 6). This phylogeny is based on complete small subunit rDNA and partial (D1–D3) large subunit rDNA sequences of 170 taxa from 77 nominal families. This is the largest set of taxa and characters yet assembled for the Digenea and therefore may be more reliable and informative than previous analyses. The phylogeny used in the present analyses, and on which the classification of Olson *et al.* (2003) is based, is based on Bayesian inference from both the large and small nuclear ribosomal genes (their Figure 3).

Perhaps the key feature of the phylogeny and classification of Olson *et al.* (2003) is its recognition of a clear basal dichotomy between two superorders named the Diplostomida (new name) and the Plagiorchiida; there is no small basal taxon although the Plagiorchiida is far larger than the Diplostomida. Fourteen orders (eight new) are recognized within the superorders. All Diplostomida are incorporated in three superfamilies in the Order Diplostomata. Of the thirteen orders recognized within the Plagiorchiida, nine contain only single superfamilies. The four more inclusive orders are the Hemiurata (Azygioidea + Hemiuroidea), Bucephalata (Bucephaloidea + Gymnophalloidea), Paramphistomata (Paramphistomoidea + Pronocephaloidea) and Xiphidiata (Gorgoderoidea + Microphalloidea + Allocreadioidea + Plagiorchioidea). The superorders, orders and superfamilies are the primary groups that we have used for considering life cycle evolution, although the trees show the superfamily Schistosomatoidea divided into its four constituent families, the Clinostomidae, Sanguinicolidae, Schistosomatidae and Spirorchiidae. This is done partly because this superfamily is so complex and interesting, and partly because it is paraphyletic with respect to the three blood-dwelling groups. In referring to digenean taxa we refer to the most inclusive taxon available. Thus, lists of taxa that share a particular character might include families, superfamilies and orders. Where an order contains just one superfamily we refer to the superfamily.

The number of taxa incorporated in the trees used has been reduced as far as possible for simplicity. For example, multiple taxa from the Cryptogonimidae, Heterophyidae and Opisthorchiidae are always represented by a single higher taxon, the Opisthorchioidea. This reflects the fact that, for the purposes of the present analysis, there is no known difference in the life cycles represented in this superfamily. A brief description of the key life cycle characteristics found in each of the 22 superfamilies recognized by Olson *et al.* (2003) is given in the Appendix. The complete phylogeny and higher classification referred to in this analysis is shown in Figure 3. Two

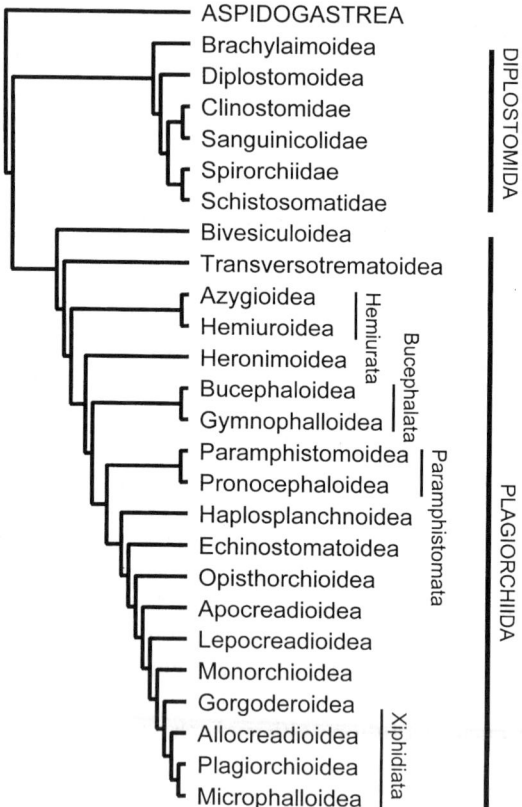

Figure 3 Relationships and higher classification of the superfamilies of the Digenea (based on Olson *et al.*, 2003) used in this analysis.

key elements of the classification remain unconfirmed. The type families of the Allocreadioidea and the Gymnophalloidea were not included in the molecular analyses, and so the composition of these superfamilies is partly speculative; see Olson et al. (2003) for a complete list of exemplar taxa analysed.

Readers should be aware that some elements of the classification proposed by Olson et al. (2003) and used here conflict with that proposed in volume one of Keys to the Trematoda (Gibson et al., 2002) and subsequent volumes. For example, we recognise the Clinostomidae as part of the Schistosomatoidea which invalidates the Clinostomoidea as recognised by Kanev et al. (2002) to include the Clinostomidae and Liolopidae. This change was necessitated by the close relationship identified between the Clinostomidae and the Sanguinicolidae. We see this change as an example of the fact that the classification of the Digenea is presently evolving quite rapidly. We gladly acknowledge, however, that all aspects of the classification remain hypotheses. Certainly the changes that we propose in no way affect the utility of Keys to the Trematoda in facilitating the identification of digeneans.

4. MAPPING AND INTERPRETING LIFE CYCLE TRAITS

The variety in the life cycle of digeneans is reduced here to distinctions in the identity of the hosts (first intermediate, second intermediate, definitive), the processes that lead to the infection of these hosts, and the morphology of the life cycle stages. It is unclear, at least *a priori*, to what extent these characters are connected. Our approach, therefore, is to derive hypotheses for these traits independently and then attempt to draw the separate hypotheses together.

4.1. First Intermediate Hosts

Figure 4 shows the distribution of the four major groups of first intermediate hosts: gastropods, bivalves, scaphopods and polychaetes. Gastropods are clearly the most common group. However, common does not necessarily equate with primitive; therefore it is necessary to analyse the data with respect to phylogeny. The outgroup, the Aspidogastrea, infect bivalves and gastropods. This allows four interpretations of the original molluscan hosts of trematodes: (a) gastropods were the basal host group; (b) bivalves were the basal host group; (c) the group as a whole coevolved with

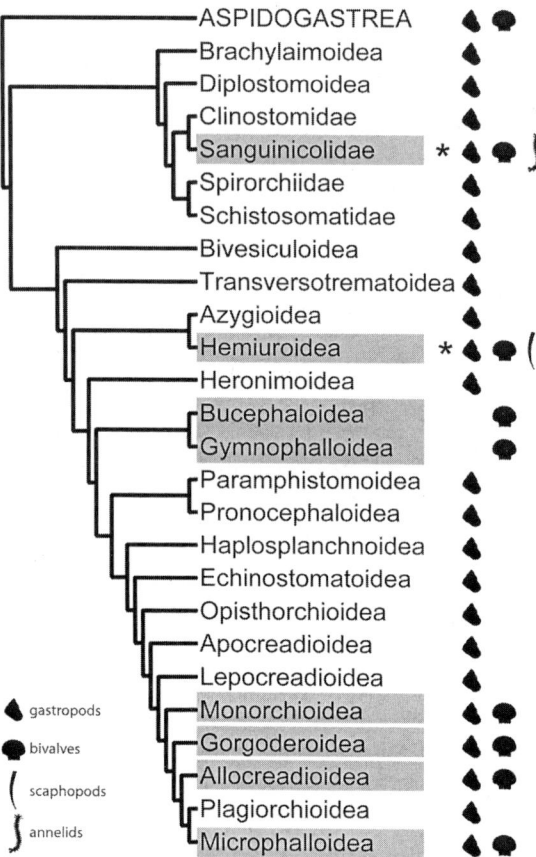

Figure 4 Digenean life cycle variation – first intermediate hosts (gastropods, bivalves, scaphopods and annelids). Shaded taxa are inferred to have host-switched from gastropods. Taxa marked with an asterisk (*) are inferred to have made two switches.

molluscs as a whole; and (d) the infection of molluscs by aspidogastreans and digeneans arose independently. We first reject (d) because the Trematoda as a whole appear to be defined in part by their shared parasitism of molluscs. We reject (c) because the association of trematodes with the other classes of molluscs is either very rare (Scaphopoda) or unknown (Aplacophora, Polyplacophora, Cephalopoda). We can distinguish between (a) and (b) comparing the number of evolutionary changes that each hypothesis requires.

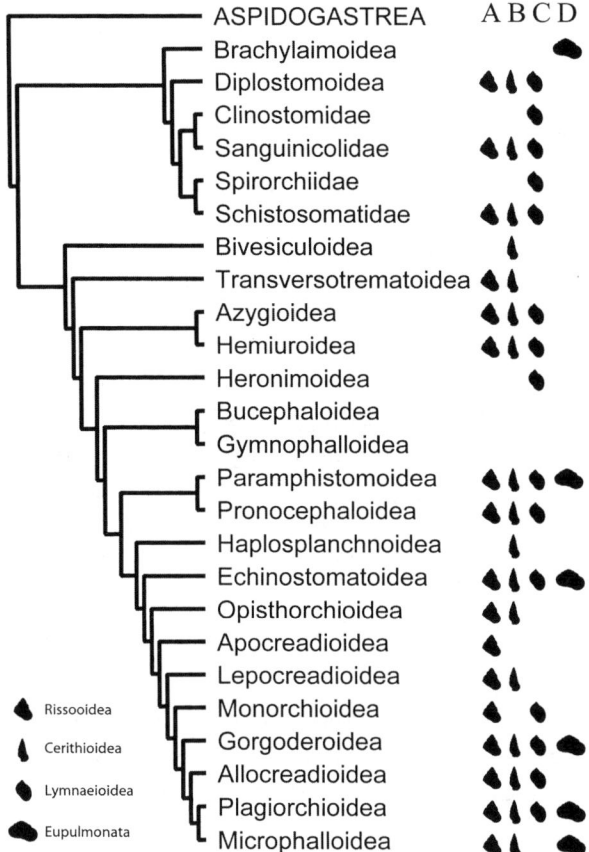

Figure 5 Digenean life cycle variation – first intermediate hosts – selected gastropod taxa. A, Rissooidea; B, Cerithioidea; C, Lymnaeoidea; D, Eupulmonata. Intermediate hosts of Aspidogastrea are not shown.

If the original host of trematodes was a bivalve, then the parasitism of gastropods, polychaetes and scaphopods shown in Figure 4 requires 24 changes of host (if no reversal of host is allowed). If reversals are allowed, then the host distribution can be explained in eleven steps. In contrast, if the gastropods were the first host group then the host distribution can be explained in just ten steps of host change and no host reversal need be invoked. The eight taxa in which such changes are hypothesized to have occurred are highlighted on Figure 4 and the two in which at least two host changes are necessary are further marked with an asterisk. Hypotheses that polychaetes or scaphopods were the first hosts are even less parsimonious. We thus conclude that gastropods were the basal host group for the Digenea.

We are also interested in the detail of the individual host associations. Scaphopods are known hosts only for a ptychogonimid (Palombi, 1942) (the family presently has only two species) and for one lecithasterid hemiuroid (Køie et al., 2002), whereas other lecithasterids use gastropods exclusively. The family Ptychogonimidae is of interest in this respect as it is likely to be one of the most basal taxa in the Hemiuroidea; unfortunately, this family was not represented in the present phylogeny. The Lecithasteridae is relatively derived within the Hemiuroidea. Polychaetes are even more restricted, occurring as hosts only for some sanguinicolids. Thus, the conclusion that scaphopods and polychaetes were acquired by host-switching is plausible as well as parsimonious. Remarkably, sanguinicolids are found in significant numbers in gastropods, bivalves and polychaetes as first intermediate hosts, but not enough is known about relationships within the Sanguinicolidae or about the life cycles of marine taxa for us to draw conclusions on the origins of this diversity.

Although the majority of digeneans infect gastropods, bivalves nevertheless form a significant intermediate host group (Figure 4). In the Diplostomida, bivalves are hosts only rarely (some sanguinicolids), but in the Plagiorchiida they are host to the Allocreadiidae, Bucephalata, Faustulidae, Gorgoderidae and Monorchiidae. The topology of relationships within the Digenea suggests that, as proposed by Hall et al. (1999) and others, parasitism of bivalves has arisen as the result of host-switching from gastropods and that it arose at least seven times (this inference assumes that the Allocreadiidae does indeed form a clade with the Opecoelidae, Brachycladiidae and Acanthocolpidae as predicted by Olson et al., 2003). The Bucephalata is by far the largest taxon that is restricted to bivalves as first intermediate hosts; no other whole superfamily is found only in bivalves.

If it is accepted that bivalves, scaphopods and polychaetes are infected as the result of host-switching from gastropods, then it is still possible that those taxa associated with gastropods will show evidence of host–parasite coevolution via congruence of the host and parasite phylogenies. However, the evidence is strongly against such an interpretation. The simplest level of phylogenetic comparison possible is that between the Diplostomida and the Plagiorchiida, the two primary clades of digeneans. About 23% of records of first intermediate hosts in our database relate to the Diplostomida and the remainder to the Plagiorchiida. Few distinctions are noticeable in the distribution of the gastropods they infect (Table 1). For both groups, there are substantial numbers of records from

Table 1 Major gastropod taxa as first intermediate hosts of Diplostomida and Plagiorchiida.

Subclass	Order	Superfamily	Diplostomida	Plagiorchiida
Eogastropoda	Patellogastropoda	Acmaeoidea		•
		Patelloidea		•
Opisthobranchia	Cephalaspidea	Haminoeoidea	•	
		Philinoidea		•
Orthogastropoda		Ampullarioidea	•	•
		Calyptraeoidea		•
		Cerithioidea	•	•
		Conoidea		•
		Fissurelloidea		•
		Littorinoidea	•	•
		Muricoidea	•	•
		Naticoidea		•
		Neritoidea	•	•
		Pyramidelloidea		•
		Rissooidea	•	•
		Trochoidea		•
		Valvatoidea	•	•
		Velutinoidea		•
		Vermetoidea		•
Pulmonata	Basommatophora	Amphiboloidea		•
		Lymnaeoidea	•	•
		Siphonarioidea		•
	Eupulmonata	Achatinoidea	•	•
		Arionoidea	•	
		Helicoidea	•	•
		Limacoidea	•	•
		Partuloidea		•
		Polygyroidea	•	•
		Pupilloidea	•	•
		Rhytidoidea	•	
		Succineoidea	•	•

"prosobranchs" (especially Rissooidea and Cerithioidea) and freshwater pulmonates (especially Lymnaeoidea), and the terrestrial pulmonates (Eupulmonata) used by the two groups are also similar. The distributions of these four gastropod taxa are shown in Figure 5, where they are scattered throughout the phylogeny. Thus, we find no reflection of the basal dichotomy of the Digenea in the first intermediate hosts. The implication of this is that there has been no deep level coevolution between the major clades of digeneans and their molluscan hosts or, if it has occurred, it is rendered unrecognizable by host-switching. If such coevolution has occurred at all it may have been within smaller clades. The Hemiuroidea, Sanguinicolidae and Bucephalidae each have a wide range of intermediate hosts and the possibility that they coevolved with their hosts has not yet been explored.

4.2. Second Intermediate Hosts

At least 17 of the 22 digenean superfamilies incorporate second intermediate hosts in their life cycles. Figure 6 shows the distribution of taxa exploited as second intermediate hosts. Table 2 shows counts for host phyla extracted from our database. Two features are immediately striking. There is a great diversity of intermediate hosts (nine phyla) and the distributions of different types of hosts are discontinuous on the phylogeny. These observations lead to the conclusion that three-host life cycles have been adopted repeatedly. The fact that three of the most basal superfamilies (Bivesiculoidea, Azygioidea and Transversotrematoidea) usually lack second intermediate hosts indicates that three-host cycles have been derived from two-host forms. Outgroup comparison with the Aspidogastrea also indicates that a two-host life cycle was plesiomorphic for the Digenea. We can conclude that at least the Diplostomida, Hemiuroidea, Bucephalata and the Echinostomatoidea, plus all remaining Plagiorchiida, must each have separately adopted second intermediate hosts.

An important component of the adoption of second intermediate hosts is the identity of host groups. Most clades have distinctive host distributions and several show perfect or almost perfect fealty to a single host group. Thus, the Azygioidea, Bucephaloidea, Opisthorchioidea and the majority of Diplostomoidea are restricted to vertebrates and the Brachylaimoidea are entirely restricted to molluscs. As each taxon with this restricted relationship with a group of second intermediate hosts is isolated from any other taxon solely using the same host group, it is arguable that each represents either an independent adoption or involved a complete switch of host group. Cribb et al. (2002) suggested that it was a hallmark of many digenean taxa to be in part defined by their range of hosts in this way. In contrast, the Echinostomatoidea occur almost equally in vertebrates and molluscs and the Gymnophalloidea and Lepocreadioidea occur in at least six and seven different phyla, respectively.

Just two clades of digeneans are frequent parasites of arthropods – the Hemiuroidea and the Xiphidiata (Table 2, Figure 6). The Hemiuroidea

LIFE CYCLE EVOLUTION IN THE DIGENEA 215

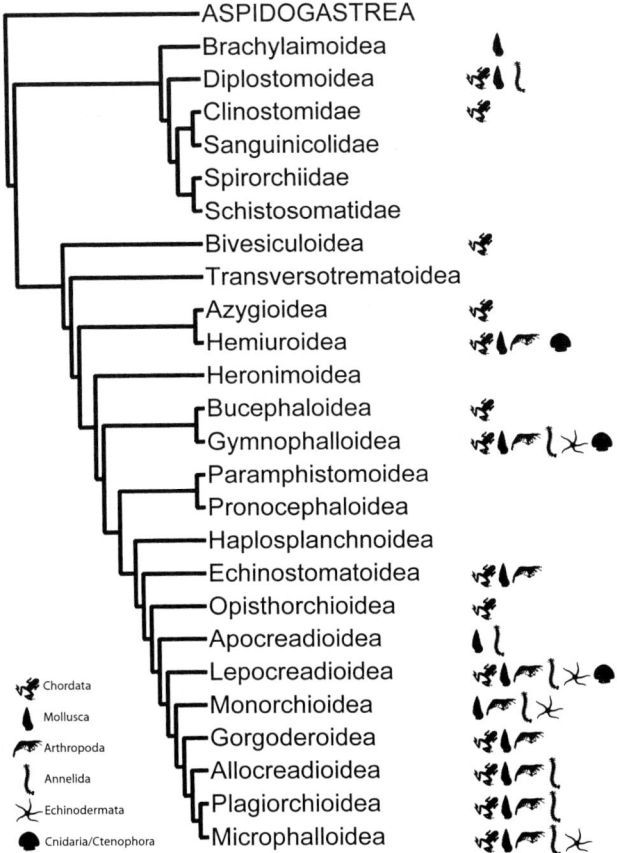

Figure 6 Digenean life cycle variation – phyla exploited as second intermediate hosts. Lepocreadioidea infects all six shown phyla – Chordata, Mollusca, Arthropoda, Annelida, Echinodermata and Cnidaria (coded with Ctenophora). Platyhelminthes and Chaetognatha are not shown.

infect crustaceans and insects by way of their specialized cystophorous cercaria which penetrates the gut of the intermediate host and injects the cercarial body into the haemocoel (e.g., see Køie, 1990). The few records of metacercariae of hemiuroids in chaetognaths, molluscs and vertebrates are likely to be reports of third intermediate hosts that have consumed infected arthropods (e.g., Zelmer and Esch, 1998); hemiuroid evolution may thus be completely coupled to arthropod second intermediate hosts. In contrast to the Hemiuroidea, the Xiphidiata have adopted external penetration of

Table 2 Second intermediate hosts. Counts are of phyla recorded as second intermediate hosts for individual species. A species might be counted more than once if shown to infect, e.g., a mollusc and a vertebrate, but not if infecting multiple species of molluscs. Superfamilies with no or negligible use of second intermediate hosts are excluded.

Superfamily	Annelida	Arthropoda	Chaetognatha	Cnidaria	Ctenophora	Echinodermata	Mollusca	Platyhelminthes	Vertebrata	Total
DIPLOSTOMIDA										
Brachylaimoidea							20			20
Diplostomoidea	4						5		76	85
Schistosomatoidea									11	11
PLAGIORCHIIDA										
Apocreadioidea	1						2			3
Echinostomatoidea		1					29		19	49
Hemiuroidea		24	4		1		1		1	31
Lepocreadioidea	3	1		4		2	4	1	3	18
Monorchioidea	1	1				1	7	2		12
Opisthorchioidea									63	63
Bucephalata										
Bucephaloidea									16	16
Gymnophalloidea	1			1		3	10		1	16
Xiphidiata										
Allocreadioidea	1	27					5		8	41
Gorgoderoidea		18					4		5	27
Microphalloidea	2	50				1	3		1	57
Plagiorchioidea	4	20					12		17	53
Total	17	142	4	5	1	7	102	3	221	556

arthropods. This is the taxon most frequently reported from arthropods, but at least 13 species from this clade are reported to be capable of infecting more than one phylum of intermediate hosts and some species are known from three phyla (Angel, 1967; Jue Sue and Platt, 1998). Approximately one-third of our records of metacercariae of Xiphidiata are from non-arthropods. We conclude that evolution of the stylet conferred the capacity to infect arthropods by penetrating their cuticle or arthrodial membranes, but that, in contrast to the Hemiuroidea, this was not at the expense of the ability to infect other groups of animals.

We conclude that many digenean taxa that have three-host life cycles are linked strongly to a phylogenetic group of intermediate hosts and that three-host cycles have arisen repeatedly. Hypotheses that consider the evolutionary history of individual digenean taxa should thus incorporate reference to the identity of the second intermediate host.

4.3. Definitive Hosts

We have mapped the distribution of definitive hosts on the phylogeny of the Digenea at the simplest level possible – chondrichthyans, teleosts and tetrapods (Figure 7). In this context the Aspidogastrea are uninformative as they occur in all three groups; they do, however, present a strong contrast to the Digenea in that three of their four families occur in chondrichthyans. Determination of the basal host group is made more complicated by problems with the interpretation of the evolutionary status of the definitive host of the blood fluke families (see Section 4.9, p. 234). If for this reason the Diplostomida are ignored, the analysis of the Plagiorchiida alone shows that the basal host group was unambiguously teleosts. This has been well accepted; most hypotheses for the evolution of the Digenea argue that the subclass arose in association with marine teleost fishes (Cable, 1974).

It is noteworthy that so few clades of digeneans occur in chondrichthyans, and that most of these have apparently host-switched from teleosts or are accidental infections (Bray and Cribb, 2003). In these categories are the Acanthocolpidae, Bucephalidae, Faustulidae, Gorgoderidae, Opecoelidae and Zoogonidae; each of these families occurs mainly in teleosts. There are three families that are not so easily explained because of their near-basal position in the case of the Ptychogonimidae and Azygiidae or the complexity of their host distribution in the case of the Sanguinicolidae, but all three require further study as discussed by Bray and Cribb (2003). Thus, the available data support an origin for the Digenea in association with teleosts followed by host-switching into chondrichthyans.

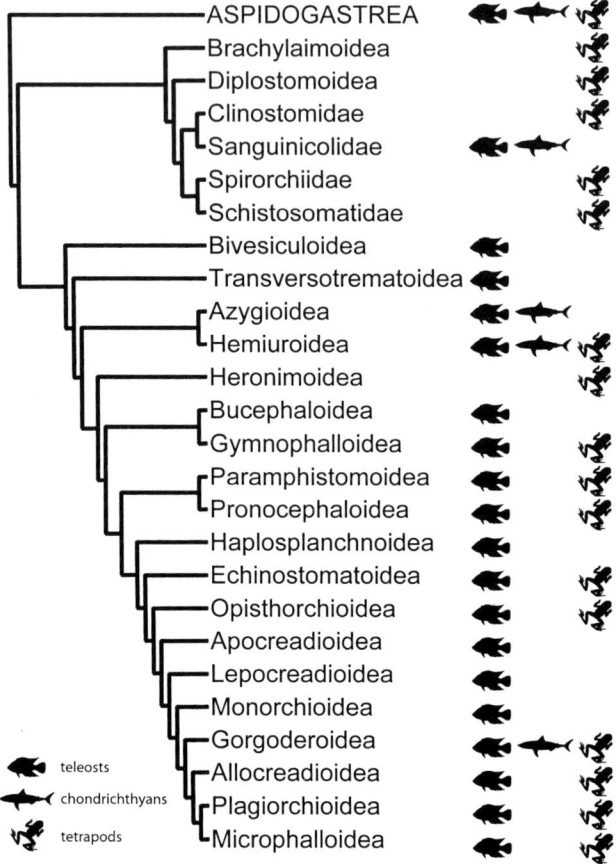

Figure 7 Digenean life cycle variation – definitive hosts coded as teleosts, chondrichthyans and tetrapods.

Adoption of parasitism in tetrapods can be considered, conveniently, in the Plagiorchiida and Diplostomida as separate sets of events. Parasitism of tetrapods is rare among the more basal Plagiorchiida (Figure 7). The two most basal taxa within the Plagiorchiida (Bivesiculoidea and Transversotrematoidea) are entirely parasites of fishes. Parasitism of tetrapods has arisen within one superfamily of the Hemiurata, the Hemiuroidea, which, although overwhelmingly parasites of teleosts, have adopted occasional parasitism of amphibians and reptiles. The monotypic Heronimoidea is the only plagiorchiidan superfamily to be restricted to tetrapods; it is restricted to freshwater turtles. The Gymnophalloidea comprise four families of parasites of teleosts and the Gymnophallidae which parasitize birds (occasionally mammals). Parasitism of tetrapods tends to be more significant in the more derived clades although the Haplosplanchnoidea, Apocreadioidea, Lepocreadioidea and Monorchioidea are all found mainly or entirely in fishes.

The Echinostomatoidea comprises principally parasites of tetrapods. The Paramphistomoidea + Pronocephaloidea are overwhelmingly parasites of tetrapods, although the Paramphistomoidea are well represented in freshwater fishes. The Opisthorchioidea is a major clade occurring in fishes, reptiles, birds and mammals. The life cycle within this family is highly uniform, depending upon a metacercaria in a fish being eaten by the definitive host. The distribution of opisthorchioids, essentially defined by a fish-eating host diet, suggests that the appearance of this clade in tetrapods is the result of a host-switch from fish. The final clade of Plagiorchiida that has radiated significantly among tetrapods is the Xiphidiata. This clade comprises four superfamilies. All four include substantial clades (families) of both fish and tetrapod parasites. The Plagiorchioidea are overwhelmingly parasites of tetrapods, except for the Macroderoididae of freshwater fishes, and it is possible that this family represents a secondary colonization of fishes. The Gorgoderoidea includes such major taxa of fish parasites as the Gorgoderidae as well as quintessentially terrestrial tetrapod parasites, the Dicrocoeliidae. Of particular interest within this clade is the Gorgoderidae itself, members of which infect teleosts, chondrichthyans and tetrapods. The Allocreadioidea are apparently restricted to fishes except for the Brachycladiidae, which are parasites of marine mammals. The Microphalloidea is heavily concentrated among tetrapods (e.g. Lecithodendriidae, Microphallidae, Pleurogenidae, Prosthogonimidae and Renicolidae) but has one major radiation within fishes, the Zoogonidae + Faustulidae. The topology of relationships within the Xiphidiata might suggest either that parasitism of tetrapods arose several times within the Xiphidiata or that they have switched back into fishes more than once.

In contrast to the Plagiorchiida, the hosts of the Diplostomida present a puzzle. All but one of the six resolved clades shown in Figure 7 are parasites of tetrapods. The only clade with adults of some species in fishes is the Sanguinicolidae, and this clade is apparently as derived as any within the Diplostomida. Bray *et al.* (1999) reported urotrematids from Chinese freshwater fishes. It is possible that these trematodes belong in the Diplostomida, but this has not yet been demonstrated. This host distribution implies that the clade of the Diplostomida, having presumably arisen with fishes, has either coevolved away from or, with the exception of the Sanguinicolidae, host-switched out of fishes. The distribution becomes even more surprising when it is realized that blood fluke life cycles are usually considered to be abbreviated three-host life cycles in which a gastrointestinal adult has been lost. Indeed, this has been the dominant and consistent interpretation of the group to the exclusion of any other interpretation of which we are aware (La Rue, 1951; Pearson, 1972; Cable, 1974; Brooks *et al.*, 1985; Shoop, 1988; Kearn, 1998). This interpretation means that effectively no extant member of the Diplostomida is primarily a parasite of a fish. What can account for this unexpected host distribution? This question is revisited when we attempt to resolve an evolutionary history for the Diplostomida as a whole.

4.4. Infection Processes – Miracidial Behaviour

All digeneans have a sexual generation that produces eggs. These eggs embryonate, either within or outside the host, to produce a miracidium. Digenean miracidia vary significantly in their morphology, but here we consider just one important behavioural trait – whether the miracidium hatches and infects the mollusc independently or is eaten by the mollusc and penetrates internally. Occurrences of ingestion of the egg are mapped in Figure 8 (shaded taxa); eggs are eaten in all Brachylaimoidea, Hemiurata, Pronocephaloidea, Opisthorchioidea and Monorchioidea and in several families of the Xiphidiata. Hatching of the miracidium and active host-finding is inferred to be plesiomorphic. Passive ingestion of eggs of the Brachylaimoidea and Dicrocoeliidae (Xiphidiata) are key adaptations in association with completely terrestrial life cycles.

4.5. Rediae and Sporocysts

All digeneans undergo asexual reproduction within a first intermediate host. The basic pattern appears to be a mother sporocyst (the adult of the miracidium) which produces a second generation of sporocysts or rediae. Life cycles in which there is reported to be only a single asexual generation (Heronimidae, Bucephalidae) are, by the position of their appearance in the phylogeny, almost certainly secondary reductions from patterns with two generations. Both heronimids and bucephalids have enormous branching sporocysts so that the need for a second generation may have been lost in both cases. Another interpretation is possible. The mother sporocyst generation may simply be so reduced as to be almost unrecognizable. The miracidium in the Heronimidae contains germinal balls, which are initially enclosed by a fine membrane. Gibson (1987) and

LIFE CYCLE EVOLUTION IN THE DIGENEA

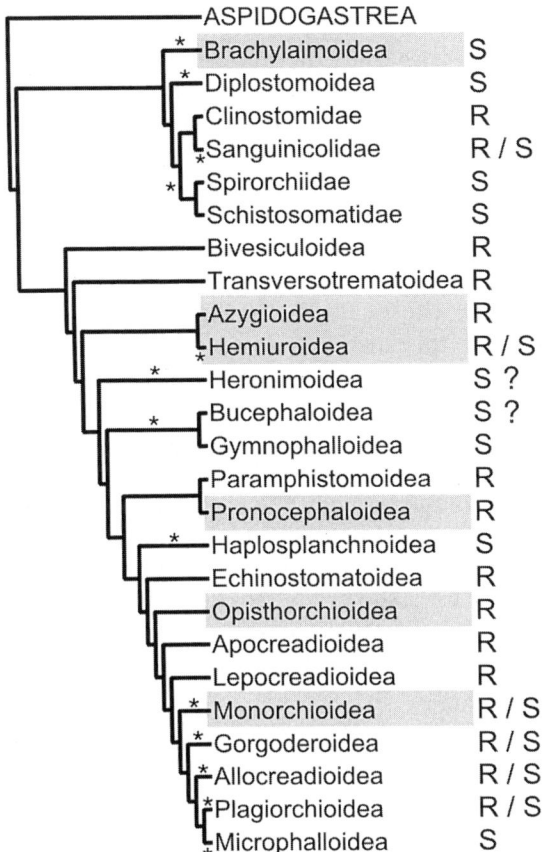

Figure 8 Digenean life cycle variation – behaviour of miracidia; cercaria-producing generation. Shaded taxa are those in which the ingested egg is the strategy adopted by the entire superfamily. R, rediae; S, sporocyst. Taxa marked with an asterisk (*) are inferred to have derived sporocysts. The query (?) next to Heronimoidea and Bucephaloidea reflects uncertainty about which generation is present (mother or daughter sporocyst).

others have suggested that this membrane may represent a highly reduced single second generation. Further indications of such reduction are seen in taxa in which the miracidium contains only a single redia (e.g. all Cyclocoelidae and the paramphistome *Stichorchis*). The possibility of the presence of only a single intra-molluscan generation is not considered further here.

Apart from the issue of the number of generations, perhaps the major observation to be made about the intra-molluscan stages is in the distinction between rediae (possessing a mouth, a pharynx and a short saccular gut) and sporocysts (which have no organs at all). Mapping of this character (Figure 8) shows that both sporocysts and rediae occur widely in the Digenea. Analysis of this distribution indicates that rediae have given rise to sporocysts. This may have happened as many as four times in the Diplostomida and at least nine times in the Plagiorchiida (clades in which sporocysts are inferred to have evolved are marked with an asterisk). The frequency of this evolutionary change suggests that this derivation occurs relatively easily. It is intriguing that the sporocyst should be the derived condition in the light of findings which have shown that in inter-specific interactions within a mollusc, rediae often eat and dominate sporocysts (Lim and Heyneman, 1972).

Our compilation of life cycle data revealed a surprising correlation. Of the taxa that occur in bivalves as first intermediate hosts (Allocreadiidae, Bucephalata, Faustulidae, Gorgoderidae, Monorchiidae and some Sanguinicolidae) only the Allocreadiidae have rediae. Even the single hemiuroid from a bivalve, identified as such by the cystophorous cercaria (Wardle, 1975), has a sporocyst. In the Allocreadiidae, the redia has a pharynx, but the gut is often described as being very small or even absent (Caira, 1989). Thus, we can infer cautiously that parasitism of bivalves has promoted the derivation of sporocysts from rediae. It is not clear why the infection of bivalves and the derivation of sporocysts should be correlated.

The Schistosomatoidea is particularly interesting in the distribution of sporocysts and rediae (Figure 8). Within the clade Clinostomidae + Sanguinicolidae, rediae are found in some sanguinicolids in polychaetes and in all Clinostomidae whereas those sanguinicolids that infect bivalves and gastropods always have sporocysts. The other clade (Schistosomatidae + Spirorchiidae) also always has sporocysts. The Sanguinicolidae is not only the only digenean family to use both of the major categories of first intermediate hosts extensively but it is one of few in which either rediae or sporocysts can be the second generation.

Whereas rediae are dominant in the Hemiuroidea, sporocysts are not uncommon and have been reported from four families of which three are also reported to have rediae. Pearson (1972) reviewed evidence that showed that the ontogeny of at least some of the species in this superfamily includes a redia that loses traces of its gut to become a sporocyst. Thus, at least in part, the Hemiuroidea may illustrate the process that led to the development of sporocysts in general.

4.6. Cercarial Tails

Figure 9 shows the distribution of the two basic kinds of cercarial tails – forked and simple (shaded). The distribution is remarkably free of ambiguity. The fork-tailed cercaria is clearly the plesiomorphic form, a conclusion

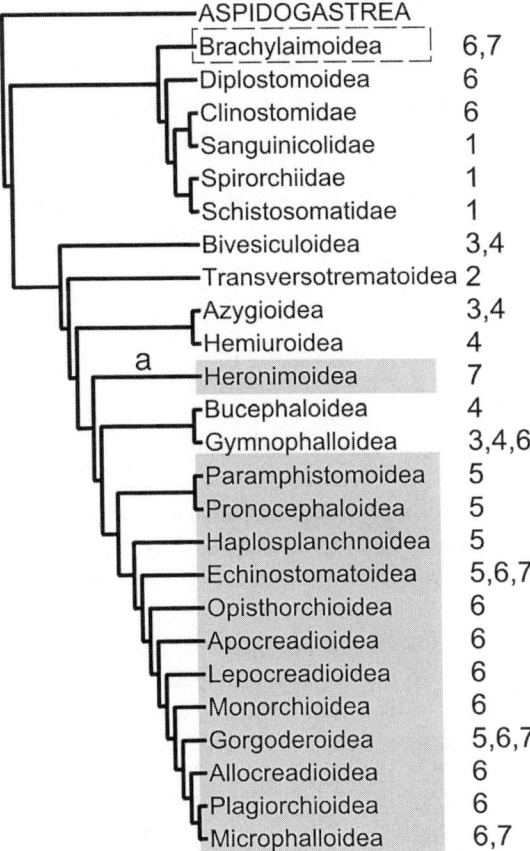

Figure 9 Digenean life cycle variation – cercarial tail; behaviour of cercaria. Shaded taxa have simple tails; the dashed 'box' indicates that only some taxa within the clade have simple tails. The only clade at the superfamily level in which the cercaria does not emerge from the first intermediate host is marked 'a'. 1. Cercaria penetrates definitive host directly. 2. Cercaria attaches directly to definitive host. 3. Cercaria eaten directly by definitive host. 4. Cercaria eaten by second intermediate host. 5. Cercaria emerges from mollusc and encysts in the open. 6. Cercaria emerges from mollusc and penetrates second intermediate host externally. 7. Cercaria remains in first intermediate host which is eaten directly by definitive host.

reached by Cable (1974). Modification to a simple tail has apparently occurred perhaps only three times for major clades. In the Diplostomida, it is modified into a simple, almost absent, tail in the Brachylaimoidea that use terrestrial molluscs; the aquatic Leucochloridiomorphidae retain a typical forked-tail. In the Plagiorchiida, separate derivations have apparently occurred in the Heronimoidea and in the huge clade comprising the Paramphistomata + all other more derived Plagiorchiida. There is also evidence for derivation of a simple tail within some otherwise forked-tailed taxa (Køie, 1979). There is no evidence that a forked-tail has ever been acquired secondarily. The topology of the phylogeny of the Digenea given by Cribb *et al.* (2001) led them to suggest that the forked-tailed cercaria of the Bucephalidae had been developed secondarily but the new topology suggests that this is not the case. Cribb *et al.* (2001) suggested that the derivation of the simple cercarial tail correlated with the abandoning of direct pursuit of the host. This hypothesis requires that highly derived taxa such as the Opisthorchioidea and Xiphidiata, which penetrate intermediate hosts but have a simple tail, were derived via forms that encysted without an intermediate host. This interpretation is supported here.

4.7. Infection Processes – Cercarial Behaviour

The digenean cercaria can behave in many distinct ways that ultimately lead to the infection of the vertebrate definitive host. We recognize seven distinct behaviours that are mapped in Figure 9.

(i) The cercaria may penetrate the definitive host directly. This behaviour occurs only in some Schistosomatoidea (Schistosomatidae, Spirorchiidae and Sanguinicolidae).
(ii) The cercaria attaches to the surface of the definitive host. This behaviour occurs only in the Transversotrematidae.
(iii) The cercaria is eaten directly by the definitive host. This behaviour occurs in several clades – Bivesiculidae, Azygiidae, Fellodistomidae and Tandanicolidae (but not in all individuals of these clades).
(iv) The cercaria is eaten by a second intermediate host, and then frequently penetrates the gut. A metacercaria forms and waits for the definitive host to eat the second intermediate host. This behaviour is probably general in the Hemiuroidea but occurs notably elsewhere in some Fellodistomidae and Gorgoderidae and rarely in the Bivesiculidae and Azygiidae.
(v) The cercaria emerges from the mollusc, encysts in the open as a metacercaria on a potential food source of the definitive host, and

waits there to be eaten. This behaviour is found in several clades (Paramphistomata, Haplosplanchnoidea, some Echinostomatoidea and Haploporidae from within the Gorgoderoidea).

(vi) The cercaria emerges from the mollusc and penetrates a second intermediate host externally. A metacercaria forms and waits for the definitive host to eat the second intermediate host. This is the most common life cycle form in the Digenea, both in terms of numbers of species and numbers of families. It occurs in most of the Diplostomida, Bucephaloidea, Opisthorchioidea, Echinostomatoidea and Lepocreadioidea, and overwhelmingly in the four superfamilies of the Xiphidiata.

(vii) The cercaria remains in the first intermediate host, which is eaten directly by the definitive host. This behaviour is found in the Heronimidae, Cyclocoelidae (Echinostomatoidea), Eucotylidae (Microphalloidea) and Hasstilesiidae and Leucochloridiidae (Brachylaimoidea) and sporadically elsewhere.

In our view, all other cercarial behaviours within the Digenea are either derivations or reductions of these seven basic types. Derivations include extension of the life cycle to four hosts as seen in some Hemiuroidea (Madhavi, 1978; Gibson and Bray, 1986; Goater et al., 1990) and Diplostomidae (Pearson, 1956) and numerous secondary abbreviations of the life cycle (Poulin and Cribb, 2002). The extent to which the seven behaviours may be derivable from each other is discussed here. Determining which of the behaviours might be basal for the Digenea is made complex because there are so many of them. To reiterate, our approach has been to use a combination of outgroup comparison (including the key assumption that two-host cycles preceded three-host cycles) and parsimonious inference from within the phylogeny of the Digenea.

The six behaviours in which the cercaria emerges (behaviours i–vi) numerically overwhelm that in which it remains within the mollusc (behaviour vii). Behaviour (vii) is characteristic only of the monotypic Heronimoidea (node a, Figure 9), but it also appears within the Brachylaimoidea (Hasstilesiidae and Leucochloridiidae), the Echinostomatoidea (Cyclocoelidae) and the Microphalloidea (Eucotylidae) and in some taxa within many other families. The relatively derived and usually nested positions of taxa that exhibit this behaviour allow us to infer that this behaviour has evolved many times. Thus, retention of the cercaria within the mollusc, to be eaten with the mollusc by the definitive host (life cycle pattern either analogous or homologous to that of the Aspidogastrea), is a derived state in the Digenea. Emergence of the cercaria is a synapomorphy for the Digenea relative to the Aspidogastrea and a plesiomorphic trait within the Digenea. Thus, as argued by Pearson (1972), an evolutionary hypothesis for the Digenea must account for the emerging cercaria as a plesiomorphic character for the Digenea, unless extinct forms are invoked.

We can now use this inference, the topology of the tree and the distribution of cercarial emergence to consider which emergent cercarial behaviour is plesiomorphic. We have presented evidence that a two-host life cycle of some sort preceded the three-host life cycle; outgroup comparison and simple logic both suggest that two-host life cycles preceded three-host cycles. We can thus conclude that behaviours (vi) and (iv) are not plesiomorphic. Behaviour (vi), external penetration of a second intermediate host, appears first in relatively highly derived taxa (Bucephalata and higher taxa, see Figure 9), and so cannot be inferred to be plesiomorphic. Behaviour (iv), ingestion of the cercaria by a second intermediate host, appears in some of the relatively basal Azygioidea and Gymnophalloidea, and generally in the Hemiuroidea and in the highly derived Gorgoderidae. This three-host life cycle may be derived from ingestion of the free-living cercaria by the definitive host in the case of the Gymnophalloidea and the Hemiuroidea and from the more common standard penetrating three-host cycle in the case of the Gorgoderidae.

There are four behaviours involving two-host cycles in which the cercaria emerges from the mollusc, which may be plesiomorphic. Behaviour (i), direct penetration of the definitive host, occurs only in derived Diplostomida (Figure 9). This behaviour cannot therefore be considered plesiomorphic for the Digenea (but see Section 4.9, p. 234). Behaviour (ii), attachment of the cercaria, is seen only in the relatively basal Transversotrematoidea. This is the only life cycle within the Plagiorchiida that leads to infection other than, ultimately, by ingestion. It is not parsimonious to infer that this life cycle strategy was basal to all others in this clade. Behaviour (v), encystment in the open on a food source, is found in the relatively derived taxa – Paramphistomata, Echinostomatoidea, Haplosplanchnoidea and the Haploporidae. Encystment in the open followed by ingestion is easily derived from ingestion of a free-swimming cercaria whereas the reverse is not necessarily the case. Again, it is not parsimonious to infer that this life cycle strategy was basal to all others in this clade.

Behaviour (iii), ingestion of the free-swimming cercaria, is found in the most basal taxon of the Plagiorchiida, the Bivesiculoidea, and in at least some representatives of two other relatively basal clades, the Azygioidea and Gymnophalloidea. The behaviour is simple and could have arisen easily by the ingestion of a free-swimming cercaria. The intuitive objection to this hypothesis is that it seems improbable that the cercaria could survive the digestive tract, especially the stomach, of its host. The evidence, however, is that this happens in several digenean taxa. Thus, we conclude that ingestion of a free-swimming cercaria is both parsimoniously and plausibly plesiomorphic for the Digenea. This conclusion may also have the implication that the first host in the digenean life cycle was a mollusc, that the ancestor of digeneans had a free-swimming adult, and that the vertebrates were added secondarily. Problems associated with these possibilities are discussed in Section 5.2 (p. 241).

4.8. A Hypothesis of Evolution within the Plagiorchiida

The individual hypotheses that have been derived above can now be combined into an overall hypothesis for life cycle evolution. This is done separately for the Plagiorchiida and the Diplostomida (see Section 4.9, p. 234) because there remain substantial difficulties in the interpretation of the Diplostomida, but both sets of hypotheses refer to the same set of inferences regarding plesiomorphic life cycle traits for the Digenea. A parsimonious hypothesis for the evolution of major life cycle traits within the Plagiorchiida is shown in Figure 10 using just 16 key taxa from the group. Most of the life cycle diversity within the group can be explained by 20 key evolutionary changes that are described below. The numbers below refer to clades labelled in Figure 10.

Plesiomorphic life cycle characters for the Digenea (1). The analyses above suggest that the common ancestor to extant Digenea had a miracidium that hatched from the egg and penetrated a gastropod. The second asexual generation developed into a redia that produced fork-tailed cercariae, which emerged from the gastropod to be eaten passively by a teleost definitive host. There was no second intermediate host. This combination of characters is found without modification in some Bivesiculidae, the most basal taxon within the Plagiorchiida.

Although apparently plesiomorphic in its life cycle, the Bivesiculidae is not necessarily primitive in all respects. A striking feature of the family is its lack of oral and ventral suckers. Absence of suckers has been commented on previously as perhaps consistent with a basal position for this family and plesiomorphic absence of suckers in the Digenea, but their wide occurrence in both the Diplostomida and Plagiorchiida leads to the parsimonious inference that oral and ventral suckers are indeed plesiomorphic for the Digenea. These parasites occur in the intestines of their host and there is no ready explanation for the absence of suckers if it has occurred by their loss. Thus, absences of suckers in this taxon may be apomorphies. A study of the homology of oral and ventral suckers in the Diplostomida and Plagiorchiida might shed fresh light on this matter. Whereas most evidence about

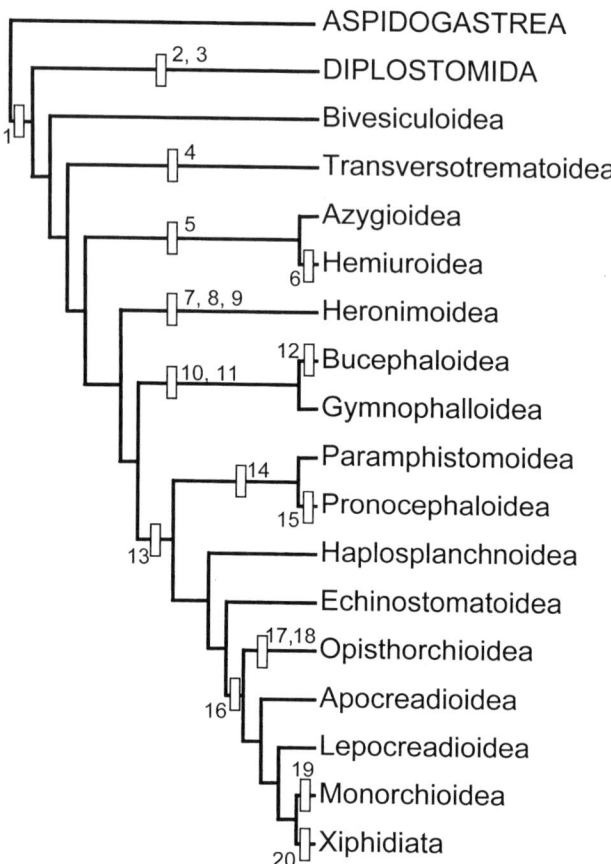

Figure 10 A hypothesis for the evolution of life cycles within the Plagiorchiida. Numbered evolutionary changes are described in the text.

bivesiculids is consistent with a two-host cycle, Cribb *et al.* (1998) showed that at least one species has incorporated (perhaps facultatively) a second intermediate host. A mother sporocyst has never been described in the Bivesiculidae. The basal position of this taxon suggests this may be informative.

Cercaria penetrates; infection of tetrapods (2,3). Penetration of a host by the cercaria is a possible synapomorphy for the Diplostomida; it is clearly independent of cercarial penetration in other taxa in the higher Plagiorchiida. The clade may also be defined by the adoption of parasitism of tetrapods as definitive hosts (see further discussion Section 4.9, p. 234).

No synapomorphy. This hypothesis proposes no life cycle synapomorphy for the Plagiorchiida.

Cercaria attaches externally to scaled fish (4). This character is an autapomorphy for the Transversotrematidae. We argued above (p. 226) that the two-host life cycle of the Transversotrematidae, which incorporates direct attachment of the cercaria, could not be inferred to be basal to that of the remainder of the Plagiorchiida. The topology of the tree means that it is thus most parsimonious to interpret it as a direct modification of the bivesiculid-type life cycle. We have mapped this as a single change. However, at least three plausible hypotheses can account for this life cycle: a two-host (ingestion) life cycle in which the cercaria has exchanged passive ingestion for active attachment (as mapped, Figure 10); an abbreviated three-host life cycle from which the third (definitive) host has been omitted; or a two-host life cycle derived independently of other digenean life cycles.

Counting changes in life cycle traits in the competing hypotheses (not shown) suggests that derivation of the cycle from a two-host ingestion cycle is the most parsimonious, if modification of the life cycle from ingestion to attachment can be construed as a single step. This may be overly optimistic as it implies complex changes. The hypothesis that the life cycle is abbreviated was proposed by Brooks *et al.* (1989), who suggested that the life cycle had been abbreviated from three to two hosts. Although this scenario is plausible, it has neither supporting evidence nor is it parsimonious in the context of the new topology of the Digenea. The hypothesis that the life cycle was derived independently by a free-swimming cercaria becoming associated with the surface of fishes is simple and requires just one step of derivation for the Transversotrematidae itself, but the topology of the remainder of the tree would then require that vertebrate parasitism was adopted at least three times in the Plagiorchiida. This is significant. The discussion in Sections 4.3 and 4.7 assumed that parasitism of vertebrates had arisen only once. This interpretation is suggested initially by both outgroup comparison and parsimony. However, the assumption may be false. We concluded that basal digeneans infected vertebrates by ingestion of a free-swimming cercaria. This interpretation may mean that outgroup comparison (with the Aspidogastrea) is uninformative (or indeed misleading) in this case. If this is the case, then separate adoptions of vertebrate parasitism within the Digenea may need to be considered. In this context, it is noteworthy that the cercaria of the Transversotrematidae is one of the most remarkable in the Digenea. It has unique arms at the base of the tail and it may be sexually mature with sperm in the seminal vesicle. Sexual maturity of the cercaria may be an echo of a free-swimming sexual adult. However, the plasticity of cercariae seen elsewhere shows that we must be highly cautious in such inferences. The origin of the transversotrematid life cycle certainly requires further investigation.

Eggs eaten (5). This life cycle character (and the associated morphological modification of the miracidium) unites the Azygioidea and the Hemiuroidea. The two-host life cycles of azygiids differ from the basal form as represented by the Bivesiculidae only in the behaviour of the miracidium, which must be eaten. The sexual adults of azygiids are quite distinct, however, as they have oral and ventral suckers. Some, perhaps most, azygiids have expanded their life cycles to three hosts but this is interpreted as a within-clade derivation that does not affect the apparently plesiomorphic status of the two-host life cycle.

Cercaria eaten by intercalated arthropod second intermediate host (6). The Hemiuroidea is a huge taxon comprising over 300 genera (including the Didymozoidae). The ingestion of the cercaria by an arthropod intermediate host is here interpreted as the intercalation of an extra host in a life cycle in which the plesiomorphic condition was the cercaria being eaten directly by the definitive host. The Hemiuroidea, some of the Gymnophalloidea and the Gorgoderidae are the only clades in which the cercaria is ingested by the second intermediate host. All other three-host life cycles are characterized by external cercarial penetration.

Cercaria remains in, and is transmitted by the ingestion of the first intermediate host; cercarial tail reduced to simple; hosts tetrapods (7–9). The two-host cycle of the Heronimoidea is here interpreted as the result of the abbreviation of a cycle in which the cercaria previously emerged. The nature of the unabbreviated life cycle cannot be deduced, but the simplest abbreviation would have been of a life cycle in which the free-living cercaria was eaten directly. The fact that this clade is represented by a single species infecting the unusual site of the lungs of a freshwater turtle suggests that this clade is a relict *sensu* Brooks and Bandoni (1988). Certainly the Heronimoidea appears to have no significant bearing on understanding the overall evolution of digenean life cycles as it did when Brooks *et al.* (1985) inferred that it might be the sister-group to the remainder of the Digenea. However, it does represent the first appearance of a simple tail within the Plagiorchiida.

Host-switch to bivalves; derivation of sporocysts (10–11). The shared parasitism of bivalves as first intermediate hosts and the derivation of sporocysts unite the Bucephaloidea and Gymnophalloidea as a clade. Some two-host fellodistomid and tandanicolid life cycles differ from the basal form only in the parasitism of bivalves and the derivation of sporocysts; as noted earlier, these characters may be linked. Some, perhaps most, fellodistomids have expanded their life cycles to three hosts but this is again interpreted as a within-clade derivation that does not affect the inferred plesiomorphic status of the two-host life cycle.

Cercaria penetrates intercalated second intermediate fish host (12). The Bucephalidae are parasitic in the intestines of fishes. The life cycle is uniformly three-host, involving the penetration of a fish and subsequent formation of a metacercaria in its tissues. There is no particular clue from living bucephalids to suggest how the three-host cycle might have arisen. We can only speculate that the capacity of cercariae to survive in association with the surface of fishes, that they once contacted accidentally, improved the overall chances of the parasite finally to be ingested by the definitive host. The origin of the remarkable cercarial tail of the Bucephalidae (furcate, but without a tail-stem) remains unexplained but it is noteworthy that Køie (1979) has shown that the same condition can also exist in the related Fellodistomidae.

Cercarial tail simple (+ encystment in open?) (13). This character unites all remaining Plagiorchiida: Haplosplanchnoidea + Echinostomatoidea + Paramphistomata + Opisthorchioidea + Lepocreadioidea + Xiphidiata.

Cribb *et al.* (2001) suggested that derivation of simple tails may relate to the adoption of external encystment. They argued that cercariae that have forked tails typically swim faster than those that have simple tails. Presumably this is energetically expensive. Thus, slower, more efficient swimming by a cercaria that will encyst in the open rather than pursue a host may have been selected for. We hypothesize that encystment in the open and presence of a simple tail are linked characters. Encystment in the open can be derived from the ingestion of a free-swimming cercaria. It requires only that the cercaria finds a suitable substrate and secretes some kind of protective cyst. This behaviour is characteristic of the Paramphistomata, the Haplosplanchnoidea and the Echinostomatoidea (although three host cycles do arise within this last clade). These taxa combine to represent a considerable number of species, pointing to the substantial advantage conferred by cercarial encystment which prolongs the effective life of the cercaria.

Encystment on vegetation – specialization for herbivory? (14). This character is used here rather speculatively to define the Paramphistomata (Pronocephaloidea + Paramphistomoidea). These trematodes have two-host life cycles in which the cercaria encysts in the open. Paramphistomata have radiated widely among herbivores and their key innovation may have been encystment on vegetation, although pronocephaloids often encyst on mollusc shells and infect carnivorous hosts as well. This life cycle strategy has apparently never expanded to a true three-host cycle, defined here as growth of the metacercaria at the expense of the second intermediate host.

An association between this clade and herbivory may also be suggested by the morphology of the adults. Both the Pronocephaloidea and the Paramphistomoidea lack an oral sucker (or pharynx) and the Pronocephaloidea and some Paramphistomoidea lack a ventral sucker. Both modifications undoubtedly affect feeding behaviour and these losses may reflect a move to ingestion of gut contents instead of browsing on the mucosa. These characteristics require further analysis but, for the present, we propose that they can be usefully combined as a single character – "specialization for herbivory".

Egg eaten (15). Ingestion of the egg appears to distinguish the Pronocephaloidea from the Paramphistomoidea, although the behaviour has not been sufficiently studied in a wide range of taxa.

No synapomorphies: Haplosplanchnoidea and Echinostomatoidea. In contrast to the Paramphistomata, these superfamilies are not here defined by life cycle synapomorphies. The simplest life cycles in these clades are those in the Haplosplanchnidae and, in the Echinostomatoidea, some Philophthalmidae, Psilostomidae, Fasciolidae and Echinostomatidae in which the cercaria encysts in the open, usually in association with the surface of potential prey items (especially molluscs) or vegetation eaten by the definitive host. The life cycle is associated with herbivory only in the Haplosplanchnidae of marine fishes and the derived Fasciolidae of terrestrial mammals. Neither superfamily has modified feeding or attachment structures as are seen in the Paramphistomata. The Echinostomatidae, by far the largest group within the Echinostomatoidea, usually encyst in association with animals. The derivation of three-host cycles within the clade is seen clearly in the Echinostomatidae and Psilostomidae where a range of levels of association with intermediate hosts leads to a true three-host life cycle. In addition, within this clade the Cyclocoelidae is characterized by an apparently abbreviated life cycle in which the cercaria encysts within the first intermediate host.

Cercaria penetrates intercalated second intermediate (invertebrate?) host (16). This character appears to define the clade comprising all remaining plagiorchiidan taxa which, except by apparent secondary modification (e.g. Eucotylidae, Haploporidae, some Monorchiidae), uniformly use second intermediate hosts that are penetrated externally with the assistance of penetration glands. It seems likely that three-host cycles appeared in the same way that is shown by the range of extant Echinostomatoidea – external association leading to progressively more intimate association and penetration.

Second intermediate hosts exclusively vertebrates; eggs eaten (17 and 18). The Opisthorchioidea represents a substantial clade of trematodes that uniformly have a three-host life cycle in which fishes and occasionally amphibians are exploited as second intermediate hosts. The position in the phylogeny occupied by the Opisthorchioidea implies that the superfamily adopted fish as intermediate hosts independently of any other clade of digeneans. The opisthorchioid cercaria is highly distinctive, uniform and apparently specialised for this behaviour. This is the third clade of the Plagiorchiida in which eggs are always eaten by the molluscan intermediate host. The fact that this is an independent development in each clade is shown by the morphological distinctions between the eggs and miracidia in each clade.

No synapomorphies: *Lepocreadioidea and Apocreadioidea*. These superfamilies each lack a clear life cycle synapomorphy. These clades include many large families most of which are essentially intestinal parasites of teleost fishes, have gastropods as first intermediate hosts and have cercariae produced in rediae. All have cercariae that penetrate a wide range of invertebrate and vertebrate hosts with the conspicuous exception of any significant presence in arthropods. The lepocreadioid clade comprising the Gorgocephalidae, Enenteridae and Gyliauchenidae (Olson et al., 2003) remains without an elucidated life cycle. All these families are concentrated in herbivorous fishes and it is possible that their life cycles are secondarily reduced and have metacercariae that are associated with algae.

Egg eaten (19). The Monorchioidea appear to be united by the necessity for the egg to be ingested by the first intermediate host. Otherwise life cycles in this superfamily broadly resemble those of the Apocreadioidea and Lepocreadioidea.

Stylet in oral sucker; arthropods adopted as main second intermediate hosts (20). The Xiphidiata are united by the presence of a stylet in the oral sucker. This structure appears to have made possible the penetration of arthropod cuticle or arthrodial membranes. Arthropods are more heavily exploited by the Xiphidiata than by any other clade of digeneans. By contrast, the Lepocreadioidea, Monorchioidea and Apocreadioidea, the three taxa immediately basal to the Xiphidiata, are either rare in or absent from arthropods. This clade includes the Haploporidae, parasites of fishes with a two-host life cycle in which the cercaria, lacking a stylet, encysts in association with algae. Membership of this clade is either an indication that this two-host cycle has evolved by the loss of a second intermediate host (and the stylet) or that it has been misplaced in the present phylogeny.

Overall the Plagiorchiida present a comprehensible system with origins in fishes and simple two-host life cycles followed by developing life cycle complexity, the adoption of second intermediate hosts, and repeated expansion with and into tetrapods.

4.9. Hypotheses of Evolution within the Diplostomida

It was a major finding of Olson *et al.* (2003) that there is a basal dichotomy in the Digenea between the Plagiorchiida and the Diplostomida. The depth and significance of this dichotomy is reflected in differences in the life cycles seen in the two clades. Inference of the pattern of evolution within the Diplostomida is considerably more difficult than for the Plagiorchiida because the basal taxon, the Brachylaimoidea, have three-host life cycles and parasitise tetrapods. As noted above, outgroup comparison and simple logic dictate that a two-host life cycle precedes a three-host cycle. In addition, parsimony analysis shows that the basal hosts for the Digenea were teleost fishes. Thus, in several important respects the most basal Diplostomida are evidently significantly derived in comparison to basal plagiorchiidans. Two equally parsimonious hypotheses for the evolution of life cycle characters in the Diplostomida are shown in Figure 11; numbers below refer to clades labelled in this figure. The hypothesis in Figure 11a proposes the following sequence*:

Plesiomorphic life cycle characters for the Digenea (1–3). As for the Plagiorchiida, this hypothesis commences with the inferences that plesiomorphically the miracidium hatches from eggs outside the host and the fork-tailed cercaria emerges from gastropods to be eaten passively by a teleost definitive host.

Definitive hosts tetrapods (4). In contrast to the Plagiorchiida, all Diplostomida except the blood flukes of fishes (Sanguinicolidae) are parasites of tetrapods.

Molluscan second intermediate host intercalated by cercarial entry into natural pores (5). All Brachylaimoidea have molluscs as second intermediate hosts except where, by abbreviation, the metacercariae form in the first intermediate host (Hasstilesiidae and Leucochloridiidae).

Egg eaten (6). A life cycle apomorphy for the Brachylaimoidea.

Vertebrate second intermediate host intercalated by cercarial penetration (7). Whereas in the Brachylaimoidea the cercaria typically enters molluscs via natural pores, in the remainder of the Diplostomida vertebrates (or rarely other taxa) are penetrated actively and penetration glands are present.

Vertebrate definitive host lost (8). This step of life cycle abbreviation gives rise to the life cycle seen in all three blood fluke families.

---

*The steps shown here exclude the evolution of the form of the second asexual generation, which is inferred to incorporate several derivations of sporocysts.

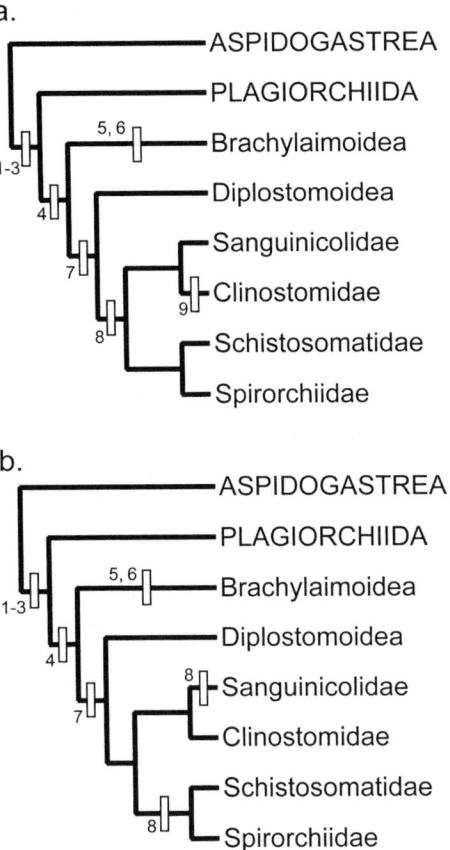

Figure 11 Two hypotheses for the evolution of life cycles within the Diplostomida assuming plesiomorphic ingestion of the cercaria by the definitive host. Numbered evolutionary changes are described in the text.

Vertebrate definitive host added (secondarily) (9). This step hypothesizes the re-extension to a three host life cycle in the Clinostomidae by the secondary addition of a vertebrate definitive host.

A second equally parsimonious hypothesis for the evolution of life cycles within the Diplostomida (Figure 11b) differs from the first by proposing the independent abbreviation of the life cycle by the Sanguinicolidae and the Schistosomatidae + Spirorchiidae (i.e. number 8 appears twice) and no secondary adoption of vertebrate parasitism by the Clinostomidae. The nature of the association between clinostomids and their hosts may allow discrimination between these two hypotheses. Whereas the other (three-host) Diplostomida are parasites of the intestine, clinostomids are found in the upper digestive tract, typically the oesophagus. This unusual site of infection could be interpreted as being associated with an independent adoption of vertebrates.

There is one substantial difficulty with these hypotheses – the vertebrate host distribution of the Diplostomida. These hypotheses construe the original vertebrate hosts of the Diplostomida to be those in which they presently develop to sexual maturity in three-host life cycles. These hosts are entirely tetrapods, animals associated with the terrestrial habitat in their evolutionary origin, if not necessarily in their present-day ecology. The hosts of the blood flukes, which include fishes, are excluded as original vertebrate hosts because these hypotheses require that they are definitive hosts only by abbreviation of a three-host cycle. It is difficult to infer what such original definitive hosts might have been. Thus, this hypothesis suggests that no extant member of the Diplostomida plesiomorphically develops to adulthood in a fish. Given that we have inferred that the original hosts of digeneans were fishes, this absence must be explained by the extinction of all clades associated with fishes so that only those that coevolved or host-switched into tetrapods remain. Such scenarios are possible, but they seem at best surprising when parasitism of the guts of fishes has proven so successful in the Plagiorchiida. We were able to infer that plagiorchiidans adopted parasitism of tetrapods repeatedly because they have left behind so many traces of these adoptions; in ten of the 11 superfamilies of the Plagiorchiida that occur in tetrapods, there are also representatives in teleost fishes. There are no such traces for the Diplostomida. The Diplostomida might have arisen only as tetrapods appeared but this hypothesis does not seem plausible because the rise of tetrapods is associated with the interface between freshwater and terrestrial environments. This context is inconsistent with the present-day distribution of sanguinicolids which includes marine teleosts, elasmobranchs and holocephalans (Smith, 1997a, b).

An alternative set of hypotheses is suggested by reconsideration of the definitive hosts of the Diplostomida. One taxon, the Sanguinicolidae does occur in fishes and indeed is found in teleosts, elasmobranchs and holocephalans. What are the implications of making the radical assumption that direct penetration as seen in this family was the plesiomorphic cercarial behaviour for the Diplostomida? We here test the hypothesis that the diplostomidan cercaria originally penetrated its definitive host. The hypothesis in Figure 12 proposes the following sequence (the numbers below refer to clades labeled in Figure 12):

Plesiomorphic life cycle characters for the Digenea (1–2). Miracidium hatches from egg outside host (1); cercaria emerges from gastropod (2).

LIFE CYCLE EVOLUTION IN THE DIGENEA

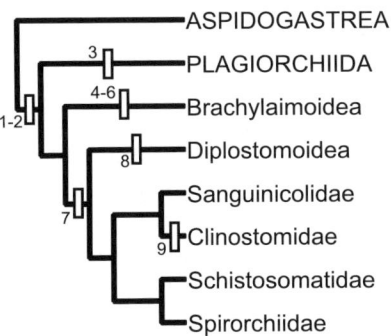

Figure 12 A hypothesis for the evolution of life cycles within the Diplostomida assuming plesiomorphic penetration of the definitive host by the cercaria. Numbered evolutionary changes are described in the text.

(A key distinction relative to the hypotheses in Figure 11 is that ingestion of the cercaria is no longer considered plesiomorphic for the Digenea as a whole but instead a synapomorphy for the Plagiorchiida.)

Cercaria eaten by teleost definitive host (3). This apomorphy serves to define the Plagiorchiida.

Cercaria forms external association with, and enters mollusc; egg eaten by mollusc; tetrapod host added by ingestion of mollusc (4–6). These three apomorphies serves to define the Brachylaimoidea and suggest that they adopted parasitism of vertebrates independently of any other digenean taxon.

Cercaria forms external association with and penetrates vertebrate host (7). This step serves to define the clade of the Diplostomoidea + Schistosomatoidea. Whereas in the Brachylaimoidea the cercaria typically enters the second intermediate host via natural pores, in the remainder of the Diplostomida (Diplostomoidea and Schistosomatoidea) the host is actively penetrated and penetration glands are present. This step is sufficient to establish the nature of blood fluke life cycles.

Vertebrate host added (8–9). The topology here requires the independent extension to three-host life cycles in the Diplostomoidea and the Clinostomidae. The comments made above about the distinctiveness of the association between clinostomids and their vertebrate hosts apply here also.

This hypothesis proposes that the blood fluke life cycle evolved by the direct association of an initially free-living cercaria with the definitive host. The three-host life cycle then evolved by the addition of a new definitive host. This host was adopted by the process of a vertebrate ingesting the initial definitive host (='host-succession extension' of Sprent (1983) and 'terminal addition' of O'Grady (1985)); throughout the Diplostomida where the life cycle has three hosts the definitive host ingests an intermediate host. The hypothesis notably requires three separate adoptions of vertebrates by the Digenea. Although other equally parsimonious hypotheses are possible (e.g. a single adoption of vertebrate parasitism by the Diplostomoidea + Schistosomatoidea and two abbreviations to blood fluke life cycles) this suffices to illustrate the possibilities of dramatically different interpretation.

This hypothesis requires nine steps of modification, the same as those shown in Figure 11. The main attraction of this hypothesis is that it provides a superior explanation of the apparent restriction of diplostomidans to tetrapods by suggesting that the former are not restricted to the latter at all. The hosts penetrated by diplostomidan cercariae are aquatic. The Leucochloridiomorphidae (Brachylaimoidea) have aquatic molluscs as intermediate hosts and the Diplostomoidea and Schistosomatoidea penetrate aquatic vertebrates; a few Diplostomoidea penetrate Annelida and Mollusca. The hosts of Schistosomatidae may not be strictly aquatic, but certainly they are infected while they are in water. The only exception to the penetration of aquatic hosts by cercariae is that part of the clade of the Brachylaimoidea that has become entirely terrestrial. So, if the host penetrated by the cercaria is interpreted as the original definitive host (rather than the actual present-day definitive host), then the range of hosts is plesiomorphically fully aquatic. Such a distribution is exactly what would be predicted for one of two basal clades of the Digenea, a group that is inferred to be a fundamentally aquatic group of parasites and in strong contrast to the tetrapod-only distribution implied by the hypotheses shown in Figure 11.

Although this "cercaria penetrates" hypothesis accounts for some aspects of the host distribution of the Diplostomida, it has three substantial difficulties in terms of intuitive plausibility. First, the hosts penetrated by the Brachylaimoidea are molluscs, whereas the rest of the clade infects mainly vertebrates. This hypothesis suggests that molluscs were once the definitive hosts of this clade. Is such an idea plausible? Jamieson (1966) discussed hypotheses proposing exactly this evolutionary scenario, but certainly no such life cycles are now extant (except perhaps by secondary reduction).

The second difficulty associated with the "cercaria penetrates" hypothesis is that it requires that a free-living adult (the proto-cercaria of Pearson, 1972) evolved into one that, for the Diplostomoidea + Schistosomatoidea, penetrated a vertebrate host and developed to a sexual adult capable of releasing its eggs while leaving no evolutionary trace of the process. This involves significant conceptual complexity. The evolution of penetration in the Brachylaimoidea is relatively easily envisaged. Their cercariae typically enter the excretory pore of another mollusc and then develop as unencysted metacercariae in the excretory vesicle or other cavities to which they migrate. Such behaviour might have arisen gradually following the formation of an external association with the mollusc. Further, this form of parasitism does not require penetration glands or the evolution of a mechanism for the escape of the eggs of the parasite. Because the Brachylaimoidea is basal to the remainder of the Diplostomida, this simple behaviour might have given rise to the more sophisticated penetration behaviours of the remainder of the clade. However, there remains a major gulf between the unsophisticated behaviour of the brachylaimoids and the sophisticated penetration of the remainder of the Diplostomida. In addition, the use of nearly completely exclusive second intermediate host groups (molluscs for Brachylaimoidea and vertebrates for Diplostomoidea + Schistosomatoidea) implies either a dramatic host-switch or independent derivation (as shown in Figure 12).

The third difficulty returns to the host distribution for the Diplostomida. If two-host blood fluke life cycles gave rise to three-host cycles, then why are none of these additions found in fishes?

The hypothesis in Figure 12 is radical in proposing that digeneans adopted vertebrate parasitism three times. Other hypotheses are possible that require fewer vertebrate host adoptions by proposing host switches (especially of the second intermediate host). It should be noted, however, that the interpretation of the evolution of the Plagiorchiida also raised the possibility that the Transversotrematidae had an adoption of vertebrate hosts separate from the remainder of that clade. The significance of this is that one of our underlying inferences from outgroup comparison with the Aspidogastrea may have been false. That is, although aspidogastreans and digeneans both share molluscs and vertebrates as alternating hosts, the two life cycles may be only partly homologous. This potential difficulty was predicted, in part, when we drew attention to the number of differences between the nature of the aspidogastrean and digenean life cycles. Finally in this context we must note that the hypotheses in Figures 11 and 12 are equally parsimonious only if the analysis is restricted to the Digenea; if the Aspidogastrea are incorporated then strict parsimony favours the hypotheses in Figure 11. We return to a general consideration of such issues in Section 5.2 (p. 241).

The evolution of life cycles within the Diplostomida thus remains poorly understood. The most basal taxa have complex, highly derived life cycles and several equally parsimonious hypotheses are presently tenable to explain them. We cannot yet discriminate satisfactorily between them.

5. PROBLEMS

5.1. Shortcomings of the Parsimony Approach

Whereas we are confident that the parsimony approach we have taken above has conceptual rigour and in some cases provides excellent insight, we are certainly aware that it is not infallible. There are two conceptual problems. The first relates to the underlying phylogeny. Even if we assume that the phylogeny is broadly correct, we can be certain that many highly informative taxa are extinct. The problem is illustrated in Figure 13(a) which shows the "true" relationships between a hypothetical outgroup (X) and four ingroup taxa (A–D). The mapping of four character states (1–4) on the five taxa allows an unambiguous interpretation of the pattern of evolution of the four character states. Figure 13(b) shows the "true" relationships of five descendants of taxon D and proposes that taxa A–C are extinct. If we

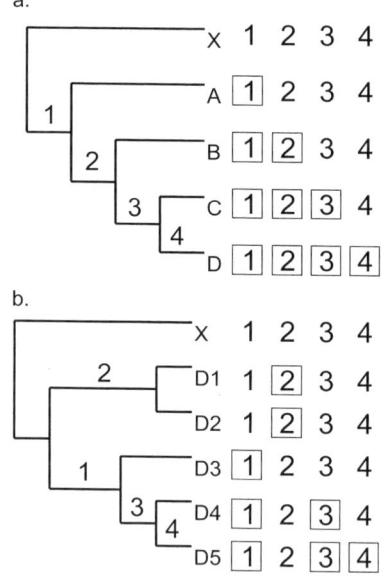

Figure 13 Problems of false inference in parsimony analysis. a. 'True' relationships of a hypothetical outgroup (X) and an ingroup (A–D) allowing inference of the pattern of derivation of four characters (1–4; derived states boxed). b. 'True' relationships of outgroup (X) and ingroup (D1–D5) following extinction of taxa A–C. Parsimonious analysis falsely suggests that all four derived character states arose within D.

refer to the "true" pattern of evolution of the characters, then we see that all four have been subject to some secondary loss in the clade of taxon D's descendants. However, without knowledge of taxa A–C this is unknowable. Relying on parsimony to interpret the evolution of these characters leads to mistakes in every case; we would conclude that all four derived character states arose within the clade of D. This type of problem is potentially significant in analysis of the Digenea. All digeneans have a highly complex life cycle that has many complex features not found in the sister taxon (Aspidogastrea; see below). It is reasonable to assume that forms that were in some way once intermediate between the Digenea and Aspidogastrea are now extinct. If such forms were incorporated in the phylogeny the possibility of false inferences would be reduced. In a sense, however, this observation simply restates the underlying problem: if all ancestors and extinct taxa were available for study, then the path of evolution would be clear. This exercise emphasizes the facts that objective parsimony analysis may easily support false hypotheses and that the evolutionary history of particular life cycle traits may be unknowable.

A second problem flows from the first. If there is no guarantee that a parsimoniously derived hypothesis is necessarily true, is there a way in which we can discover the truth? If parsimony analysis leaves us with a hypothesis that is actually false, we suspect that the error can be detected only if it is first deemed to be intuitively unsatisfactory and is thus subjected to further scrutiny. In our view, the inference of a parsimonious but unsatisfying hypothesis should quite reasonably be taken as the spur for further consideration of the underlying data, ideally with the goal of either finding further objective mappable evidence or exploring the homology of the existing characters. This is logical, but methodologically dangerous. The decision that a parsimoniously derived hypothesis is "unsatisfying" carries the implication that the analyst has preconceived ideas of the "true" answer. Such preconceptions (insights or bias?) can lead to the "adjustment" or "development" of the data and repeated analysis until the "correct" result is found. Such an approach has the capacity to lead to the support of a false hypothesis just as easily as can the uncritical acceptance of parsimonious hypotheses. In essence, it is difficult for the analyst to remain both interested and disinterested and passionate and dispassionate.

5.2. Conflict between Hypotheses

Analysis of the evolution of life cycles in the Digenea creates exactly the kind of problems that we have described above. In our analysis we largely avoided addressing a substantial problem that bears on much of the analysis above – how did the basic digenean life cycle evolve? The problem that we have avoided is easily described. The topology of the Platyhelminthes as a whole, and the position of the Neodermata (Figure 2) allow the parsimonious interpretation that neodermatans as a whole adopted parasitism of vertebrates. This hypothesis suggests that, subsequently, the Trematoda adopted molluscs as second intermediate hosts. It is parsimonious to propose then that the initial digenean two-host cycle resembled that of present-day aspidogastreans as described earlier.

Two difficulties are posed by this hypothesis. First, it implies that vertebrates were adopted just once by the ancestors of the Neodermata; parsimony suggests that all present-day associations can be explained by coevolution or host-switching, not the separate adoption of vertebrates *de novo*. However, the hypotheses developed above may imply separate adoption of vertebrates, at least by the Diplostomida and the Plagiorchiida. More extravagant hypotheses, which are not without support, might suggest even more adoptions. First in the list of candidates for independent adoption of vertebrates might well be the Transversotrematidae.

The second problem relates to the evolution of the cercaria. Parsimonious interpretation of the topology of the Neodermata suggests that the digenean life cycle arose from an aspidogastrean-style cycle. The difficulty with this is that, in the Aspidogastrea, infection of the definitive host occurs by ingestion of the mollusc. By contrast, our analysis of cercarial behaviour allows the unambiguous conclusion that for extant Digenea the cercaria plesiomorphically emerged from the mollusc. An overall analysis of the Neodermata suggests that an aspidogastrean-style life cycle gave rise to a digenean cycle in which the cercaria emerged from the mollusc. However, there is no compelling narrative to describe how this might have occurred. Hypotheses for the Digenea such as that of Pearson (1972) suggested that cercarial emergence occurred in the context of a mollusc-only life cycle. This idea makes it relatively easy to postulate hypotheses in which some of the distinctive vertebrate infection strategies of the Diplostomida and the Plagiorchiida were derived independently.

Thus, objective parsimony analysis of the major taxa of Neodermata suggests one set of hypotheses whereas analysis of the Trematoda in isolation suggests quite different hypotheses. In principle, the resolution of the problem should occur by way of an expanded analysis that incorporates more taxa and more objectively mappable characters, if extant taxa have sufficiently retained the evolutionary markers. In practice, however, the task is formidable; present analyses rely on more than a century of accumulated study of digenean life cycles. The truth here may well be unknowable.

5.3. Ten Questions

We believe that substantial progress is being made in the understanding of the evolution of the digenean life cycle. However, many questions remain to be answered. Below we pose ten questions that we consider both important and unanswered, but potentially answerable when incorporated in the approach espoused herein.

(i) What, if anything, is homologous between the aspidogastrean and digenean life cycles? An answer to this question should address two related issues: Is the infection of molluscs in the two groups truly homologous? Did the Trematoda as a whole inherit parasitism of vertebrates from a single neodermatan adoption of vertebrate parasitism?
(ii) Has any major clade of digeneans coevolved (in the sense of strict cophyly with little or no host-switching) with molluscan first intermediate hosts?
(iii) How did life cycles within the Diplostomida arise, and are their origins linked to the appearance of tetrapods?
(iv) Why are the Sanguinicolidae uniquely found in gastropods, bivalves and polychaetes as first intermediate hosts and holocephalans, chondrichthyans and teleosts as definitive hosts?
(v) Why are digeneans conspicuously rare in chondrichthyans?
(vi) Why do the Bivesiculidae lack oral and ventral suckers?
(vii) How did the life cycle of the Transversotrematidae arise?
(viii) Why have the Paramphistomata lost their oral suckers?
(ix) What is the significance of the correlation between the infection of bivalves and the presence of sporocysts?
(x) What was the chronology of the evolution of the digenean life cycle?

If these questions are answered with evidence that is both objective and compelling, the evolution of the digenean life cycle may be considered well-understood. Answers to some of the questions have been inferred herein, but should certainly be subject to further scrutiny. Other questions require further descriptive work. We are by no means beyond the age of information gathering; for a number of intriguing digenean families there is no life cycle information at all and for many it is inadequate. For some of the questions we may already have all the information that can be gathered. Some questions may prove unanswerable based on evidence stemming from extant taxa.

Analysis of the evolution of life cycles should occur at many different levels within a phylogenetic framework. For example, we note inconsistencies when the phylogeny of the Neodermata and the Digenea are considered separately. Such inconsistencies are informative in that they demand further work and illumination. Detailed analyses of smaller clades (e.g. superfamilies) will minimize the need to make assumptions about more inclusive clades (e.g. orders) and will provide a better understanding of both plasticity and conservation in the evolution of digenean life cycles. Moreover, new phylogenies with even greater representation of digenean diversity will extend our ability to infer the sequence, acquisition and loss of life cycle traits. We hope that our efforts will go some way in guiding workers who wish to elucidate the evolution of parasite life cycles within other taxa.

APPENDIX

Summary of key life cycle characteristics of major taxa of the Digenea discussed in this review. Taxa are listed in the order shown in the figures

Order Diplostomida

Brachylaimoidea. This superfamily comprises mainly families that have entirely terrestrial life cycles (Brachylaimidae, Hasstilesiidae and Leucochloridiidae) and the Leucochloriodiomorphidae which has an aquatic cycle. In the Leucochloriodiomorphidae fork-tailed cercariae emerge from freshwater snails and enter other molluscs in which they form metacercariae; the molluscs are eaten to complete the life cycle (Allison, 1945). In the Brachylaimidae almost tail-less cercariae leave the terrestrial gastropod and enter another gastropod where they form metacercariae (Mas-Coma and Montoliu, 1986). In the Hasstilesiidae and Leucochloridiidae (Bakke, 1980) the life cycle is two-host, the cercariae forming metacercariae in the snail in which they are produced. The egg must be ingested by the gastropod.

Diplostomoidea. This superfamily comprises six families (Diplostomidae, Strigeidae, Brauninidae, Bolbocephalodidae, Cyathocotylidae, Proterodiplostomidae) (Niewiadomska, 2002) all of which occur as adults in tetrapods only. The life cycle is highly uniform. Fork-tailed cercariae are produced in sporocysts in gastropods. These penetrate and form metacercariae especially in fishes and amphibians but also in molluscs and annelids (Pearson, 1961). In some Diplostomidae the life cycle is expanded to incorporate four hosts by the intercalation of a mesocercaria (a form between the cercaria and the metacercaria) (Pearson, 1956). Eggs typically hatch and penetrate the first intermediate host.

Schistosomatoidea. This superfamily comprises all the blood flukes (Sanguinicolidae, Schistosomatidae and Spirorchiidae) and the Clinostomidae. Kanev et al. (2002) include the Clinostomidae and Liolopidae in a separate superfamily, the Clinostomoidea. The relationship of clinostomids and liolopids has not been confirmed by molecular evidence and almost nothing is known of liolopid life cycles. In the Schistosomatoidea as conceived here fork-tailed cercariae are produced in rediae or sporocysts in gastropods, bivalves or polychaetes and penetrate their definitive hosts directly in the case of the blood flukes (Wall, 1951; Køie, 1982; Blair and Islam, 1983) or enter fishes or amphibians in which metacercariae form in the case of the Clinostomidae (Dönges, 1974). Eggs typically hatch and penetrate the first intermediate host.

Order PLAGIORCHIIDA

Bivesiculoidea. This superfamily includes only the Bivesiculidae, and is the most basal clade in the Plagiorchiida. It comprises about 30 species of digeneans that occur as sexual adults in the small intestine of fishes. Forktailed cercariae are produced in rediae in gastropods. After emergence from the snail the body is withdrawn into the base of the tail; the cercaria then awaits consumption by the definitive host (Cable and Nahhas, 1962; Pearson, 1968; Cribb et al., 1998). In one cycle it has been shown that a fish intermediate host may be intercalated (Cribb et al., 1998). Eggs typically hatch and penetrate the first intermediate host.

Transversotrematoidea. This superfamily includes only the Transversotrematidae, a tiny family of about 10 species found under the scales of marine and freshwater fish in the Indo-Pacific region. Their unique and highly modified fork-tailed cercariae are produced in rediae in gastropods and attach directly to the definitive host (Cribb, 1988). Eggs typically hatch and penetrate the first intermediate host.

Azygioidea. This superfamily includes only the Azygiidae which are parasitic in marine and freshwater fishes. Their life cycle is essentially identical to that of the Bivesiculidae in that fork-tailed cercariae are produced in rediae in gastropods, the cercarial body is withdrawn into the tail after emergence, and the definitive host is infected by ingesting the cercaria directly (Sillman, 1962). Some cycles are known in which cercariae emerge with eggs already formed in the uterus (Dickerman, 1946). In many cycles (especially those leading to infection of elasmobranchs) a second intermediate host is apparently intercalated in the life cycle (Brinkmann, 1988). The egg must be ingested by the gastropod.

Hemiuroidea. This superfamily comprises 12 families (Gibson, 2002) and the Didymozoidae and it is one of the major clades of digeneans parasitic in marine and freshwater fishes (a few have entered tetrapods). Their life cycles feature a specialized fork-tailed cercaria known as a cystophorous cercaria. These are produced in sporocysts or rediae in gastropods and very rarely in scaphopods or bivalves. When cystophorous cercariae are eaten by the second intermediate host (which may well always be an arthropod), a specialized structure, the delivery tube, everts, penetrates the host's gut and injects the cercarial body into the haemocoel of the arthropod where an unencysted metacercaria forms (Køie, 1990). These are then transmitted to the definitive host by ingestion. A number of hemiuroid life cycles have been extended to four hosts by the intercalation of an extra host between the second intermediate and definitive hosts (Madhavi, 1978; Goater et al., 1990). The life cycle may also be abbreviated; Jamieson (1966) reported a cycle in which all stages were found within the gastropod. The egg must be ingested by the gastropod.

Heronimoidea. This superfamily includes only the Heronimidae which is itself represented by a single species, *Heronimus chelydrae*, which lives in the lungs of freshwater turtles in North America. Simple-tailed cercariae are produced in a massive branched sporocyst in gastropods. The cercariae do not emerge from the snail host, but are eaten with the snail by the turtle (Crandall, 1960). The egg hatches and the miracidium penetrates the first intermediate host.

Bucephaloidea. This superfamily contains only one family, the Bucephalidae, which is a major clade parasitic almost exclusively in the intestines of teleost fishes. Distinctive fork-tailed cercariae emerge from branching sporocysts in bivalves and penetrate fishes in which they form metacercariae which are transmitted by ingestion to other fishes (Matthews, R.A., 1974). Miracidia typically hatch and penetrate the first intermediate host.

Gymnophalloidea. This superfamily comprises the Gymnophallidae of birds and the Botulisaccidae, Callodistomidae, Fellodistomidae and Tandanicolidae and fishes. Nothing is known of the life cycles of the Botulisaccidae and Callodistomidae. All known gymnophalloids use bivalves as first intermediate hosts and have fork-tailed cercariae that emerge from sporocysts. The simplest cycles are seen in the Tandanicolidae and some Fellodistomidae in which the cercariae are eaten directly by the definitive host (Angel, 1971; Køie, 1980). In some Fellodistomidae and all known Gymnophallidae the cercariae are eaten by or penetrate a range of second intermediate hosts in which a metacercaria forms (Campbell, 1985). Miracidia typically hatch and penetrate the first intermediate host.

Paramphistomoidea. This superfamily is a major radiation of up to 12 families found principally in teleosts, amphibians, reptiles and mammals. Cercariae are simple-tailed and produced in rediae in gastropods. Cercariae emerge from the gastropod and encyst in the open. Encystment is typically on vegetation or on hard surfaces of food (e.g. gastropods) that is eaten by definitive host (Durie, 1956). The life cycle never involves a true second intermediate host. Miracidia typically hatch and penetrate the first intermediate host.

Pronocephaloidea. This superfamily is a major radiation of six families found in all classes of vertebrates except chondrichthyans. Cercariae are simple-tailed and produced in redia in gastropods. They emerge to encyst in the open, typically on vegetation or on hard surfaces of food (e.g. gastropods) that is eaten by definitive host (Martin, 1956). The life cycle never involves a true second intermediate host. The life cycle differs materially from that of the Paramphistomoidea only in the miracidium, which must be eaten by the gastropod host whereupon a mechanical structure injects it into the body cavity of the host.

Haplosplanchnoidea. This superfamily contains only its type-family, trematodes parasitic in the intestines of marine herbivorous fishes. Few life cycles have been elucidated for the family but it appears that simple-tailed cercariae are produced in sporocysts in gastropods and emerge to encyst in association with algae to be eaten subsequently by the definitive host (Cable, 1954). Miracidia typically hatch and penetrate the first intermediate host externally.

Echinostomatoidea. This superfamily is a major clade comprising mainly families parasitic in tetrapods. Cercariae are simple-tailed and produced in rediae in gastropods. The cercariae may encyst in the first intermediate host (Cyclocoelidae) (Johnston and Simpson, 1940), encyst in the open (Fasciolidae, some Philophthalmidae, Psilostomidae and Echinostomatidae) (Cable, 1954; Howell and Bearup, 1967) or in association with or inside second intermediate hosts (usually molluscs or vertebrates – some Psilostomidae and Echinostomatidae) (Johnston and Angel, 1942). Miracidia typically hatch and penetrate the first intermediate host.

Opisthorchioidea. This superfamily comprises three main families, the Cryptogonimidae, Opisthorchiidae and Heterophyidae. They are parasites of teleosts, reptiles, birds and mammals. Simple-tailed cercariae are produced in rediae in gastropods and emerge to penetrate fishes (rarely amphibians) in which a metacercaria develops (Bearup, 1961; Cribb, 1986). The egg must be ingested by the gastropod.

Apocreadioidea. This superfamily contains only the Apocreadiidae which are parasites of the intestines of marine and freshwater teleost fishes. Few life cycles are known. Typically, simple-tailed cercariae are produced in rediae with in gastropods. These cercariae emerge to penetrate and encyst in molluscs and annelids which are eaten by the definitive host (Stunkard, 1964). Miracidia typically hatch and penetrate the first intermediate host externally.

Lepocreadioidea. This superfamily comprises a major assemblage of perhaps nine families of trematodes overwhelmingly parasitic in teleosts. Cercariae are always simple-tailed and almost always infect a second intermediate host. First intermediate hosts are gastropods in which rediae occur. Cercariae penetrate a wide range of phyla of invertebrates and rarely fishes in which a metacercaria develops (Køie, 1975; Watson, 1984; Køie, 1985). Life cycles are completely unknown for Gorgocephalidae, Enenteridae and Gyliauchenidae. Miracidia typically hatch and penetrate the first intermediate host.

Monorchioidea. This superfamily includes parasites of marine and freshwater fishes in the families Monorchiidae and Lissorchiidae. Monorchiidae produce cercariae in sporocysts in bivalves (Bartoli et al., 2000) whereas Lissorchiidae produce cercariae in both rediae and sporocysts in gastropods (Stunkard, 1959). Second intermediate hosts are a wide range of invertebrates. It appears that typically eggs must be eaten by the first intermediate host.

Xiphidiata. The Xiphidiata is a huge clade parasitic in all classes of vertebrates except chondrichthyans. The classification proposed by Olson *et al.* (2003) recognizes four superfamilies: Allocreadioidea, Gorgoderoidea, Microphalloidea and Plagiorchioidea. Overall, life cycles within this clade are characterized by simple-tailed cercariae that have a stylet, emerge from gastropods, and penetrate arthropod second intermediate hosts; this life cycle is found in all four superfamilies (Yamaguti, 1943; Johnston and Angel, 1951; Stunkard, 1968; Prévôt et al., 1976). There are many important variations. First intermediate hosts are bivalves for the Allocreadiidae, Gorgoderidae and Faustulidae (Thomas, 1958; Chun and Kim, 1982). Both rediae and sporocysts are common in the Gorgoderoidea whereas sporocysts are overwhelmingly dominant in the other three superfamilies. Some families, notably the Haploporidae which also has no second intermediate host, lack stylets (Shameem and Madhavi, 1991). Metacercaria are formed most frequently in arthropods but occur in almost any animal. The extent to which these variations are genuine within-clade variation or may reflect weaknesses in the current classification and phylogeny is not clear. Cercariae may encyst within first intermediate hosts (all Eucotylidae, some Microphallidae,

Opecoelidae and others). Miracidia may hatch or await ingestion of the egg.

ACKNOWLEDGEMENTS

T.H.C. is supported by grants from the Australian Research Council and the Australian Biological Resources Study. P.D.O. and D.T.J.L. are supported by a Wellcome Trust Senior Research Fellowship (043965) to D.T.J.L. We thank Trudy Wright for gathering and assembling much of the life cycle data analysed here.

REFERENCES

Allison, L.N. (1945). *Leucochloridiomorpha constantiae* (Mueller) (Brachylaimidae), its life cycle and taxonomic relationships among digenetic trematodes. *Transactions of the American Microscopical Society* **62**, 127–168.
Angel, L.M. (1967). The life-cycle of *Echinoparyphium hydromyos* sp. nov. (Digenea: Echinostomatidae) from the Australian water-rat. *Parasitology* **57**, 19–30.
Angel, L.M. (1971). *Burnellus* gen. nov. (Digenea: Fellodistomatidae), the life history of the type-species, *B. trichofurcatus* (Johnston & Angel, 1940), and a note on a related species, *Tandanicola bancrofti* Johnston, 1927, both from the Australian freshwater catfish, *Tandanus tandanus*. *Parasitology* **62**, 375–384.
Bakke, T.A. (1980). A revision of the family Leucochloridiidae Poche (Digenea) and studies on the morphology of *Leucochloridium paradoxum* Carus, 1835. *Systematic Parasitology* **1**, 189–202.
Barker, S.C. and Cribb, T.H. (1993). Sporocysts of *Mesostephanus haliasturis* (Digenea) produce miracidia. *International Journal for Parasitology* **23**, 137–139.
Bartoli, P., Jousson, O. and Russell-Pinto, F. (2000). The life cycle of *Monorchis parvus* (Digenea: Monorchiidae) demonstrated by developmental and molecular data. *Journal of Parasitology* **86**, 479–489.
Bearup, A.J. (1961). Observations on the life cycle of *Stictodora lari* (Trematoda: Heterophyidae). *Proceedings of the Linnean Society of New South Wales* **86**, 251–257.
Blair, D. and Islam, K.S. (1983). The life-cycle and morphology of *Trichobilharzia australis* n. sp. (Digenea: Schistosomatidae) from the nasal blood vessels of the black duck (*Anas superciliosa*) in Australia, with a review of the genus *Trichobilharzia*. *Systematic Parasitology* **5**, 89–117.
Bray, R.A. & Cribb, T.H. (2003). The digeneans of elasmobranchs – distribution and evolutionary significance. In: *Taxonomie, écologie et évolution des métazoaires parasites* (C. Combes and J. Jourdane, eds). Tome I, 67–96. Perpignan: Presses Universitaire de Perpignan.

Bray, R.A., Gibson, D.I. and Zhang, J. (1999). Urotrematidae Poche, 1926 (Platyhelminthes: Digenea) in Chinese freshwater fishes. *Systematic Parasitology* **44**, 193–200.

Brinkmann, A. (1988). Presence of *Otodistomum* sp. metacercariae in Norwegian marine fishes. *Sarsia* **73**, 79–82.

Brooks, D.R. and Bandoni, S.M. (1988). Coevolution and relicts. *Systematic Zoology* **37**, 19–33.

Brooks, D.R., O'Grady, R.T. and Glen, D.R. (1985). Phylogenetic analysis of the Digenea (Platyhelminthes: Cercomeria) with comments on their adaptive radiation. *Canadian Journal of Zoology* **63**, 411–443.

Brooks, D.R., Bandoni, S.M., MacDonald, C.A. and O'Grady, R.T. (1989). Aspects of the phylogeny of the Trematoda Rudolphi, 1808 (Platyhelminthes: Cercomeria). *Canadian Journal of Zoology* **67**, 2609–2624.

Cable, R.M. (1954). Studies on marine digenetic trematodes of Puerto Rico. The life cycle in the family Haplosplanchnidae. *Journal of Parasitology* **40**, 71–76.

Cable, R.M. (1974). Phylogeny and taxonomy of trematodes with reference to marine species. In: *Symbiosis in the Sea* (W.B. Vernberg, ed.), pp. 173–193. Columbia, South Carolina: University of South Carolina Press.

Cable, R.M. (1982). Phylogeny and taxonomy of the malacobothrean flukes. In: *Parasites – their world and ours* (D.F. Mettrick and S.S. Desser, eds), pp. 194–197. Amsterdam: Elsevier Biomedical Press.

Cable, R.M. and Nahhas, F.M. (1962). *Bivesicula caribbensis* sp. n. (Trematoda: Digenea) and its life history. *Journal of Parasitology* **48**, 536–538.

Caira, J.N. (1989). A revision of the North American papillose Allocreadiidae (Digenea) with independent cladistic analyses of larval and adult forms. *Bulletin of The University of Nebraska State Museum* **11**, 1–91.

Campbell, D. (1985). The life cycle of *Gymnophallus rebecqui* (Digenea: Gymnophallidae) and the response of the bivalve *Abra tenuis* to its metacercariae. *Journal of the Marine Biological Association of the United Kingdom* **65**, 589–601.

Chun, S.K. and Kim, Y.G. (1982). [Studies on the life history of the trematode parasitic in *Meretrix lusoria* Röding]. *Bulletin of National Fisheries University of Busan* **22**, 31–44.

Crandall, R.B. (1960). The life history and affinities of the turtle lung fluke, *Heronimus chelydrae* MacCallum, 1902. *Journal of Parasitology* **46**, 289–307.

Cribb, T.H. (1986). The life cycle and morphology of *Stemmatostoma pearsoni*, gen. et sp. nov., with notes on the morphology of *Telogaster opisthorchis* Macfarlane (Digenea: Cryptogonimidae). *Australian Journal of Zoology* **34**, 279–304.

Cribb, T.H. (1988). Life cycle and biology of *Prototransversotrema steeri* Angel, 1969 (Digenea: Transversotrematidae). *Australian Journal of Zoology* **36**, 111–129.

Cribb, T.H., Anderson, G.R., Adlard, R.D. and Bray, R.A. (1998). A DNA-based demonstration of a three-host life-cycle for the Bivesiculidae (Platyhelminthes: Digenea). *International Journal for Parasitology* **28**, 1791–1795.

Cribb, T.H., Bray, R.A., Littlewood, D.T.J., Pichelin, S. and Herniou, E.A. (2001). The Digenea. In: *Interrelationships of the Platyhelminthes* (D.T.J. Littlewood and R.A. Bray, eds), pp. 168–185. London: Taylor & Francis.

Cribb, T.H., Chisholm, L.A. and Bray, R.A. (2002). Diversity in the Monogenea and Digenea: does lifestyle matter? *International Journal for Parasitology* **32**, 321–328.

Dickerman, E.E. (1946). Studies on the trematode family Azygiidae. III. The morphology and life cycle of *Proterometra sagittaria* n. sp. *Transactions of the American Microscopical Society* **65**, 37–44.

Dönges, J. (1971). The potential number of redial generations in echinostomatids (Trematoda). *International Journal for Parasitology* **1**, 51–59.
Dönges, J. (1974). The life cycle of *Euclinostomum heterostomum* (Rudolphi, 1809) (Trematoda: Clinostomatidae). *International Journal for Parasitology* **4**, 79–90.
Durie, P.H. (1956). The paramphistomes (Trematoda) of Australian ruminants III. The life-history of *Calicophoron calicophorum* (Fischoeder) Nasmark. *Australian Journal of Zoology* **4**, 152–157.
Ferguson, M.A., Cribb, T.H. and Smales, L.R. (1999). Life-cycle and biology of *Sychnocotyle kholo* n. g., n. sp. (Trematoda: Aspidogastrea) in *Emydura macquarii* (Pleurodira: Chelidae) from southern Queensland, Australia. *Systematic Parasitology* **43**, 41–48.
Gibson, D.I. (1987). Questions in digenean systematics and evolution. *Parasitology* **95**, 429–460.
Gibson, D.I. (2002). Superfamily Hemiuroidea Looss, 1899. In: *Keys to the Trematoda* (D.I. Gibson, A. Jones and R.A. Bray, eds), Vol. 1, pp. 299–413. Wallingford and New York: CABI Publishing.
Gibson, D.I. and Bray, R.A. (1986). The Hemiuridae (Digenea) of fishes from the north-east Atlantic. *Bulletin of the British Museum (Natural History) Zoology* **51**, 1–125.
Gibson, D.I., Jones, A. and Bray, R.A., eds (2002). *Keys to the Trematoda*. Vol. 1. Wallingford and New York: CABI Publishing.
Goater, T.M., Browne, C.L. and Esch, G.W. (1990). On the life history and functional morphology of *Halipegus occidualis* (Trematoda: Hemiuridae), with emphasis on the cystophorous cercaria stage. *International Journal for Parasitology* **20**, 923–934.
Hall, K.A., Cribb, T.H. and Barker, S.C. (1999). V4 region of small subunit rDNA indicates polyphyly of the Fellodistomidae (Digenea) which is supported by morphology and life-cycle data. *Systematic Parasitology* **43**, 81–92.
Howell, M.J. and Bearup, A.J. (1967). The life histories of two bird trematodes of the family Philophthalmidae. *Proceedings of the Linnaean Society of New South Wales* **92**, 182–194.
Huelsenbeck, J.P., Ronquist, F., Nielsen, R. and Bollback, J.P. (2001). Bayesian inference of phylogeny and its impact on evolutionary biology. *Science* **294**, 2310–2314.
Jamieson, B.G.M. (1966). Larval stages of the progenetic trematode *Parahemiurus bennettae* Jamieson, 1966 (Digenea, Hemiuridae) and the evolutionary origin of cercariae. *Proceedings of the Royal Society of Queensland* **77**, 81–91.
Johnston, T.H. and Angel, L.M. (1942). The life history of the trematode, *Paryphostomum tenuicollis* (S. J. Johnston). *Transactions of the Royal Society of South Australia* **66**, 119–123.
Johnston, T.H. and Angel, L.M. (1951). The life history of *Plagiorchis jaenschi*, a new trematode from the Australian water rat. *Transactions of the Royal Society of South Australia* **74**, 49–58.
Johnston, T.H. and Simpson, E.R. (1940). The anatomy and life history of the trematode *Cyclocoelium jaenschi* n. sp. *Transactions of the Royal Society of South Australia* **64**, 273–278.
Jue Sue, L. and Platt, T.R. (1998). Description and life-cycle of two new species of *Choanocotyle* n. g. (Trematoda: Plagiorchiida), parasites of Australian freshwater turtles, and the erection of the family Choanocotylidae. *Systematic Parasitology* **41**, 47–61.

Kanev, I., Radev, V. and Fried, B. (2002). Superfamily Clinostomoidea Lühe, 1901. In: *Keys to the Trematoda* (D.I. Gibson, A. Jones and R.A. Bray, eds). Vol. 1, pp. 111–125. Wallingford and New York: CABI Publishing.

Kearn, G.C. (1998). *Parasitism and the Platyhelminths.* London & New York: Chapman & Hall.

Køie, M. (1975). On the morphology and life-history of *Opechona bacillaris* (Molin, 1859) Looss, 1907 (Trematoda, Lepocreadiidae). *Ophelia* **13**, 63–86.

Køie, M. (1979). On the morphology and life-history of *Monascus* [= *Haplocladus*] *filiformis* (Rudolphi, 1819) Looss, 1907 and *Steringophorus furciger* (Olsson, 1868) Odhner, 1905 (Trematoda, Fellodistomidae). *Ophelia* **18**, 113–132.

Køie, M. (1980). On the morphology and life-history of *Steringotrema pagelli* (Van Beneden, 1871) Odhner, 1911 and *Fellodistomum fellis* (Olsson, 1868) Nicoll, 1909 [syn. *S. ovacautum* (Lebour, 1908) Yamaguti, 1953] (Trematoda, Fellodistomidae). *Ophelia* **19**, 215–236.

Køie, M. (1982). The redia, cercaria and early stages of *Aporocotyle simplex* Odhner, 1900 (Sanguinicolidae) – a digenetic trematode which has a polychaete annelid as the only intermediate host. *Ophelia* **21**, 115–145.

Køie, M. (1985). On the morphology and life-history of *Lepidapedon elongatum* (Lebour, 1908) Nicoll, 1910 (Trematoda, Lepocreadiidae). *Ophelia* **24**, 135–153.

Køie, M. (1990). On the morphology and life-history of *Hemiurus luehei* Odhner, 1905 (Digenea: Hemiuridae). *Journal of Helminthology* **64**, 193–202.

Køie, M., Karlsbakk, E. and Nylund, A. (2002). A cystophorous cercaria and metacercaria in *Antalis entalis* (L.) (Mollusca, Scaphopoda) in Norwegian waters, the larval stage of *Lecithophyllum botryophorum* (Olsson, 1868) (Digenea, Lecithasteridae). *Sarsia* **87**, 302–311.

La Rue, G.R. (1951). Host-parasite relations among the digenetic trematodes. *Journal of Parasitology* **37**, 333–342.

Lim, H.K. and Heyneman, D. (1972). Intramolluscan inter-trematode antagonism: a review of factors influencing the host-parasite system and its possible role in biological control. *Advances in Parasitology* **10**, 191–268.

Littlewood, D.T.J., Rohde, K., Bray, R.A. and Herniou, E.A. (1999). Phylogeny of the Platyhelminthes and the evolution of parasitism. *Biological Journal of the Linnean Society* **68**, 257–287.

Lockyer, A.E., Olson, P.D. and Littlewood, D.T.J. (2003). Utility of complete large and small subunit rRNA genes in resolving the phylogeny of the Neodermata (Platyhelminthes): implications and a review of the cercomer theory. *Biological Journal of the Linnean Society* **78**, 155–171.

Maddison, W.P. and Maddison, D.R. (2000). *MacClade. Version 4.* Sunderland, Massachusetts: Sinauer Associates.

Madhavi, R. (1978). Life history of *Genarchopsis goppo* Ozaki, 1925 (Trematoda: Hemiuridae) from the freshwater fish *Channa punctata. Journal of Helminthology* **52**, 251–259.

Martin, W.E. (1956). The life cycle of *Catatropis johnstoni* n. sp. (Trematoda: Notocotylidae). *Transactions of the American Microscopical Society* **75**, 117–128.

Martins, E.P., ed. (1996). *Phylogenies and the comparative method in animal behavior.* Oxford: Oxford University Press.

Mas-Coma, S. and Montoliu, I. (1986). The life cycle of *Brachylaima ruminae* n. sp. (Trematoda: Brachylaimidae), a parasite of rodents. *Zeitschrift für Parasitenkunde* **72**, 739–753.

Matthews, B.F. (1980). *Cercaria vaullegeardi* Pelseneer, 1906 (Digenea: Hemiuridae); the daughter sporocyst and emergence of the cercaria. *Parasitology* **81**, 61–69.
Matthews, R.A. (1974). The life-cycle of *Bucephaloides gracilescens* (Rudolphi, 1819) Hopkins, 1954 (Digenea: Gasterostomata). *Parasitology* **68**, 1–12.
Murrills, R.J., Reader, T.A.J. and Southgate, V.R. (1985). Studies on the invasion of *Notocotylus attenuatus* (Notocotylidae: Digenea) into its snail host *Lymnaea peregra*: the contents of the fully embryonated egg. *Parasitology* **91**, 397–405.
Niewiadomska, K. (2002). Superfamily Diplostomoidea Poirier, 1886. In: *Keys to the Trematoda* (D.I. Gibson, A. Jones and R.A. Bray, eds), Vol. 1, pp. 159–166. Wallingford and New York: CABI Publishing.
O'Grady, R.T. (1985). Ontogenetic sequences and the phylogenetics of parasitic flatworm life cycles. *Cladistics* **1**, 159–170.
Olson, P.D., Cribb, T.H., Tkach, V.V., Bray, R.A. and Littlewood, D.T.J. (2003). Phylogeny and classification of the Digenea (Platyhelminthes: Trematoda). *International Journal for Parasitology* **33**, 733–755.
Palombi, A. (1942). Il ciclo biologico di *Ptychogonimus megastoma* (Rud.). Osservazioni sulla morfologia e fisiologia delle forme larvali e considerazioni filogenetiche. *Rivista di Parassitologia* **6**, 117–172.
Pearson, J.C. (1956). Studies on the life cycles and morphology of the larval stages of *Alaria arisaemoides* Augustine and Uribe, 1927 and *Alaria canis* LaRue and Fallis, 1936 (Trematoda: Diplostomidae). *Canadian Journal of Zoology* **34**, 295–387.
Pearson, J.C. (1961). Observations on the morphology and life cycle of *Neodiplostomum intermedium* (Trematoda: Diplostomatidae). *Parasitology* **51**, 133–172.
Pearson, J.C. (1968). Observations on the morphology and life-cycle of *Paucivitellosus fragilis* Coil, Reid & Kuntz, 1965 (Trematoda: Bivesiculidae). *Parasitology* **58**, 769–788.
Pearson, J.C. (1972). A phylogeny of life-cycle patterns of the Digenea. *Advances in Parasitology* **10**, 153–189.
Pearson, J.C. (1992). On the position of the digenean family Heronimidae: an inquiry into a cladistic classification of the Digenea. *Systematic Parasitology* **21**, 81–166.
Poulin, R. and Cribb, T.H. (2002). Trematode life cycles: short is sweet? *Trends in Parasitology* **18**, 176–183.
Prévôt, G., Bartoli, P. and Deblock, S. (1976). Cycle biologique de *Maritrema misenensis* (A. Palombi, 1940), n. comb. (Trematoda: Microphallidae, Travassos, 1920) du midi de la France. *Annales de Parasitologie Humaine et Comparée* **51**, 433–446.
Rohde, K. (1972). The Aspidogastrea, especially *Multicotyle purvisi* Dawes, 1941. *Advances in Parasitology* **10**, 78–151.
Rohde, K. (1973). Structure and development of *Lobatostoma manteri* sp. nov. (Trematoda: Aspidogastrea) from the Great Barrier Reef, Australia. *Parasitology* **66**, 63–83.
Rohde, K. (1975). Early development and pathogenesis of *Lobatostoma manteri* Rohde (Trematoda: Aspidogastrea). *International Journal for Parasitology* **5**, 597–607.
Rohde, K. (2001). The Aspidogastrea: an archaic group of Platyhelminthes. In: *Interrelationships of the Platyhelminthes* (D.T.J. Littlewood and R.A. Bray, eds), pp. 159–167. London: Taylor & Francis.

Rohde, K. and Sandland, R. (1973). Host-parasite relations in *Lobatostoma manteri* Rohde (Trematoda: Aspidogastrea). *Zeitschrift für Parasitenkunde* **42**, 115–136.
Sewell, R.B.S. (1922). Cercariae indicae. *Indian Journal of Medical Research* **10** (Supplement), 1–370.
Shameem, U. and Madhavi, R. (1991). Observations on the life-cycles of two haploporid trematodes, *Carassotrema bengalense* Rekharani & Madhavi, 1985 and *Saccocoelioides martini* Madhavi, 1979. *Systematic Parasitology* **20**, 97–107.
Shoop, W.L. (1988). Trematode transmission patterns. *Journal of Parasitology* **74**, 46–59.
Sillman, E.I. (1962). The life history of *Azygia longa* (Leidy, 1851) (Trematoda: Digenea), and notes on *A. acuminata* Goldberger, 1911. *Transactions of the American Microscopical Society* **81**, 43–65.
Smith, J.W. (1997a). The blood flukes (Digenea: Sanguinicolidae and Spirorchidae) of cold-blooded vertebrates: Part 1. A review of the literature published since 1971, and bibliography. *Helminthological Abstracts* **66**, 255–294.
Smith, J.W. (1997b). The blood flukes (Digenea: Sanguinicolidae and Spirorchidae) of cold-blooded vertebrates: Part 2. Appendix I: Comprehensive parasite-host list; Appendix II: Comprehensive host-parasite list. *Helminthological Abstracts* **66**, 329–344.
Sprent, J.F.A. (1983). Observations on the systematics of ascaridoid nematodes. In: *Concepts in nematode systematics* (A.R. Stone, H.M. Platt and L.F. Khalil, eds), pp. 303–319. London: Academic Press.
Stunkard, H.W. (1959). The morphology and life-history of the digenetic trematode, *Asymphylodora amnicolae* n. sp.; the possible significance of progenesis for the phylogeny of the Digenea. *Biological Bulletin* **117**, 562–581.
Stunkard, H.W. (1964). The morphology, life history and systematics of the digenetic trematode, *Homalometron pallidum* Stafford, 1904. *Biological Bulletin* **126**, 163–173.
Stunkard, H.W. (1968). The asexual generations, life-cycle, and systematic relations of *Microphallus limuli* Stunkard, 1951 (Trematoda: Digenea). *Biological Bulletin* **134**, 332–343.
Thomas, J.D. (1958). Studies on the structure, life history and ecology of the trematode *Phyllodistomum simile* Nybelin, 1926 (Gorgoderidae: Gorgoderinae) from the urinary bladder of brown trout, *Salmo trutta* L. *Proceedings of the Zoological Society of London, Series B* **130**, 397–435.
Wall, L.D. (1951). The life history of *Vasotrema robustum* (Stunkard 1928), Trematoda: Spirorchiidae. *Transactions of the American Microscopical Society* **70**, 173–184.
Wardle, W.J. (1975). *Cercaria anadarae* sp. n. parasitizing a bivalve mollusc, *Anadara brasiliana* (Lamarck), from the northwest Gulf of Mexico. *Journal of Parasitology* **61**, 1048–1049.
Watson, R.A. (1984). The life cycle and morphology of *Tetracerasta blepta*, gen. et. sp. nov., and *Stegodexamene callista*, sp. nov. (Trematoda: Lepocreadiidae) from the long-finned eel, *Anguilla reinhardtii* Steindachner. *Australian Journal of Zoology* **32**, 177–204.
Yamaguti, S. (1943). On the morphology of the larval forms of *Paragonimus westermanii* with special reference to their excretory system. *Japanese Journal of Zoology* **10**, 461–467.
Zelmer, D.A. and Esch, G.W. (1998). Bridging the gap: the odonate naiad as a paratenic host for *Halipegus occidualis* (Trematoda: Hemiuridae). *Journal of Parasitology* **84**, 94–96.

Progress in Malaria Research: the Case for Phylogenetics

Stephen M. Rich[1] and Francisco J. Ayala[2]

[1]*Division of Infectious Disease, Tufts University School of Veterinary Medicine, 200 Westboro Road, North Grafton, MA 01536, USA*
[2]*Department of Ecology and Evolutionary Biology, University of California, Irvine, 321 Steinhaus Hall, Irvine, CA 92697-2525, USA*

Abstract	255
1. The Malaria Phylum: Apicomplexa	256
1.1. Origin of the Apicomplexan Plastid	256
1.2. Evolution of Apicomplexan Life Cycles	257
2. Morphology, Phylogenetics and *Plasmodium* Systematics	258
2.1. Host Affiliation of Malaria Parasites	260
2.2. Affiliation of Malaria Parasites with their Arthropod Vectors	266
3. Evolution and Extant Distribution of Malignant Human Malaria: *P. falciparum*	266
3.1. Malaria's Eve Hypothesis	268
3.2. Constraints on Synonymous Substitutions	272
4. Concluding Remarks	275
References	275

ABSTRACT

Malaria, from the Italian for "bad air", is a term used to describe a human disease caused by any of four parasites of the genus, *Plasmodium*. There are in fact over 200 described species of *Plasmodium* that parasitize reptiles, birds, and mammals, and may or may not cause disease in these various hosts. In this chapter, we highlight important evolutionary studies that have been undertaken to determine the relatedness among these species and their place in the taxonomic hierarchy. We begin by providing an overview of our present understanding of the phylum to which malaria parasites belong—Apicomplexa. The unique characteristics of these parasites reflect both their adaptation to the parasitic life style as well as some vestigial remnants of their pre-parasitic evolutionary past. Phylogenetic analyses provide the means for discerning the means by which these characteristics have come into existence. We next discuss the systematics of the genus *Plasmodium*. Morphology, genomic structure and content as well as host affiliation of these parasites are all traits that have been used for establishing taxonomic arrangements. Molecular phylogenetics has proven to be an invaluable tool in this regard and so we discuss the current phylogenetic picture of the genus as well as the correspondence among the various datasets (morphology, molecules, and host-preference). Lastly, we present a detailed account of our current understanding of the evolutionary past of the most deadly of the human malaria species—*P. falciparum*.

1. THE MALARIA PHYLUM: APICOMPLEXA

The genus *Plasmodium* belongs to the Apicomplexa, a large and complex phylum with > 5000 known species and many more to be described (Corliss, 1994). All Apicomplexa are parasitic and are characterized by their eponymous structure, the apical complex, which plays an important role in the parasite's penetration of host cells. Apicomplexa other than *Plasmodium* that cause disease in humans include *Toxoplasma* spp., *Cryptosporidium* spp., and *Babesia* spp., although any one of these accounts for a very small fraction of the risk associated with *Plasmodium*. Since there is no fossil record of apicomplexans (Margulis *et al.*, 1993), molecular investigations provide the primary insight to the origins of the phylum. These studies have indicated that the Apicomplexa are very ancient, perhaps as old as the multicellular kingdoms of plants, fungi, and animals, and thus somewhat older than one billion years (Ayala *et al.*, 1998).

1.1. Origin of the Apicomplexan Plastid

A common structural feature of many apicomplexans is the occurrence of an organelle called the plastid or apicoplast, which is a non-photosynthetic homologue of the chloroplast found in plants and algae (Lang-Unnasch *et al.*, 1998). The function of the plastid organelle in apicomplexans is unknown, but recent studies have indicated that it shares a similar secondary endosymbiotic origin to that of dinoflagellates (Fast *et al.*, 2001). Interestingly, apicomplexan parasites have retained a metabolic mechanism referred to as the shikimate pathway, which is essential in the production of various aromatic compounds including the synthesis of the amino acids tryptophan, phenylalanine, and tyrosine. This vital functionality in plants is associated with the plastid organelle, and so the apicoplast may also be the site of this activity. Since this pathway is absent from the animal hosts of apicomplexans, it provides ideal targets for inhibitory agents such as glyphosate, the active ingredient of the common herbicide Roundup™ (Sigma) (Roberts *et al.*, 1998).

The plastid genomes of apicomplexans are highly divergent among the various members of the phylum in terms of both content and nucleotide sequence. Indeed, the degree of divergence is high to the extent that the plastid genome provides few informative phylogenetic markers for deep evolutionary inquiries. Moreover, it appears that horizontal transfer of genes has occurred between the plastid and the other apicomplexan organelle, the mitochondrion (Obornik *et al.*, 2002). This paralogous situation would make determination of phylogeny a perilous task. Fast *et al.* (2001) were able to derive robust species trees by focusing their study on more conserved genes in the nuclear genome that encode proteins localizing to specific plastid targets. The results of this study indicate that the ancestors of apicomplexans were probably photosynthetic, and that the modern descendents have conserved the plastid as a relic that may have co-opted some new functional role for the parasites that have retained it (Figure 1). What is clear is that the plastid is essential to parasite survival insofar as it serves a number of metabolic and house-keeping functions of the cell (Roos *et al.*, 2002). Hence, the discovery of this evolutionary relic has presented new opportunities for drug therapies that target plastid genes in the parasites (Surolia *et al.*, 2002; Touze *et al.*, 2002).

1.2. Evolution of Apicomplexan Life Cycles

Apicomplexans have complex life cycles that are partitioned into distinct developmental stages that permit the parasite to invade different host environments. In some instances, this complex life cycle may involve more than one host species. Parasites that have a single unique host species are called *monogenetic*; for example, *Cryptosporidium parvum* that infects animal gut epithelia requires only one host species for its transmission. However, all *Plasmodium* spp. are *digenetic* parasites, requiring two separate host species to complete their life cycle. The two host species are

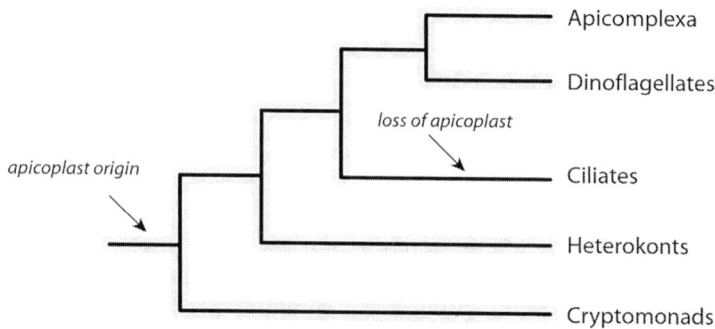

Figure 1 Phylogeny of protozoal phyla, showing the origin of the plastid-like apicoplast organelle. The most parsimonious explanation for the observed pattern is that the apicoplast is derived from a very early endosymbiosis and that this organelle was subsequently lost in the ciliates (Fast *et al.*, 2001).

distinguished as being either definitive or intermediate, with the definitive host being the site in which sexual reproduction of the parasite occurs. A long-standing debate regarding *Plasmodium* evolution is whether these parasites evolved directly from monogenetic parasites of ancient marine invertebrate ancestors of modern chordates, or whether they descended from other digenetic parasites (Huff, 1938; Manwell, 1955; Garnham, 1966; Barta, 1989). To date, molecular phylogenetic comparisons have not given complete resolution to this issue, but it is apparent from analyses of ribosomal DNA and other genetic loci that the digenetic life style has multiple independent origins among apicomplexans (Barta, 1989; Escalante and Ayala, 1995; Fast *et al.*, 2002).

Vertebrates are the typical intermediate hosts of *Plasmodium*, while invertebrate species are the definitive hosts or vectors. *Plasmodium* are intracellular parasites occupying the blood cells of the intermediate host for a large part of their life cycle; accordingly, their invertebrate vectors are blood-feeding organisms, most typically mosquitoes, although in the case of some of the reptilian malariae, sand flies serve as the vector (Kimsey, 1992).

2. MORPHOLOGY, PHYLOGENETICS AND *PLASMODIUM* SYSTEMATICS

Before the widespread application of molecular phylogenetics, parasite systematists determined relationships of malaria parasites by comparison of discernible traits such as host- and vector-preference, morphology, geographical distribution and life history. Many of these parasite characteristics retain importance for diagnostic purposes and assessment of risk, but their patterns of similarity among *Plasmodium* species do not necessarily reflect phyletic relationships. For example, the four species traditionally considered parasitic to humans: *P. falciparum*, *P. malariae*, *P. ovale*, and *P. vivax* are – despite the commonality of their infectivity to humans – markedly different in their respective geographic distribution, pathogenicity, and life cycle. Analyses of nucleotide sequences of rDNA indicate that the common ancestry of these four species probably dates back to the origin of the genus-wide radiation approximately 130 million years (MY) ago. Moreover, each of these parasites has much more recent common ancestry with distinct non-human primate malarias, suggesting that *Plasmodium* have found preference for primate hosts, including humans, multiple times in their evolutionary history (Figure 2) (Escalante and Ayala, 1995; Ayala, *et al.*, 1998).

In other instances, malaria parasites that appear very similar turn out to be quite different when examined molecularly. For example, *P. azurophilum* is a lizard malaria parasite infecting Anoles in the Caribbean.

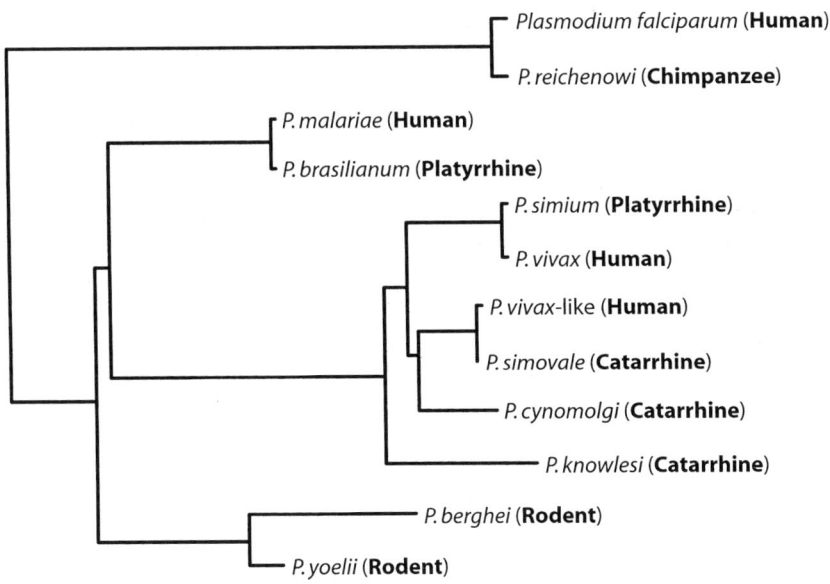

Figure 2 Phylogeny of primate malaria species based on the circumsporozoite gene (Ayala *et al.*, 1998).

Perkins (2000) found that these parasites actually comprise a complex of at least two cryptic species, one infecting lymphocytes and the other parasitizing erythrocytes. The two species are virtually indistinguishable microscopically, but they bear distinct mitochondrial DNA (mtDNA) haplotypes that suggest they diverged prior to their association with their Anole hosts (Perkins, 2000).

Accordingly, we should evaluate cautiously traditional taxonomy and systematic arrangements of *Plasmodium* and other Apicomplexa species. Biological similarity (or difference) is, of course, the stuff from which comparative studies are made; however, it is imperative that we determine whether these similarities are attributable to common ancestry or to convergence on favorable conditions of transmission. If, for example, particular pathogenicity factors appear to follow phylogenetic trends, then it may be possible to anticipate future threats from organisms that are closely related and hence more likely to possess these traits. However, if such traits appear to have evolved in multiple instances among species, it may be the case that the plasticity of the trait is such that even benign organisms may take on pathogenic characters under the right, or for our sake, wrong conditions.

2.1. Host Affiliation of Malaria Parasites

If phylogenetics is an endeavor to determine evolutionary relationships among species, then systematics is the discipline of assigning taxonomic designations to these organisms. While it should be the case that the best taxonomy will reflect phylogenetic relatedness, we know that this is not always the case. Among parasite systematics, this disparity is often a consequence of a long held notion that parasites evolve with their hosts. This assumption implies that different host species should have different species parasitizing them, and the phylogeny of host and parasite should be concordant. As we have indicated above, this is not always so, as demonstrated by the discordant phylogeny of humans and their malaria parasites. This is where molecular phylogenetics has proven extremely useful: by providing a means of objectively evaluating these co-evolutionary hypotheses using a set of less ambiguous characters, i.e., nucleotide and amino acid sequences. The comparison of molecular phylogenies of hosts and parasites, allows us to determine whether alterations in host preference – *host shifts* – have occurred.

Among avian and reptilian malaria parasites, host-shifts appear to have been a common occurrence in the evolution of the genus *Plasmodium* (Bensch *et al.*, 2000; Ricklefs and Fallon, 2002). The same can be said for three of the four species of human *Plasmodium* species, whose evolutionary divergence greatly predates the origin of the primates themselves. It follows that their parasitic associations with humans are phylogenetically independent; that is to say that some of these species have been laterally transmitted to the human ancestral lineage from other distinct host species. In this instance, the conclusion is consistent with the diversity of physiological and epidemiological characteristics of the human *Plasmodium* species (Gilles and Warrell, 1993).

Molecular phylogenetics has proven invaluable for determining the state of affairs among the human malaria parasites. Analysis of the nucleotide sequence of the circumsporozoite protein (*Csp*) gene of two of these extremely divergent human parasites *P. malariae* and *P. vivax*, found that they are virtually indistinguishable from two New World monkey parasites, *P. brasilianum* and *P. simium*, respectively (Figure 2) (Ayala et al., 1998). From this, we inferred that a lateral transfer between hosts has occurred in recent times, either from monkeys to humans or vice versa. The *Csp* genetic distance between *P. malariae* and *P. brasilianum* is 0.002 ± 0.002, not greater than the distance among the various isolates of *P. malariae* ($n=2$), *P. vivax* ($n=4$), or *P. falciparum* ($n=8$) (Ayala et al., 1998). This suggests that *P. malariae* (isolated from humans) and *P. brasilianum* (isolated from New World monkey) have not long diverged from one another and might be considered a single species exhibiting "host polymorphism" (Escalante and Ayala, 1994), i.e., able to parasitize more than one host species. A similar situation appears in the case of the human agent of benign tertian malaria *P. vivax*, and the New World primate malaria species, *P. simium*; both of which are also genetically indistinguishable.

Whether or not the two species in each of the human–primate parasite pairs (*P. vivax–P. simium* and *P. malariae–P. brasilianum*) should be considered the same or distinct species is merely a matter of nomenclatural convenience, and hence is not biologically substantive. More important is the conclusion that two of the four known human malaria parasites have nearly identical platyrrhine (New World monkey) parasite relatives. This is a strong indication that a host-switch has occurred in recent times, or perhaps that it continues to occur, as would be the case if there were a demonstrated host-polymorphism.

Determining the direction of the host-switch between human and platyrrhine – either from monkey to human, or human to monkey – holds great biological relevance for understanding the evolution of the genus and understanding the origin of disease. Human and platyrrhine monkeys are distantly related and have come to be geographically coupled following the first human colonization of the Americas, which occurred within the last 15,000 years. The host-switch may have occurred at that time or, alternatively, it may not have happened until after the second influx of humans to South America, i.e. when Europeans began to colonize the Americas in the 16th century. Whether 500 or 15,000 years have passed since the host-switch, it would be a mere moment in evolutionary time and so it is not surprising that the human and platyrrhine parasites are so little diverged. Both *P. simium* and *P. brasilianum* are known to be infectious to humans (Gilles and Warrell, 1993). Epidemiological serosurveys of humans and monkeys in French Guiana indicate that platyrrhines may actually serve as zoonotic reservoirs for human disease (Fandeur *et al.*, 2000), thus lending support to the host-polymorphism hypothesis.

Unlike the Old World primate parasite, *P. reichenowi*, that thrives exclusively in chimpanzees, the platyrrhine malaria parasites are quite capacious in their host preference, and so these New World parasites appear quite susceptible to host-switches. *P. simium* infects at least three, and *P. brasilianum* has been identified in as many as 26 species of New World monkeys (Gysin, 1998). We have argued in the past on the grounds of evolutionary parsimony, that the host-switches observed in *vivax/simium* and *malariae/brasilianum* were most likely to have occurred from primates to humans (Escalante and Ayala, 1995; Ayala *et al.*, 1998). Based on the observed host distribution, the alternative explanation of a switch from humans to primates seems less plausible since it would require that multiple independent switches have occurred. For example, in the case of the *malariae-brasilianum* host switch, it is improbable that no less than 26 human to platyrrhine (or platyrrhine to platyrrhine) host-switches would have occurred in the 15,000-year history of human habitation of South America.

P. vivax and *P. malariae* have widespread global distributions, while the complementary *P. simium* and *P. brasilianum* are restricted to South America. This is not inconsistent with a host-switch of the platyrrhine parasites to human hosts since humans have been remarkably vagile in the past several centuries and could have carried their parasites wherever they may have traveled. For example, a survey of ribosomal DNA sequences revealed the occurrence of a *P. brasilianum* isolate in Myanmar (Kawamoto *et al.*, 2002). The most plausible explanation for New World monkey malaria in Southeast Asia is that an infected human carried it there. This pattern of host transmission may become more evident as additional molecular genotypes of malaria isolates collected from around the globe become available.

Geographical distribution records are somewhat ambiguous with respect to determining the direction of host-switch in the case of *P. vivax* and *P. simium*. *P. vivax* is the most cosmopolitan of the human malarias, but it is notably absent from sub-Saharan Africa. Absence of *vivax*-malaria in the region has been attributed to the widespread occurrence of a genetic mutation of the Duffy blood group proteins in indigenous sub-Saharan peoples. Duffy proteins are expressed on the surface of erythrocytes, and are necessary for receptor–ligand mediated invasion of *P. vivax* into red blood cells (Miller et al., 1976). Duffy receptors do not play a role in infection by the main African malaria parasite, *P. falciparum*. Individuals with an Fy^{a-b-} mutation do not express the Duffy receptor, suggesting an adaptive response for resistance to *P. vivax*. This adaptation is found primarily in particular parts of Africa, and its occurrence is inconsistent with a recent introduction of *vivax*-malaria to humans, suggesting that the occurrence of this human mutation was in response to an ancient exposure to *P. vivax*, which has since been nearly extirpated from the continent. The possibility that the Duffy mutation may have arisen in response to some other selection pressure cannot be eliminated. Indeed, Duffy negative individuals are resistant to *P. knowlesi* (Mason et al., 1977), which is an Old World monkey parasite, and hence one with which human ancestors have shared a common geographical range for millions of years. The Duffy mutations may have reached high frequency in sub-Saharan Africa as a counter response to risk of exposure to this zoonotic malaria known to be infectious to humans lacking the Duffy negative genotype.

Historical documentation of non-malignant (i.e. non-*P. falciparum*) malaria in humans is similarly equivocal. Firstly, there is no record of quartan malaria in South America prior to European colonization. This would be consistent with the interpretation that *P. vivax* (as well as *P. malariae* and *P. falciparum*) was introduced to the New World by the European colonizers and their African slaves. The weakness of this argument is that it relies on negative evidence, particularly unreliable when there are few records or studies that would have likely manifested the presence of malaria in the New World before the year 1500, even if it had indeed been present.

In the Old World, historical records are more complete. Chinese medical writings (dated 2700 B.C.), cuneiform clay tablets from Mesopotamia (~2000 B.C.), the Ebere Egyptian Papyrus (ca. 1570 B.C.), and Vedic-period Indian writings (1500–800 B.C.) mention severe periodic fevers, spleen enlargement and other symptoms suggestive of malaria (Sherman, 1998). Spleen enlargement and the malaria antigen have been detected in Egyptian mummies more than 3000 years old (Miller et al., 1994; Sherman, 1998). Hippocrates' (460–370 B.C.) discussion of tertian and quartan fevers, "leaves little doubt that by the fifth century B.C. *Plasmodium malariae* and *P. vivax* were present in Greece" (Sherman, 1998, p. 3). If this interpretation is correct, the association of *P. malariae* and *P. vivax* with humans could not be attributed to a host switch from monkeys to humans that would have occurred after the European colonization of the Americas.

This would be definitive evidence, so long as one accepts the interpretation that the fevers described by Hippocrates were indeed caused by the two particular species *P. vivax* and *P. malariae* (rather than, say, *P. ovale*).

The matter will be resolved by comparing the genetic diversity of the human and primate parasites. Genetic diversity will be greater in the donor host than in the recipient host of the switch. If the transfer has been from human to monkeys, the amount of genetic diversity, particularly at silent nucleotide sites and other neutral polymorphisms, will be much greater in *P. vivax* than in *P. simium*, and in *P. malariae* than in *P. brasilianum* (including in each comparison the polymorphisms present in the various monkey host species). A transfer from monkey to humans should yield much lower polymorphism in the human than in the monkey parasites. Due to their lesser role in human mortality and morbidity, these malaria species have not garnered the attention that has been lavished on *P. falciparum*, and so very little genetic diversity data are available for *P. vivax* and *P. malariae*, and far, far less for *P. brasilianum* and *P. simium*. Acquisition of this kind of data will be of great benefit in evaluating the origins of human malaria and in determining whether animals may serve as disease reservoirs.

In addition to expanding the breadth of genomic sequence data from well-characterized malaria species, it is also imperative that more taxa be included in phylogenetic surveys. Exclusion of taxa may lead to biased results in determining relationships among species. For example, Waters *et al.* (1991) found that the human parasite *P. falciparum* is most closely related to the avian *Plasmodium* parasites and hence concluded that "infection by *P. falciparum* is a recent acquisition of humans and possibly coincident with the onset of agriculture-based life style". In a follow-up study utilizing the same dataset, however, with the inclusion of a sequence from the chimpanzee parasite *P. reichenowi*, Escalante and Ayala (1994) found that the closest relative of *P. falciparum* is *P. reichenowi* and not the avian species. The latter study demonstrated conclusively that *P. falciparum* has been associated with the human lineage since a time that predates the split of humans and chimpanzees.

One of the most comprehensive molecular phylogenetic surveys of *Plasmodium* species to date is that of Perkins and Schall (2002), who examined the mtDNA *cytochrome-b* (*cyt-B*) gene from 52 species of mammalian, avian and saurian parasite species of the genus *Plasmodium* and three related genera (*Haemoproteus*, *Hepatocystis*, and *Leucocytozoon*). The resulting phylogeny (Figure 3) revealed two important conclusions, (1) *Plasmodium* spp. are paraphyletic with respect to *Haemoproteus* and *Hepatocystis* parasites, and (2) *P. falciparum* and *P. reichenowi* belong to a monophyletic group of mammalian parasites at the exclusion of avian and

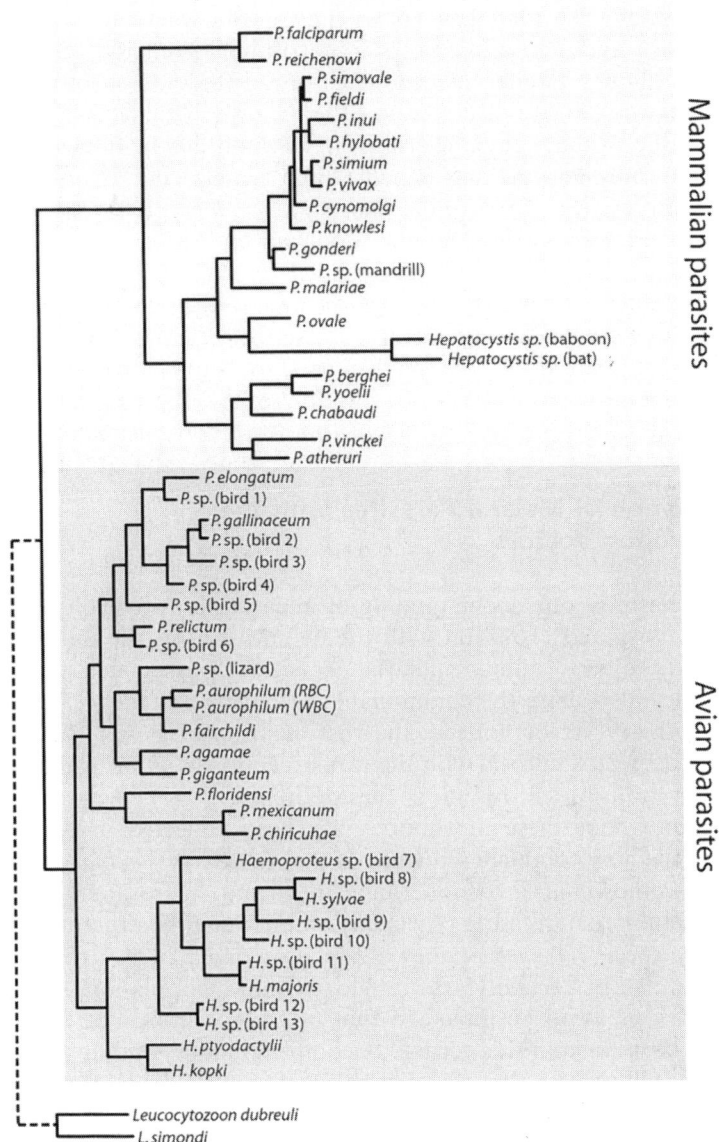

Figure 3 Phylogeny of several species of parasites representing four genera (*Plasmodium*, *Haemoproteus*, *Hepatocystis*, and *Leucocytozoon*) (Perkins and Schall, 2002). The dash line connecting the Leucocytozoon outgroup indicates that this branch is not drawn to scale with genetic distances (the actual branch would be much longer).

saurian species. The phylogeny also suggests that the clade consisting of *P. reichenowi* and *P. falciparum* is quite distinct from the other primate malarias, although likelihood ratio tests (using the method of Shimodaira and Hasegawa (1999)) failed to reject the hypothesis that these two species belong to the primate clade (Perkins and Schall, 2002). It should also be noted that while Perkins and Schall (2002) criticize the taxon-bias shortcomings of previous efforts to determine *Plasmodium* phylogeny, their own dataset excludes *P. brasilianum*, an important and quite possibly a human parasite whose position is controversial as explained in preceding paragraphs. Of course, no single study will be able to include all known-species let alone those species that are yet to be described. Limitations due to taxon bias can be minimized by using phylogenetic tools to test hypotheses and in doing so, making certain that the assumptions of the inquiry and the scientific design of the experiment are sufficient to achieve a robust conclusion.

2.2. Affiliation of Malaria Parasites with their Arthropod Vectors

Another chasm in our understanding of malaria parasite evolution is with regard to its shared evolution with the definitive host, the mosquito. Not only are the diverse human malaria parasites all associated with human disease, they also share the commonality of having an *Anopheles* mosquito as their primary vector. Indeed, the vast majority of *Plasmodium* spp. are transmitted by mosquitoes, with the rare exception of a few lizard malarias which are transmitted by sand flies (Kimsey, 1992). A great deal of information about these mosquitoes has been collected in recent years, including the now complete sequence of the *Anopheles gambiae* (Holt et al., 2002). Chromosomal inversion, allozymes, and microsatellites have all proven useful in determining phylogeny and population structure of several anopheline species (Powell et al., 1999; Walton et al., 2000; Krzywinski and Besansky, 2002). Certainly, the phylogeny and population structure of malaria parasites will be linked to that of their vectors, but to date little work has been done in this regard. A complete understanding of evolution of malaria will require that this gap be filled.

3. EVOLUTION AND EXTANT DISTRIBUTION OF MALIGNANT HUMAN MALARIA: *P. FALCIPARUM*

The classical designation of *P. falciparum* as the agent of "malignant" malaria is due to the severity of disease it causes relative to the other three human parasites. It is among the greatest threats to global human health, particularly in the developing world. With as many as 500 million malaria cases per year (among which >2 million will result in death of the patient), the burden on these developing countries is staggering (Trigg and Kondrachine, 1998). Despite more than a century of biomedical research, which saw unprecedented examples of international collaboration to eradicate the disease, the situation only seems to be worsening as drug-resistant parasites and pesticide-resistant mosquitoes dominate the landscape.

Indeed, *P. falciparum* has demonstrated remarkable adaptive potential in overcoming every effort to thwart its transmission. Novel strategies are currently in development. These include the innovation of new therapeutic modalities (Macreadie *et al.*, 2000; Price, 2000), development of protective vaccines (Guerin *et al.*, 2002; Moorthy and Hill, 2002; Plebanski *et al.*, 2002), and efforts to develop refractory mosquito vectors (Atkinson and Michel, 2002; Ito *et al.*, 2002). Choosing the most effective means of reducing malaria transmission, for example selecting among the more than 40 vaccines currently in development, will require careful consideration of the parasites' ability to circumvent intervention. Accordingly, it is crucial that we look first to the evolutionary processes that have facilitated the long-term association of the parasite and its human host. Understanding this evolutionary past will help to interpret how genetic variation among extant *P. falciparum* parasites actuates to become adaptive response.

In stark contrast to the situation in the other human malaria parasites, *P. falciparum* has shared a parallel evolutionary trajectory with its chimpanzee counterpart, *P. reichenowi*. *P. falciparum* is distinct from, although closely related to *P. reichenowi*. Indeed, *P. falciparum* is more closely related to *P. reichenowi* than it is to any other *Plasmodium* species. The time of divergence between these two *Plasmodium* species is estimated at 5–7 million years (MY) ago, which is roughly consistent with the time of divergence between the two host species, human and chimpanzee (Escalante and Ayala, 1995). A parsimonious interpretation of this state of affairs is that *P. falciparum* is an ancient human parasite associated with our ancestors at least since the divergence of the hominids from the great apes, and that the divergence of *P. falciparum* and *P. reichenowi* was concurrent with the divergence of their host species, humans and chimps.

Since *P. falciparum* seemingly shares a long-term evolutionary history with its human host, investigators have hypothesized that some of the allelic forms of *P. falciparum* antigenic determinants may date back even beyond the origin of the human species (Hughes, 1993; Hughes and Hughes, 1995). This situation would be analogous to that found in human HLA loci, where extant allelic variants are older than the human species

(Ayala et al., 1994). An abundance of epidemiological information suggests that extant populations of *P. falciparum* are remarkably variable. However, most studies of genetic diversity have focused on genes encoding either antigenic determinants or drug resistance factors. Moreover, much of the observed diversity is attributable to changes that have occurred in nucleotide positions leading to changes in the encoded amino acid, i.e. nonsynonymous or replacement substitutions. Replacement polymorphisms in antigenic genes are usually under strong diversifying selection imposed by the host's immune system (Escalante et al., 1998). Since nucleotide variants that alter amino acid sequence of antigenic genes may be maintained by selection, their rates of evolution are likely to be quite erratic, and hence they do not readily lead to accurate estimation of the age of species (Rich et al., 2000). Silent-site or synonymous polymorphisms do not alter the encoded amino acid and so are thought to be largely neutral with respect to natural selection. Accordingly, nucleotide substitutions at these sites occur at steady rate through geological time periods, as a function of the mutation rate and time elapsed. The rate of such substitutions can be obtained empirically by counting differences among gene sequences from species for which the divergence time is known.

3.1. Malaria's Eve Hypothesis

We sought to determine the age of the extant distribution of *P. falciparum* by measuring the degree of synonymous polymorphisms among isolates from globally distributed locations (Rich et al., 1998). Surprisingly, we found a paucity of such polymorphism, and based on these observations we estimated that the current distribution of *P. falciparum* throughout the world's tropical regions is derived from a small ancestral population within the very recent past. We determined the upper confidence interval of the age of this recent ancestry at 8000–60,000 years (Rich et al., 1998, 2000). We referred to this conclusion as the Malaria's Eve hypothesis.

In the few short years following our first report of Malaria's Eve, the issue has created a contentious debate. Our initial conclusion was based on sequences that were then available from GenBank, and the only criteria for inclusion of genes in our dataset was that they had to be void of repetitive DNA sequences and show no evidence of being under positive selection. At that time (1998), the amount of sequence data available for the species was rather limited, but since that time this dataset has grown exponentially, culminating in the complete genome sequence of *P. falciparum* published in 2002 (Gardner et al., 2002). This has spurred other investigators to carefully scrutinize Malaria's Eve hypothesis.

One of these studies entailed a large scale sequencing survey of 25 introns located on the second chromosome, from eight *P. falciparum* isolates collected among global sites (Volkman *et al.*, 2001). The findings of this study confirmed our previous result: there is an extreme scarcity of silent site polymorphism among extant distributions of *P. falciparum*. Among some 32,000 nucleotide sites examined, Volkman *et al.* (2001) found only 3 silent single nucleotide polymorphisms (SNPs) and concluded that the age of Malaria's Eve was somewhere between 3200 and 7700 years, depending on the calibration of the substitution clock.

Conway *et al.* (2000) have presented further evidence in support of Malaria's Eve based on analysis of the *P. falciparum* mitochondrial genome. They examined the entire mitochondrial DNA (mtDNA) sequence of *P. falciparum* isolates originating from Africa (NF54), Brazil (7G8), and Thailand (K1 and T9/96), as well as the chimpanzee parasite, *P. reichenowi*. Alignment of the four complete mtDNA sequences (5965 bp) showed that 139 sites contain fixed differences between *falciparum* and *reichenowi*, whereas only 4 sites were polymorphic within *falciparum*. The corresponding estimates of divergence (K, between *P. reichenowi* and *P. falciparum*) and diversity (π, within *P. falciparum* strains), are 0.1201 and 0.0004, respectively. In short, divergence in mtDNA sequence between the two species is 300-fold greater than the diversity within the global *P. falciparum* population. If we use the rDNA-derived estimate of 8 my as divergence time between *P. falciparum* and *P. reichenowi*, then the estimated origin of the *P. falciparum* mtDNA lineages is 26,667 years (i.e. 8 my/300), which corresponds quite well with our estimate based on 10 nuclear genes (Rich *et al.*, 1998). In a subsequent survey of a total of 104 isolates from Africa ($n = 73$), Southeast Asia ($n = 11$), and South America ($n = 20$); Conway *et al.* (2000) determined that the extant global population of *P. falciparum* is derived from three mitochondrial lineages that started in Africa, and migrated subsequently (and independently) to South America and Southeast Asia. Each mitochondrial lineage is identified by a unique arrangement of the 4 polymorphic mtDNA nucleotide sites.

Arguments against the Malaria's Eve hypothesis come in two forms. The first argument is that the loci chosen in the studies described above are a biased sample and do not reflect the levels of polymorphism in the genome as a whole. The second counterargument concedes that nucleotide polymorphisms are scarce, however this is not attributable to recent origins, but rather reflects strong selection pressure against the occurrence of synonymous SNPs.

Another study reports an "ancient" origin of a *P. falciparum* is based on a survey of sequences available from the GenBank database (Hughes and Verra, 2001). As with the data in our original paper (Rich *et al.*, 1998), these

GenBank sequences are compiled from a variety of sources and many of the entries may contain sequencing errors associated with *taq* misincorporation during the PCR amplification of alleles. Moreover, some of the sequences included in the Hughes and Verra paper were not carefully examined, and the comparisons include multiple nucleotide sequences from a single clone derived in different laboratories. For example, GenBank entries AF239801 and AF282975 are both *falcipain-2* sequences from *P. falciparum* clone W2. Regardless of possible errors, the overwhelming message from their compiled data is that there is indeed a dearth of polymorphism. In fact, among the 23 loci examined, which comprised over 10,000 codons, only six contained synonymous SNPs in 4-fold degenerate codons. Nonetheless, Hughes and Verra (2001) conclude that time to most recent common ancestry of *P. falciparum* must be 300,000–400,000 years.

A most ambitious effort to quantify polymorphism in *P. falciparum* involved a survey of > 200 kb from the completed chromosome 3 (Mu *et al.*, 2002). The authors reported 31 and 62 polymorphisms among 80,415 noncoding and 192,400 synonymous nucleotide sites, respectively. Using the equation and mutation rates from our paper (Rich *et al.*, 1998), Mu *et al.* (2002) estimated that the common ancestor to be between 102,000 and 177,000 years old. At this level of polymorphism, i.e. 62 of 192,400 (or 0.03%), the error rate in PCR and sequencing becomes relevant and bears great impact on estimates of recent ancestry. Mu *et al.* (2002) re-amplified and re-sequenced 56 of the SNP (both synonymous and nonsynonymous) containing regions and in this second pass found that 2 of the SNPs were in error (an error rate of ~4.0%). Because of this, the previously described paper by Volkman *et al.* (2001) incorporated a highly redundant approach to assure integrity of the data. Their methods involved meticulous bi-directional sequencing of three clones from each of three independent PCR amplifications, or an 18-fold redundancy (Hartl *et al.*, 2002).

Another concern about calculation of the age of Malaria's Eve pertains to the estimation of mutation rates. The estimates used by Mu *et al.* (2002) are from a very small number of nucleotides (708 bp) compared between the rhoptry-associated protein gene of *P. falciparum* and *P. reichenowi* (Rich *et al.*, 1998). The neutral mutation rate may vary among chromosomal regions and its estimation is subject to sampling error. Even slight perturbations in its calculation will have exponential effects on estimation of age of the common ancestor. Reliable estimates of the mean age of the recent common ancestor are in the range of 4000 to 380,000 years (Table 1). While at first glance these differences of nearly two-orders of magnitude appear unsatisfactory, the differences are, in fact, quite small in light of the 5 MY age of the species dating back to its split with the chimpanzee parasite. This means that the global, extant distribution of *P. falciparum*,

Table 1 Summary of results from various age estimates of the most recent common ancestor (MRCA) of extant populations of *P. falciparum*.

Reference	Source basis	Mean estimate ± 95% C.I. of MRCA (10^5 years)*	
		5 MY	7 MY
Rich et al., 1998	Non-antigenic housekeeping coding regions, synonymous sites	0 (±0.38)	0 (±0.53)
Conway et al., 2000	mtDNA (complete genome)	(na) (na)	0.267** (na)
Volkman et al., 2001	Introns on chromosome 2	0.032 (na)	0.077 (na)
Hughes and Verra, 2002	Antigenic and housekeeping coding sequences, synonymous sites	2.94 (±0.9)	3.85 (±0.77)
Mu et al., 2002	Coding regions of chromosome 3	1.26 (±0.46)	1.77 (±0.33)

*age estimates are provided for two different calibrations of the substitutional clock based on the two estimated divergence times (5 and 7 MY) of the *P. falciparum* – *P. reichenowi* split (see Rich et al., 1998 for details).
**estimate based on 8 MY divergence of *P. falciparum*–*P. reichenowi* as described in text.

with its abundant diversity of antigens and drug resistance factors, originated in only a small fraction (at most ~5%) of the time since the origin of the species. This finding contrasts greatly with the previous estimates of some antigenic variation to the order of 35 MY old (Hughes and Hughes, 1995; Rich et al., 2000).

Despite discrepancies in the estimation of age of the Malaria's Eve common ancestry, it is clear that nucleotide polymorphisms are scarce in many portions of the *P. falciparum* genome (Conway and Baum, 2002; Hartl et al., 2002). A second criticism of the recent origins hypothesis concedes the paucity of synonymous site polymorphism, but attributes this to constraints on the genome itself. One proposition is that the extreme AT content of the *P. falciparum* genome may suggest that some constraint is acting upon mutations that lead to unfavorable codon sequences (Arnot, 1991; Saul and Battistutta, 1988; Saul, 1999). As we have argued elsewhere, this does not seem to be the case, since in spite of AT content as high as 84% in third positions, there appears to be an equal proportion of A and T nucleotides in third positions of four-fold degenerate codons (Rich and Ayala, 1999, 2000). Moreover, the fact that synonymous substitutions are in evidence in the divergence between *P. falciparum* and *P. reichenowi* (which has a similarly extreme AT content), indicates that mutations can and do occur (Rich and Ayala, 1999).

3.2. Constraints on Synonymous Substitutions

Hartl et al. (2002) have pointed out that genomic constraints seem unlikely given the variability of microsatellite markers among introns, intergenic regions and in some cases, coding sequences (Su and Wellems, 1996; Anderson et al., 1999, 2000; Volkman et al., 2001). Nonetheless, Forsdyke (2002) has argued that the extreme conditions of the *P. falciparum* genome present a situation where selection for genomic composition exceeds the selection on the proteins encoded by these genes. The argument is leveled not so much against the Malaria's Eve hypothesis in particular, but rather the author attempts to refute the notion that neutral evolution is even possible. This warrants further discussion.

In an attempt to assign adaptive significance to the occurrence of a simple–repetitive sequence element (the Epstein-Barr Nuclear Antigen-1, *EBNA-1*) in the genome of the Epstein-Barr virus (EBV), Forsdyke (2002) argues that the selective pressure for particular genomic content and/or arrangement supercedes the selection acting on encoded proteins (phenotype). The *EBNA-1* can be removed from the genome without any loss of function in the virus. Because EBV, like most viruses, tends to lose extraneous genetic elements nonessential to its survival, Forsdyke (2002) maintains that the *EBNA-1* must have a function other than that typically assigned to genes, i.e., to encode message. To establish this fact, he has developed several descriptive parameters that are based on the nucleotide composition and secondary-folding potential of nucleotide sequences. These parameters are termed as potential "pressures" acting on the genome to maintain a particular configuration and/or composition. Forsdyke (2002) tested whether the region in question has extraordinary values for the pressure-parameters, and found that in the EBNA-1 region, there was an excessive skew in purine content (A and G), which would limit the potential for folding of the molecule and hence reduce recombination. The potential benefit of this situation is not explained and its biological relevance remains unclear.

The analysis of the EBV provided the analytical bases of Forsdyke's claim that *P. falciparum* is under pressure for reduced nucleotide polymorphism. He chose to examine the individual sequence content of two *P. falciparum* genes coding for surface antigens, *Csp* and *Msp-2* (merozoite surface protein-2). As with the EBNA-1, he found that there was a high bias toward purines (primarily A in this case) and a strong potential for secondary folding within the repetitive regions of both *Msp-2* and *Csp*. The only conclusion drawn from this was that the high folding potential might enhance recombination in the repeat regions of both genes. The model is neither predictive nor explanatory, and even offers very little in the way of descriptive value. If it were demonstrated that these extraordinary pressure regions had significantly less (or greater) synonymous site polymorphism, and that pressure was predictive of this polymorphism, the author's claim may bear some relevance. However, neither of these claims can be made particularly because the author chose to examine two of the most highly polymorphic loci known in *P. falciparum*. What is clear is that silent site polymorphisms are in evidence in non-*falciparum* malaria species, and that synonymous substitutions have occurred in the evolution of *P. falciparum* and *P. reichenowi*. On this basis, we maintain that while substitutions may be constrained due to nucleotide composition and/or codon usage bias, these constraints do not explain the paucity of *P. falciparum* synonymous site variation. Therefore, the Malaria's Eve hypothesis remains the most likely explanation for this state of affairs.

In addition to the analyses of genetic polymorphism data, there is independent information in support of the Malaria's Eve hypothesis. Sherman (1998) notes the late introduction and low incidence of *falciparum* malaria in the Mediterranean region. Hippocrates (460–370 B.C.) describes quartan and tertian fevers, but there is no mention of severe malignant tertian fevers, which suggests that *P. falciparum* infections did not yet occur in classical Greece, as recently as 2400 years ago. Interestingly, Tishkoff *et al.* (1997) traced the origin of malaria-resistant *G-6pd* genotypes in humans to the spread of agricultural societies some 5000 years ago. The recent origin of this mutation in humans suggests a similarly recent association with widespread exposure to the malaria parasite.

How can we account for a recent demographic sweep of *P. falciparum* across the globe, given its long-term association with the hominid lineage? One likely hypothesis is that human parasitism by *P. falciparum* has long been highly restricted geographically, and has dispersed throughout the Old World continents only within the last several thousand years, perhaps within the last 10,000 years, after the Neolithic revolution (Coluzzi 1994, 1997, 1999). Three possible scenarios may explain this historically recent dispersion: (1) changes in human societies, (2) genetic changes in the host–parasite–vector association that have altered their compatibility, and (3) climatic changes that entailed demographic changes (migration, density, etc.) in the human host, the mosquito vectors, and/or the parasite.

One factor that may have impacted the widespread distribution of *P. falciparum* in human populations from a limited original focus, probably in tropical Africa, may have been changes in human living patterns, particularly the development of agricultural societies and urban centers that increased human population density (Livingston, 1958; Weisenfeld, 1967; de Zulueta *et al.*, 1973; de Zulueta, 1994; Coluzzi, 1997, 1999; Sherman, 1998). Genetic changes that have increased the affinity within the parasite-vector-host system are also a possible explanation for a recent expansion, not mutually exclusive with the previous one. Coluzzi (1997, 1999) has cogently argued that the worldwide distribution of *P. falciparum* is recent and has come about, in part, as a consequence of a recent dramatic rise in vectorial capacity due to repeated speciation events in Africa of the most anthropophilic members of the species complexes of the *Anopheles gambiae* and *A. funestus* mosquito vectors. Biological processes implied by this account may have been associated with, and even dependent on the onset of agricultural societies in Africa (scenario 1) and climatic changes (scenario 3), specifically gradual increase in ambient temperatures after the Würm glaciation, so that about 6,000 years ago climatic conditions in the Mediterranean region and the Middle East made the spread of *P. falciparum* and its vectors beyond tropical Africa possible (de Zulueta *et al.*, 1973; de Zulueta, 1994; Coluzzi, 1997, 1999). The three scenarios are likely interrelated. Once demographic and climatic conditions became suitable for propagation of *P. falciparum*, natural selection would have facilitated evolution of *Anopheles* species that were highly anthropophilic and effective *falciparum* vectors (de Zulueta *et al.*, 1973; Coluzzi, 1997, 1999).

4. CONCLUDING REMARKS

Traditional approaches to the study of infectious disease were necessarily partitioned into a diverse array of disciplinary approaches corresponding roughly to the hierarchical arrangement of biological systems ranging from examination of individual molecules to study of epidemiological patterns of pathogen transmission among host populations. This had the unfortunate consequence of creating an often-disjointed understanding of human disease. At present, it is possible to perceive the biological aspects of disease transmission in a more astute fashion by incorporating information that spans this hierarchy: from genomes to populations of pathogenic organisms and hosts. This opportunity arises as a result of the confluence of several scientific disciplines, including population genetics and molecular biology into a unified field of genomics. Phylogenetic analyses play an important role in this new science, by allowing us not only to determine relatedness of individual pathogen, but also in sorting out the interrelatedness of genome elements which themselves have arisen by phyletic processes within the lines of descent leading to the organisms that possess them.

REFERENCES

Anderson, T.J., Su, X.Z., Bockarie, M., Lagog, M. and Day, K.P. (1999). Twelve microsatellite markers for characterization of *Plasmodium falciparum* from fingerprick blood samples. *Parasitology* **119**, 113–125.

Anderson, T.J., Su, X.Z., Roddam, A. and Day, K.P. (2000). Complex mutations in a high proportion of microsatellite loci from the protozoan parasite *Plasmodium falciparum*. *Molecular Ecology* **9**, 1599–1608.

Arnot, D.E. (1991). Possible mechanisms for the maintenance of polymorphisms in *Plasmodium* populations. *Acta Leidensia* **60**, 29–35.

Atkinson, P.W. and Michel, K. (2002). What's buzzing? Mosquito genomics and transgenic mosquitoes. *Genesis* **32**, 42–48.

Ayala, F.J., Escalante, A., O'Huigin, C. and Klein, J. (1994). Molecular genetics of speciation and human origins. *Proceedings of the National Academy of Sciences, USA* **91**, 6787–6794.

Ayala, F., Escalante, A., Lal, A. and Rich, S. (1998). Evolutionary Relationships of Human Malarias. In: *Malaria: Parasite Biology, Pathogenesis, and Protection* (I.W. Sherman, ed.), pp. 285–300. Washington, D.C.: American Society of Microbiology.

Barta, J.R. (1989). Phylogenetic analysis of the class Sporozoea (phylum Apicomplexa Levine, 1970): evidence for the independent evolution of heteroxenous life cycles. *Journal Parasitology* **75**, 195–206.

Bensch, S., Stjernman, M., Hasselquist, D., Ostman, O., Hansson, B., Westerdahl, H. and Pinheiro, R.T. (2000). Host specificity in avian blood parasites: a study of *Plasmodium* and *Haemoproteus* mitochondrial DNA amplified from birds. *Proceedings of the Royal Society of London, Series B*, **267**, 1583–1589.
Coluzzi, M. (1994). Malaria and the afro-tropical ecosystems: impact of man-made environmental changes. *Parassitologia* **36**, 223–227.
Coluzzi, M. (1997). Evoluzione Biologica & i Grandi Problemi della Biologia, pp. 263–285. Accademia dei Lincei.
Coluzzi, M. (1999). The clay feet of the malaria giant and its African roots: hypotheses and inferences about origin, spread and control of *Plasmodium falciparum*. *Parassitologia* **41**, 277–283.
Conway, D.J., Fanello, C., Lloyd, J.M., Al-Joubori, B.M., Baloch, A.H., Somanath, S.D., Roper, C., Oduola, A.M., Mulder, B., Povoa, M.M., Singh, B. and Thomas, A.W. (2000). Origin of *Plasmodium falciparum* malaria is traced by mitochondrial DNA. *Molecular and Biochemical Parasitology* **111**, 163–171.
Conway, D.J. and Baum, J. (2002). In the blood–the remarkable ancestry of *Plasmodium falciparum*. *Trends in Parasitology* **18**, 351–355.
Corliss, J. (1994). An interim utilitarian ("user-friendly") hierarchical classification and characterization of the protists. *Acta Protozoologica* **33**, 1–51.
de Zulueta, J., Blazquez, J. and Maruto, J.F. (1973). Entomological aspects of receptivity to malaria in the region of Navalmoral of Mata. *Revista de Sanidad e Higiene Publica (Madrid)* **47**, 853–870.
de Zulueta, J. (1994). Malaria and ecosystems: from prehistory to posteradication. *Parassitologia* **36**, 7–15.
Escalante, A.A. and Ayala, F.J. (1994). Phylogeny of the malarial genus *Plasmodium*, derived from rRNA gene sequences. *Proceedings of the National Academy of Sciences, USA* **91**, 11373–11377.
Escalante, A.A. and Ayala, F.J. (1995). Evolutionary origin of *Plasmodium* and other Apicomplexa based on rRNA genes. *Proceedings of the National Academy of Sciences, USA* **92**, 5793–5797.
Escalante, A.A., Lal, A.A. and Ayala, F.J. (1998). Genetic polymorphism and natural selection in the malaria parasite *Plasmodium falciparum*. *Genetics* **149**, 189–202.
Fandeur, T., Volney, B., Peneau, C. and de Thoisy, B. (2000). Monkeys of the rainforest in French Guiana are natural reservoirs for *P. brasilianum/P. malariae* malaria. *Parasitology* **120**, 11–21.
Fast, N.M., Kissinger, J.C., Roos, D.S. and Keeling, P.J. (2001). Nuclear-encoded, plastid-targeted genes suggest a single common origin for apicomplexan and dinoflagellate plastids. *Molecular Biology and Evolution* **18**, 418–426.
Fast, N.M., Xue, L., Bingham, S. and Keeling, P.J. (2002). Re-examining alveolate evolution using multiple protein molecular phylogenies. *Journal of Eukaryotic Microbiology* **49**, 30–37.
Forsdyke, D. (2002). Selective pressures that decrease synonymous mutations in *Plasmodium falciparum*. *Trends in Parasitology* **18**, 411.
Gardner, M.J., Hall, N., Fung, E., White, O., Berriman, M., Hyman, R.W., Carlton, J.M., Pain, A., Nelson, K.E., Bowman, S., Paulsen, I.T., James, K., Eisen, J.A., Rutherford, K., Salzberg, S.L., Craig, A., Kyes, S., Chan, M.S., Nene, V., Shallom, S.J., Suh, B., Peterson, J., Angiuoli, S., Pertea, M., Allen, J., Selengut, J., Haft, D., Mather, M.W., Vaidya, A.B., Martin, D.M., Fairlamb, A.H., Fraunholz, M.J., Roos, D.S., Ralph, S.A., McFadden, G.I., Cummings, L.M.,

Subramanian, G.M., Mungall, C., Venter, J.C., Carucci, D.J., Hoffman, S.L., Newbold, C., Davis, R.W., Fraser, C.M. and Barrell, B. (2002). Genome sequence of the human malaria parasite *Plasmodium falciparum*. *Nature* **419**, 498–511.

Garnham, P.C.C. (1966). *Malaria Parasites and Other Haemosporidia*. Oxford, UK: Blackwell Scientific Publications.

Gilles, H.M. and Warrell, D.A. (1993). *Bruce-Chwatt's Essential Malariology*, 3rd ed. London: Edward Arnold.

Guerin, P.J., Olliaro, P., Nosten, F., Druilhe, P., Laxminarayan, R., Binka, F., Kilama, W.L., Ford, N. and White, N.J. (2002). Malaria: current status of control, diagnosis, treatment, and a proposed agenda for research and development. *Lancet Infectious Diseases* **2**, 564–573.

Gysin, J. (1998). Animal models: Primates. In: *Malaria: Parasite Biology, Pathogenesis, and Protection* (I.W. Sherman, ed.), pp. 419–441. Washington, DC: ASM Press.

Hartl, D.L., Volkman, S.K., Nielsen, K.M., Barry, A.E., Day, K.P., Wirth, D.F. and Winzeler, E.A. (2002). The paradoxical population genetics of *Plasmodium falciparum*. *Trends in Parasitology* **18**, 266–272.

Holt, R.A., Subramanian, G.M., Halpern, A., Sutton, G.G., Charlab, R., Nusskern, D.R., Wincker, P., Clark, A.G., Ribeiro, J.M., Wides, R., Salzberg, S.L., Loftus, B., Yandell, M., Majoros, W.H., Rusch, D.B., Lai, Z., Kraft, C.L., Abril, J.F., Anthouard, V., Arensburger, P., Atkinson, P.W., Baden, H., de Berardinis, V., Baldwin, D., Benes, V., Biedler, J., Blass, C., Bolanos, R., Boscus, D., Barnstead, M., Cai, S., Center, A., Chatuverdi, K., Christophides, G.K., Chrystal, M.A., Clamp, M., Cravchik, A., Curwen, V., Dana, A., Delcher, A., Dew, I., Evans, C.A., Flanigan, M., Grundschober-Freimoser, A., Friedli, L., Gu, Z., Guan, P., Guigo, R., Hillenmeyer, M.E., Hladun, S.L., Hogan, J.R., Hong, Y.S., Hoover, J., Jaillon, O., Ke, Z., Kodira, C., Kokoza, E., Koutsos, A., Letunic, I., Levitsky, A., Liang, Y., Lin, J.J., Lobo, N.F., Lopez, J.R., Malek, J.A., McIntosh, T.C., Meister, S., Miller, J., Mobarry, C., Mongin, E., Murphy, S.D., O'Brochta, D.A., Pfannkoch, C., Qi, R., Regier, M.A., Remington, K., Shao, H., Sharakhova, M.V., Sitter, C.D., Shetty, J., Smith, T.J., Strong, R., Sun, J., Thomasova, D., Ton, L.Q., Topalis, P., Tu, Z., Unger, M.F., Walenz, B., Wang, A., Wang, J., Wang, M., Wang, X., Woodford, K.J., Wortman, J.R., Wu, M., Yao, A., Zdobnov, E.M., Zhang, H., Zhao, Q., Zhao, S., Zhu, S.C., Zhimulev, I., Coluzzi, M., della Torre, A., Roth, C.W., Louis, C., Kalush, F., Mural, R.J., Myers, E.W., Adams, M.D., Smith, H.O., Broder, S., Gardner, M.J., Fraser, C.M., Birney, E., Bork, P., Brey, P.T., Venter, J.C., Weissenbach, J., Kafatos, F.C., Collins, F.H. and Hoffman, S.L. (2002). The genome sequence of the malaria mosquito *Anopheles gambiae*. *Science* **298**, 129–149.

Huff, C.G. (1938). Studies on the evolution of some disease-producing organisms. *Quarterly Reviews of Biology* **13**, 196–206.

Hughes, A.L. (1993). Coevolution of immunogenic proteins of *Plasmodium falciparum* and the host's immune system. In: *Mechanisms of Molecular Evolution* (A.G. Clark, ed.), pp. 109–127. Sunderland, MA: Sinauer Associates.

Hughes, M.K. and Hughes, A.L. (1995). Natural selection on *Plasmodium* surface proteins. *Molecular and Biochemical Parasitology* **71**, 99–113.

Hughes, A.L. and Verra, F. (2001). Very large long-term effective population size in the virulent human malaria parasite *Plasmodium falciparum*. *Proceedings of the Royal Society of London, Series B* **268**, 1855–1860.

Ito, J., Ghosh, A., Moreira, L.A., Wimmer, E.A. and Jacobs-Lorena, M. (2002). Transgenic anopheline mosquitoes impaired in transmission of a malaria parasite. *Nature* **417**, 452–455.
Kawamoto, F., Win, T.T., Mizuno, S., Lin, K., Kyaw, O., Tantulart, I.S., Mason, D.P., Kimura, M. and Wongsrichanalai, C. (2002). Unusual *Plasmodium malariae*-like parasites in southeast Asia. *Journal of Parasitology* **88**, 350–357.
Kimsey, R.B. (1992). Host association and the capacity of sand flies as vectors of lizard malaria in Panama. *International Journal for Parasitology* **22**, 657–664.
Krzywinski, J. and Besansky, N.J. (2002). Molecular systematics of *Anopheles*: from subgenera to subpopulations. *Annual Reviews in Entomology* **27**, 27.
Lang-Unnasch, N., Reith, M.E., Munholland, J. and Barta, J.R. (1998). Plastids are widespread and ancient in parasites of the phylum Apicomplexa. *International Journal for Parasitology* **28**, 1743–1754.
Livingston, F.B. (1958). Anthropological implications of sickle cell gene distribution in West Africa. *American Anthropologist* **60**, 533–560.
Macreadie, I., Ginsburg, H., Sirawaraporn, W. and Tilley, L. (2000). Antimalarial drug development and new targets. *Parasitology Today* **16**, 438–444.
Manwell, R. (1955). Some evolutionary possibilities in the history of the malaria parasites. *Indian Journal of Malariology* **9**, 247–253.
Margulis, L., McKhann, H. and Olendzenski, L. (1993). *Illustrated Guide of Protoctista*. Boston: Jones and Bartlett.
Mason, S.J., Miller, L.H., Shiroishi, T., Dvorak, J.A. and McGinniss, M.H. (1977). The Duffy blood group determinants: their role in the susceptibility of human and animal erythrocytes to *Plasmodium knowlesi* malaria. *British Journal of Haematology* **36**, 327–335.
Miller, L.H., Mason, S.J., Clyde, D.F. and McGinniss, M.H. (1976). The resistance factor to *Plasmodium vivax* in blacks. The Duffy blood-group genotype, FyFy. *New England Journal of Medicine* **295**, 302–304.
Miller, R.L., Ikram, S., Armelagos, G.J., Walker, R., Harer, W.B., Shiff, C.J., Baggett, D., Carrigan, M. and Maret, S.M. (1994). Diagnosis of *Plasmodium falciparum* infections in mummies using the rapid manual ParaSight-F test. *Transactions of the Royal Society of Tropical Medicine and Hygiene* **88**, 31–32.
Moorthy, V. and Hill, A.V. (2002). Malaria vaccines. *British Medical Bulletin* **62**, 59–72.
Mu, J., Duan, J., Makova, K.D., Joy, D.A., Huynh, C.Q., Branch, O.H., Li, W.H. and Su, X.Z. (2002). Chromosome-wide SNPs reveal an ancient origin for *Plasmodium falciparum*. *Nature* **418**, 323–326.
Obornik, M., Van de Peer, Y., Hypsa, V., Frickey, T., Slapeta, J.R., Meyer, A. and Lukes, J. (2002). Phylogenetic analyses suggest lateral gene transfer from the mitochondrion to the apicoplast. *Gene* **285**, 109–118.
Perkins, S.L. (2000). Species concepts and malaria parasites: detecting a cryptic species of *Plasmodium*. *Proceedings of the Royal Society of London, Series B* **267**, 2345–2350.
Perkins, S.L. and Schall, J.J. (2002). A molecular phylogeny of malarial parasites recovered from cytochrome b gene sequences. *Journal of Parasitology* **88**, 972–978.
Plebanski, M., Proudfoot, O., Pouniotis, D., Coppel, R.L., Apostolopoulos, V. and Flannery, G. (2002). Immunogenetics and the design of *Plasmodium falciparum* vaccines for use in malaria-endemic populations. *Journal of Clinical Investigation* **110**, 295–301.

Powell, J.R., Petrarca, V., della Torre, A., Caccone, A. and Coluzzi, M. (1999). Population structure, speciation, and introgression in the *Anopheles gambiae* complex. *Parassitologia* **41**, 101–113.

Price, R.N. (2000). Artemisinin drugs: novel antimalarial agents. *Expert Opinion on Investigational Drugs* **9**, 1815–1827.

Rich, S.M., Licht, M.C., Hudson, R.R. and Ayala, F.J. (1998). Malaria's Eve: evidence of a recent population bottleneck throughout the world populations of *Plasmodium falciparum*. *Proceedings of the National Academy of Sciences, USA* **95**, 4425–4430.

Rich, S.M. and Ayala, F.J. (1999). Reply to Saul. *Parasitology Today* **15**, 39–40.

Rich, S.M., Ferreira, M.U. and Ayala, F.J. (2000). The origin of antigenic diversity in *Plasmodium falciparum*. *Parasitology Today* **16**, 390–396.

Rich, S.M. and Ayala, F.J. (2000). Population structure and recent evolution of *Plasmodium falciparum*. *Proceedings of the National Academy of Sciences, USA* **97**, 6994–7001.

Ricklefs, R.E. and Fallon, S.M. (2002). Diversification and host switching in avian malaria parasites. *Proceedings of the Royal Society of London, Series B* **269**, 885–892.

Roberts, F., Roberts, C.W., Johnson, J.J., Kyle, D.E., Krell, T., Coggins, J.R., Coombs, G.H., Milhous, W.K., Tzipori, S., Ferguson, D.J., Chakrabarti, D. and McLeod, R. (1998). Evidence for the shikimate pathway in apicomplexan parasites. *Nature* **393**, 801–805.

Roos, D.S., Crawford, M.J., Donald, R.G., Fraunholz, M., Harb, O.S., He, C.Y., Kissinger, J.C., Shaw, M.K. and Striepen, B. (2002). Mining the *Plasmodium* genome database to define organellar function: what does the apicoplast do? *Philosophical Transactions of the Royal Society of London, Series B* **357**, 35–46.

Saul, A. and Battistutta, D. (1988). Codon usage in *Plasmodium falciparum*. *Molecular and Biochemical Parasitology* **27**, 35–42.

Saul, A. (1999). Circumsporozoite polymorphisms, silent mutations and the evolution of *Plasmodium falciparum*. *Parasitology Today* **15**, 38–39.

Sherman, I.W. (1998). A brief history of malaria and the discovery of the parasite's life cycle. In: *Malaria: Parasite Biology, Pathogenesis, and Protection* (I.W. Sherman, ed.), pp. 3–10. Washington, D.C.: American Society of Microbiology.

Shimodaira, H. and Hasegawa, M. (1999). Multiple comparisons of log-likelihoods with applications to phylogenetic inference. *Molecular Biology and Evolution* **16**, 1114–1116.

Su, X. and Wellems, T.E. (1996). Toward a high-resolution *Plasmodium falciparum* linkage map: polymorphic markers from hundreds of simple sequence repeats. *Genomics* **33**, 430–444.

Surolia, N., Ramachandra Rao, S.P. and Surolia, A. (2002). Paradigm shifts in malaria parasite biochemistry and anti-malarial chemotherapy. *Bioessays* **24**, 192–196.

Tish, K.N. and Pillans, P.I. (1997). Recrudescence of *Plasmodium falciparum* malaria contracted in Lombok, Indonesia after quinine/doxycycline and mefloquine: case report. *New Zealand Medical Journal* **110**, 255–6.

Touze, J.E., Fourcade, L., Pradines, B., Hovette, P., Paule, P. and Heno, P. (2002). Mechanism of action of antimalarials. Value of combined atovaquone/proguanil. *Médecine Tropicale* **62**, 219–224.

Trigg, P. and Kondrachine, A. (1998). The current global malaria situation. In: *Malaria: Parasite Biology, Pathogenesis, and Protection* (I.W. Sherman, ed.), pp. 11–22. Washington, DC.

Volkman, S.K., Barry, A.E., Lyons, E.J., Nielsen, K.M., Thomas, S.M., Choi, M., Thakore, S.S., Day, K.P., Wirth, D.F. and Hartl, D.L. (2001). Recent origin of *Plasmodium falciparum* from a single progenitor. *Science* **293**, 482–484.

Walton, C., Handley, J.M., Tun-Lin, W., Collins, F.H., Harbach, R.E., Baimai, V. and Butlin, R.K. (2000). Population structure and population history of *Anopheles dirus* mosquitoes in Southeast Asia. *Molecular Biology and Evolution* **17**, 962–974.

Waters, A.P., Higgins, D.G. and McCutchan, T.F. (1991). *Plasmodium falciparum* appears to have arisen as a result of lateral transfer between avian and human hosts. *Proceedings of the National Academy of Sciences, USA* **88**, 3140–3144.

Weisenfeld, S.L. (1967). Sickle-cell trait in human biological and cultural evolution. Development of agriculture causing increased malaria is bound to gene-pool changes causing malaria reduction. *Science* **157**, 1134–1140.

Phylogenies, the Comparative Method and Parasite Evolutionary Ecology

Serge Morand[1] and Robert Poulin[2]

[1]*CBGP (Centre de Biologie et de Gestion des Populations),*
Campus International de Baillarguet, CS 30 016,
34980 Montferrier sur Lez, France
[2]*Department of Zoology, University of Otago. P.O. Box 56,*
Dunedin, New Zealand

Abstract	282
1. Introduction	282
2. Phylogenetic Effects and Constraints, and the Need for Phylogenies	283
3. The Phylogenetically Independent Contrasts Method	284
4. Diversity and Diversification	287
5. The Phylogenetic Eigenvector Method	290
5.1. Example Using Mammals	290
6. The Study of Host–Parasite Co-adaptation Using The Independent Contrasts Method	292
7. The Study of Host–Parasite Co-adaptation Using PER	293
8. Scepticism About Comparative Methods: Why Bother with Phylogeny?	295
9. Phylogenetically Structured Environmental Variation	296
10. Conclusions	299
Acknowledgements	299
References	299

ABSTRACT

A growing number of comparative analyses in the field of parasite evolution and ecology have used phylogenetically based comparative methods. However, the comparative approach has not been used much by parasitologists. We present the rationale for the use of phylogenetic information in comparative studies, and we illustrate the use of several phylogenetically based comparative methods with case studies in parasite evolutionary ecology. The independent contrasts method is the most popular one, but presents some problems for studying co-adaptation between host and parasite life traits. The eigenvector method has been recently proposed as a new method to estimate and correct for phylogenetic inertia. We illustrate this method with an investigation of patterns of helminth parasite species richness across mammalian host species. This method seems to perform well in situations where host and parasite phylogenies are not perfectly congruent, but one might still want to correct for the effects of both. Finally, we present a method recently proposed for variation partitioning in a phylogenetic context, i.e. the phylogenetically structured environmental variation.

1. INTRODUCTION

One of the most powerful approaches to the study of adaptation and evolution in general involves comparing different species and searching for a predictable and consistent fit between their traits and some environmental variable of interest. This is the essence of the comparative method (Harvey and Pagel, 1991). It has now become almost globally accepted that one needs to take into account phylogenetic information in cross-species studies of adaptation (Felsenstein, 1985). The basic argument is mostly a statistical one, with phylogenetically closely related species not representing truly independent samples. Simulation studies and empirical analyses have shown that ignoring phylogenetic relationships among species included in a comparative analysis may lead to spurious conclusions due to high type I or type II errors (Gittleman and Luh, 1992; Purvis *et al.*, 1994; Diaz-Uriarte and Garland, 1996). A growing number of comparative analyses in the field of parasite evolution and ecology have used phylogenetically based comparative methods (Poulin, 1995, 1998; Morand, 2000), but their use is far from universal. Here, we present the rationale for the use of phylogenetic information in comparative studies, and we illustrate the use of several phylogenetically based comparative methods with case studies in parasite evolutionary ecology.

2. PHYLOGENETIC EFFECTS AND CONSTRAINTS, AND THE NEED FOR PHYLOGENIES

Several terms have been proposed in the literature to refer to a potential influence of evolutionary history, i.e. phylogeny, on the observed pattern of diversity: phylogenetic effects, phylogenetic constraints, phylogenetic niche conservatism or phylogenetic inertia.

Derrickson and Ricklefs (1988) placed much emphasis on the difference between phylogenetic effects and phylogenetic constraints. A phylogenetic effect is only the expression of the tendency of related species to be similar because they share a common history, whereas a phylogenetic constraint is the effect of history on the changes in diversification of a given clade (Derrickson and Ricklefs, 1988). In this sense phylogenetic effects and phylogenetic inertia are equivalent. However, according to McKitrick (1993), the definition of phylogenetic constraints, rather than placing the emphasis on the causes themselves, focuses more on the consequences of the constraints. McKitrick (1993) proposed the following definition in which a phylogenetic constraint is "any result or component of the phylogenetic history of a lineage that prevents an anticipated course of evolution in that lineage". This definition is close to the definition of the term exaptation proposed by Gould and Vrba (1982). The influence of an ancestor on its descendants has been termed phylogenetic inertia by Harvey and Pagel (1991). Finally, the concept of phylogenetic niche conservatism refers to shared attributes of phylogenetic inertia and ecological factors (Grafen, 1989; Harvey and Pagel, 1991).

So why are all these concepts so important for comparative analyses in parasite evolutionary ecology? Consider as an example the evolution of body sizes in parasitic nematodes. There is a wide range of adult body sizes among extant species of parasitic nematodes, from a couple of millimetres to several centimetres in length. Body size is the most influential determinant of parasite fecundity and therefore fitness (Skorping *et al.*, 1991; Morand 1996), and investigating what determines its evolution is thus extremely relevant. The existence of phylogenetic effects becomes apparent when one looks at the distribution of body sizes among higher taxa of nematodes. For instance, lets consider parasitic nematodes belonging to the families Oxyuridae and Ascarididae, which are common intestinal parasites of mammals. Whatever the host species in which they occur, members of the family Ascarididae are almost invariably larger-bodied than members of the family Oxyuridae (Anderson, 2000). Clearly, oxyurids must have inherited their small size from their common ancestor, whereas ascaridids have inherited their large size from their common ancestor. Species within a family are not independent of one another because they share traits simply by being related. It is essential to take phylogenetic relatedness into account in a comparative analysis of the evolution of body size in parasitic nematodes. This is true of practically all traits, in any taxon, whether parasitic or not. The additional problem that is mainly restricted to parasites and other obligate symbionts is that their traits are not only inherited from their ancestors, but their expression is also likely to be an adaptation to their environment, i.e. their hosts. Hosts have their own evolutionary history, i.e. their own phylogeny. Getting back to nematode body sizes, there is good evidence showing that adult body sizes covary with host sizes (Morand *et al.*, 1996); ignoring how host sizes themselves have evolved could lead to erroneous conclusions. Some host groups, such as bats and rodents among mammals, have similar body sizes despite having completely different phylogenetic origins, whereas more closely-related host species (e.g. among rodents, from tiny mice to the 50 kg capybara) may differ widely in body size despite a more recent phylogenetic divergence. So, in investigations of traits such as parasite body sizes, the ideal scenario might involve an analysis that takes into account *both* the phylogenies of hosts and parasites. Clearly, ignoring phylogeny altogether should not be an option anymore.

The major problem with comparative analyses of parasites has been that most of them are small and cryptic, without fossil records (except in rare exceptions), and their diversity has received much less attention than that of free-living animals (Poulin and Morand, 2000). For a long time, robust phylogenetic hypotheses were lacking for most groups, which limited the application of comparative approaches to studies of parasite ecology and evolution. However, recent developments in molecular phylogenetics have provided numerous historical frameworks that allow the investigation of parasite evolution in a proper phylogenetic context (e.g., Blaxter *et al.*, 1998; Littlewood and Bray, 2001).

3. THE PHYLOGENETICALLY INDEPENDENT CONTRASTS METHOD

The phylogenetically independent contrasts method (Felsenstein, 1985; Martins and Garland, 1991; Garland *et al.*, 1992) has been developed to resolve the problem of non-independence of data (i.e., traits measured across different species) in comparative studies. Felsenstein (1985) suggested a procedure for calculating comparisons between pairs of taxa at each bifurcation in a known phylogeny; since these bifurcations represent independent evolutionary events, contrasts between sister taxa issued from one bifurcation are thus independent from contrasts computed for other bifurcations. Since its development in 1985, the independent contrasts method has become the most widely used in comparative biology, even if sister group analyses can also be performed.

In a phylogenetic tree, the independent events (on which an analysis can be performed) correspond to ancestral (or internal) nodes that give rise to daughter branches. For each internal node, values for a given variable are obtained by averaging the values of its own daughter branches. Then the difference for each variable between the two daughter branches of each node is calculated. In the calculation of contrasts, the direction of subtraction is arbitrary. Multiple nodes (i.e. unresolved polytomies) can be treated in a way that gives a single contrast (Purvis and Garland, 1993). Pairs of sister branches that diverged a long time ago are likely to produce greater contrasts than pairs of sister branches that diverged recently. It is thus necessary to standardise each contrast through division by its standard deviation where the standard deviation of a contrast is the square root of the sum of the branch lengths issued from it (Garland et al., 1992). The main assumption of independent contrasts is a Brownian model of character evolution or random walk model. Changes in the mean phenotype are expected to occur at a constant rate and to be non-directional. The Brownian model of character evolution can be tested by regressing the absolute values of contrasts against the estimated nodal values. In the absence of information on branch length, one can assume each branch length to be equal to unity. Another method is proposed by Grafen (1989) for assigning arbitrary lengths. In this method the age of a node is assigned as the number of daughter groups descended from that node minus one. Nevertheless, Garland et al. (1992) showed that using arbitrary or real branch lengths often leads to similar results. In order to check that contrasts are properly standardised it is suggested to perform a regression of the absolute values of standardised contrasts versus their standard deviations. In case of positive relationship it is necessary to transform branch lengths before computing standard deviations (Garland et al., 1992). Contrasts can then be analysed using standard parametric tests, although all correlations between contrasts are forced through the origin. Non-parametric methods can also be used to analyse the contrasts (e.g. sign test). Several programs have been developed for performing independent contrasts analyses, e.g. CAIC (Purvis and Rambaut, 1995). Clearly, not taking into account the phylogenetic information may lead to spurious results, as illustrated by Morand and Poulin (1998) for the relationship between mammalian population density and parasite species richness. A non-phylogenetic approach (cross-species comparisons) leads to the conclusion that parasite species richness correlates negatively with mammalian population density (Figure 1A), whereas the use of the independent contrasts method showed a positive relationship (Figure 1B).

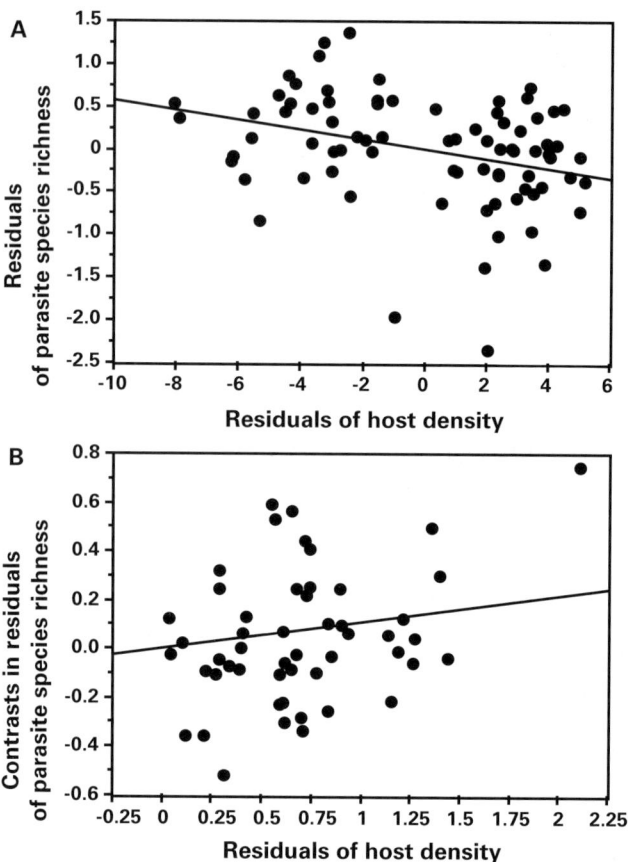

Figure 1 Relationship between host body density and parasite species richness (both variables controlled for host sampling effort) (A) using cross-species comparisons ($p < 0.01$), (B) using independent contrasts ($p < 0.05$) (redrawn after Morand and Poulin, 1998).

A robust and accurate phylogeny is the most important assumption of the independent contrasts method (Harvey and Pagel, 1991; Symonds, 2002) as well as for sister group analyses. Because of the general lack of phylogenetic information, the use of randomly generated phylogenetic trees has been proposed as an alternative (Losos, 1994; Martins, 1996; Abouheif, 1998). For instance, this procedure was used by Simkova *et al.* (2000) for the investigation of morphological adaptations of *Dactylogyrus* spp. to their Cyprinid hosts. However, a recent simulation study performed by

Symonds (2002) showed that random phylogenies may actually perform worse than analyses using raw data with no attempt to control for phylogeny.

With respect to comparative analysis of parasites, the problem is how to use the independent contrasts method when investigating the coadaptation of traits in hosts and their parasites, since ideally phylogenetic information from both hosts and parasites should be taken into account simultaneously. The independent contrasts method does not allow for this requirement. However, if the trait studied is clearly influenced much more by the phylogenetic history of parasites rather than hosts, or vice versa, then the independent contrasts method is a powerful tool for studies on parasites.

4. DIVERSITY AND DIVERSIFICATION

Instead of continuous traits, one may want to investigate differences in rates of diversification among related taxa. The study of biodiversity involves the measurement of net rates of diversification (rates of speciation minus rates of extinction). Extinction and speciation rates may differ among clades, and they may explain the relative "proliferation" of species in one clade compared to others. The search for causal links between key innovations and the diversification of platyhelminths (Brooks and McLennan, 1991) has stimulated several investigators to study diversification in parasite lineages.

Diversification rates can be tested using sister group analyses (Barraclough et al., 1999). Desdevises et al. (2001) tested whether the level of host specificity affects parasite species diversification within a monogenean family. They were the first to use the MacroCAIC program (Agapow and Isaac, 2002), derived from CAIC (Purvis and Rambaut, 1995), in a study on parasites. MacroCAIC is designed specifically for studies of diversification, and uses a basic approach based on phylogenetically independent contrasts. Desdevises et al. (2001) found no effect of host specificity on the species diversification of the group of monogeneans they investigated. Here, we investigate the effect of body size on monogenean diversification across all monogenean families using the data collected by Poulin (2002). Body size is often believed to be a key driver of diversification rates in animals in general (Orme et al. 2002a,b), and our example illustrates how this fundamental issue can be addressed with parasites in a phylogenetic context.

MacroCAIC allows one to use species richness as a variable in a comparative analysis to estimate whether other traits (here body size) are associated with high speciation rate (species richness, genus richness, or family richness within a higher clade). Species, genus, or family richness per clade cannot be used as any other continuous variable with independent contrasts because, in this case, the estimated richness value at each internal node in the phylogeny is not the average of the values of branches issued from the node, but their sum.

In our example using monogenean body size and diversification, the variables tested were:

- the natural log of the CLS/CHS ratio; where CLS is the species richness of the clade with lower mean body size, for each node of the phylogeny, and CHS is the species richness of its sister clade with larger mean body size;
- the natural log of the CLG/CHG ratio; where CLS is the number of genera in the clade with lower mean body size, for each node of the phylogeny, and CHS is the number of genera in its sister clade with larger mean body size;
- the natural log of the CLF/CHF ratio; where CLS is the number of families in the clade with lower mean body size, for each node of the phylogeny, and CHS is the number of families in its sister clade with larger mean body size.

When the ratio is 1, i.e., ln ratio $= 0$, the number of species (or the number of genera or families) in each sister clade is the same, and when the ratio is greater than one (positive ln ratio), the clade with the most species (or genera or families) is also the one with the highest mean body size. The analyses were then performed across pairs of sister clades at each internal node of the phylogeny.

The ln(CLS/CHS), the ln(CLG/CHG) and the ln(CLF/CHF) were regressed against standardised contrasts for ln(mean body size), using MacroCAIC. If the hypotheses tested are true, we should observe an increase of diversification with mean body size.

The regression equations using contrasts (forced through origin) were:

ln(CLS/CHS) $= -1.14$ ln(mean body size) ($R^2 = 0.023$, $p = 0.35$, Figure 2A).
ln(CLG/CHG) $= -0.50$ ln(mean body size) ($R^2 = 0.007$, $p = 0.62$, Figure 2B).
ln(CLF/CHF) $= -1.62$ ln(mean body size) ($R^2 = 0.13$, $p = 0.10$, Figure 2C).

Thus, no significant relationships were found between mean body size and diversification at any taxonomic level among the Monogenea, i.e. clades with higher species richness are not characterized by smaller or larger mean body size than their sister clades. Nevertheless, the methods are now available, and similar comparative studies of parasite diversification are now possible within other parasite groups with well-resolved phylogenies.

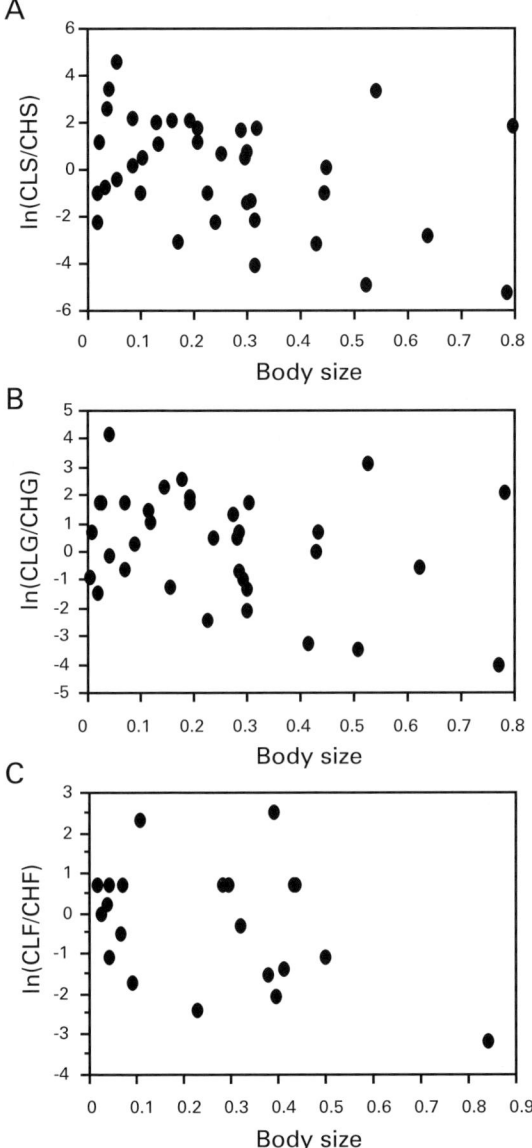

Figure 2 Relationships between the diversification ratio between sister clades of monogeneans (clade with high diversification/clade with low diversification) and mean body size (in ln), computed on independent contrasts using MacroCAIC. The analyses were performed at the level of (A) species, (B) genera and (C) families. None of the relationships were significant.

5. THE PHYLOGENETIC EIGENVECTOR METHOD

The phylogenetic eigenvector regression (PER) has been recently proposed by Diniz-Filho *et al.* (1998) as a new method to estimate and correct for phylogenetic inertia, based on an approach that is very different from that used when computing phylogenetically independent contrasts.

Diniz-Filho *et al.* (1998) only used the first principal coordinates selected with reference to a broken-stick model. All the principal coordinates extracted from the distance matrix that are significantly related to the dependent variable(s) can be used. The broken-stick model does not provide reasons to select variables of importance for the explanation of the dependent variable. As distances used might not always be Euclidean, negative eigenvalues may be produced (see Legendre and Legendre, 1998). In this case, it is possible to apply some correction methods, as presented in Gower and Legendre (1986) or Legendre and Legendre (1998). Again, a phylogenetic distance matrix can be obtained either from the raw data (e.g. sequence alignments), which avoids the reconstruction of a tree, or computed from a patristic distance matrix representing a phylogenetic tree. In simple terms, this method consists of transforming phylogenetic information into numbers that can then be used as any other variable.

A principal coordinate analysis (PCoA) is then performed on a pairwise phylogenetic distance matrix between species. Eigenvectors and eigenvalues are extracted from this analysis (Figure 3). Traits under analysis are regressed on eigenvectors retained by a broken-stick model and the residuals express the independent evolution of each species, whereas estimated values express phylogenetic trends in the data. Multiple regression analyses can also be performed using eigenvectors as predictors (see below).

5.1. Example Using Mammals

We illustrate this method with an investigation of patterns of helminth parasite species richness across mammalian host species. We used data on 79 mammal species (see Morand and Poulin (1998) for sources). Data on various life history traits were obtained from the literature (see Morand (2000) for sources). We also used recent advances in the knowledge of mammalian phylogenetic relationships (Cooper and Fortey, 1998; Murphy *et al.*, 2001) to derive a working phylogeny of the 79 species included in the analyses. In this case, we correct for host phylogeny, as we expect host features to influence the number of parasite species they harbour.

PARASITE EVOLUTIONARY ECOLOGY

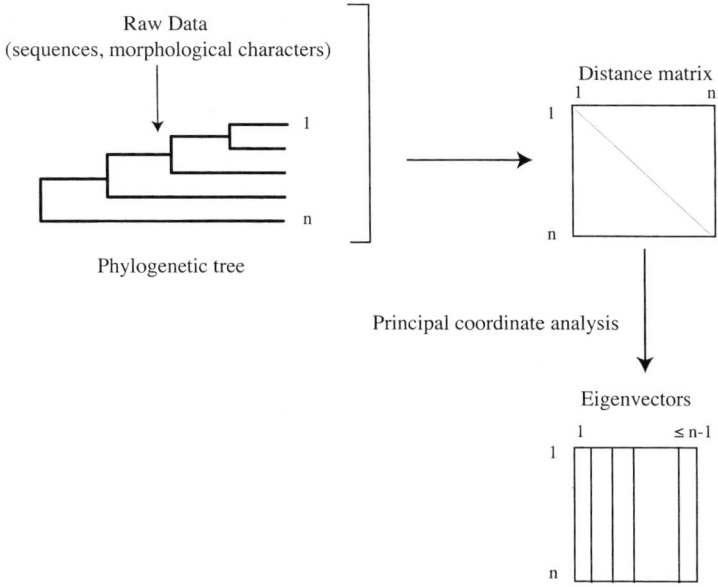

Figure 3 Principal coordinate analysis (PcoA) performed on phylogenetic distance matrix in order to extract eigenvectors representing the phylogenetic information (the maximum number of eigenvectors is $(n-1)$ but this may not always be the case) (after Desdevises, 2001).

Morand and Harvey (2000), using the independent contrasts method, have found that parasite species richness is positively correlated with the basic metabolic rate of mammal hosts and negatively correlated with mammal longevity. Basal metabolic rate (BMR) gives an indication of energy expenditure by animals, and represents the minimum energetic cost necessary for maintaining the activity of an organism. BMR, which scales with body mass, has been related to a great number of variables such as body mass, dietary habits, brain size, and reproductive strategies, and the reasons for the residual variation have been discussed (see Morand and Harvey, 2000).

We applied the PER to these data and all partial regression coefficients of the linear correlations are given in Figure 4. The subset of determinants was obtained using a stepwise regression with a backward elimination procedure.

First, we detected a phylogenetic effect on many of the host life history traits as depicted by linear correlations between these traits and the first principal coordinate of the phylogenetic matrix. It appears that the observed parasite species richness is not correlated with any of the principal eigenvectors. Second, the stepwise procedure allowed the selection of three

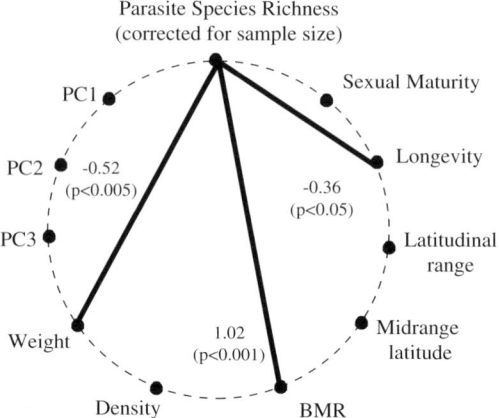

Figure 4 Diagram of the correlations between parasite species richness (corrected for host sampling size), host attributes (sexual maturity, longevity, weight, density, geographic distribution) and eigenvectors representing host phylogeny (principal coordinates: PC1, PC2, PC3), and results of a backward procedure on a multiple linear regression to select a subset of explanatory variables (determinants) of parasite species richness of mammals.

determinants of parasite species richness (corrected for sampling size), namely host body weight, host BMR and host longevity. We found a negative effect of body size and much more importantly a negative effect of host longevity ($R^2 = 0.58$, $p = 0.0011$). These results illustrate how the PER can be used in multivariate analyses including corrections for phylogenetic effects.

6. THE STUDY OF HOST–PARASITE CO-ADAPTATION USING THE INDEPENDENT CONTRASTS METHOD

Although the phylogenetically independent contrasts method does not allow for the simultaneous inclusion of two phylogenies, such as that of hosts and parasites, it can be used to control for both phylogenies in the unusual case where both phylogenies are perfectly congruent. In other words, in such cases one needs only to derive contrasts from one phylogeny to correct for the other one at the same time. This is particularly important when comparing the effect of one host trait (which may constrained by host phylogeny) on one parasite trait (which may constrained by parasite phylogeny).

Morand et al. (2000) examined the relationship between body size of pocket gophers and body size of their chewing lice (Hafner et al., 1994; Hafner and Page, 1995). Because chewing lice of mammals grasp the hair of the host in a semi-circular head groove, the head groove appears to be critically important to the louse's survival. Morand et al. (2000) hypothesised that the observed correlation between louse body size and gopher body size (Harvey and Keymer, 1991) may reflect a relationship between gopher hair-shaft diameter and louse head-groove dimensions, suggesting that there is a "lock-and-key" relationship between these two anatomical features.

They used the CAIC program to obtain independent contrast values for gopher and louse body sizes (Purvis and Rambaut, 1995). However, because the independent contrasts method compares the nodes of a single phylogeny, it was necessary that the host and parasite phylogenies being compared be topologically identical. Accordingly, Morand et al. (2000) restricted the analysis to the congruent portions of the gopher and louse phylogenies. They showed a positive relationship between gopher hair-shaft diameter and louse head-groove width and concluded that changes in body size of chewing lice may be driven by a mechanical relationship between the parasite's head-groove dimension and the diameter of the hairs of its host. Louse species living on larger host species may be larger simply because their hosts have thicker hairs, which requires that the lice have a wider head groove. This study of gopher hair-shaft diameter and louse head-groove dimensions suggests that there is a "lock-and-key" relationship between these two anatomical features. It also illustrates how the phylogenetically independent contrasts method can, in some cases, simultaneously control for *both* host and parasite phylogenies.

7. THE STUDY OF HOST–PARASITE CO-ADAPTATION USING PER

In most situations, however, host and parasite phylogenies are not perfectly congruent, but one might still want to correct for the effects of both. In these cases, PER provides a solution.

We investigate the covariation of life history traits between hosts and parasites. We reinvestigate the case of primates and their oxyurid parasites (Sorci et al., 1997). Sorci et al. (1997) tested the hypothesis of a positive covariation between parasite body length (reflecting parasite longevity) and host longevity using the independent contrasts method. However, this method can only be applied when there is a high degree of cospeciation in the host system; otherwise, it ignores the phylogenetic information of either hosts or parasites. In the present example, the phylogenies of primates and their oxyurid nematodes are not perfectly congruent (Figure 5), and it would be preferable to control for the influence of both phylogenies.

The use of eigenvalues allows this to be done. Two principal coordinate analyses (PCoA) are first performed on a pairwise phylogenetic distance matrix between species of primates and between species of oxyurid parasites (Figure 4). Eigenvectors and eigenvalues are extracted from this analysis.

We then regress life history traits on eigenvectors retained by a broken-stick model, which retains the statistically explaining eigenvectors (see Legendre and Legendre, 1998) independently for primates and for oxyurid parasites. All traits are then corrected for *all* phylogenetically confounding effects and then can be compared.

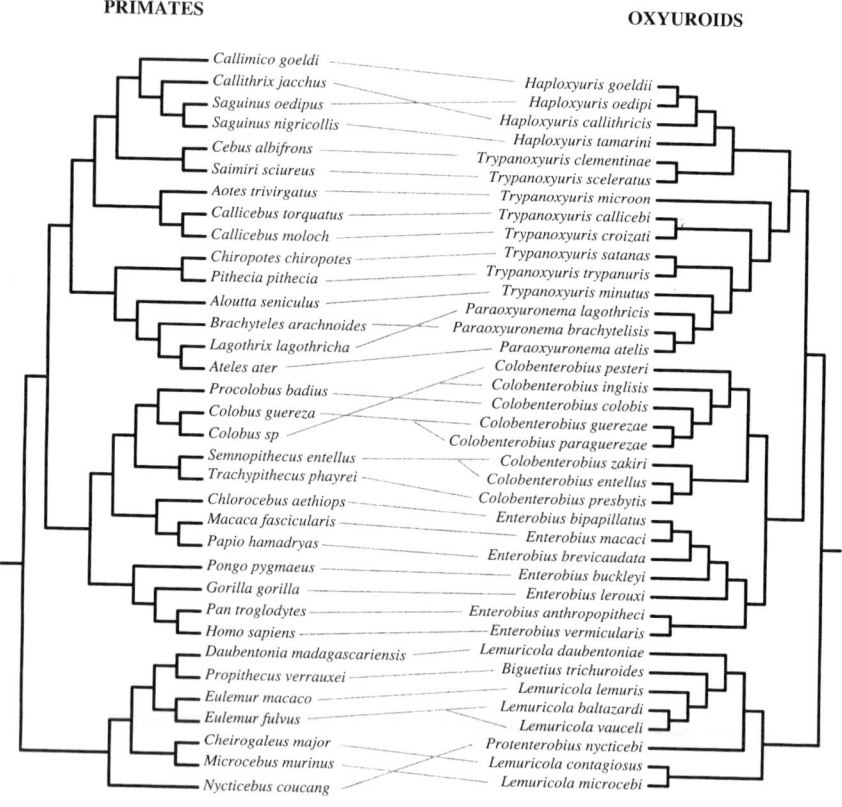

Figure 5 Tanglegram of phylogenies of oxyuroid nematodes and their primate hosts (after Hugot, 1999).

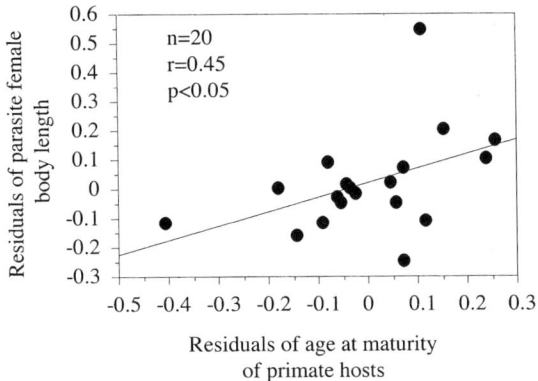

Figure 6 Relationship between female body length of oxyuroids and primate longevity (corrected for body weight). Both variables are controlled for phylogenetic confounding effects using PER (phylogenetically eigenvector regression) (see text).

We find that female parasite body length is positively correlated with host longevity (Figure 6) after correcting for host body mass and phylogeny using PER. This result confirms the previous finding of Sorci et al. (1997). However, the great advantages of this method are that it corrects for both host and parasite phylogenies and that it can be used in host–parasite system where the degree of cospeciation is very low (Desdevises et al., 2002a, 2002b) and even if the degree of host specificity is low.

8. SCEPTICISM ABOUT COMPARATIVE METHODS: WHY BOTHER WITH PHYLOGENY?

Several authors have questioned the use of phylogenetic comparative method (Ricklefs and Starck, 1996; Björklund, 1997; Price, 1997). Leroi et al. (1994) argued that comparative methods are "valuable for examining the evolutionary history of traits but they will often mislead in the study of adaptive processes". Their major concern was that we know very little about the evolutionary genetic mechanisms responsible for the distribution of traits among species. They claimed that it is very difficult to justify any evolutionary scenario without evidence of historical selection forces and, more important, the genetic relations among traits. Some of their arguments concern mainly the invocation of constraints in the explanation of either adaptation or phylogenetic conservatism. Their second criticism is about "the confounding of the causal influence of selection with that of genetic correlations". This is a more serious critique but, again, the problem applies more to inferences about the causality of the correlations than to the methods themselves. Indeed, Leroi *et al.* (1994) concluded their essay with the acknowledgement "that the methods of comparative biology and genetics might be usefully combined".

A different kind of criticism came from Westoby *et al.* (1995a, b). Their concern was that a phylogenetic correction (i.e. phylogenetically based comparative analysis) is not a correction, but rather a conceptual decision that gives priority to one interpretation over another. The comparative method, and particularly the independent contrasts method, assumes that part of the variation of a given trait is correlated with phylogeny and another part is correlated with ecology. They criticised that the procedure first removes the phylogenetic influences before estimating the influence of present-day ecological factors.

The problem is that we only know the actual phenotypes of species under the current selective regimes. We do not know what may have happened to ancestral species and what their phenotypes were. Hence, phylogenetic conservatism may be described as follows: "the ancestor of a lineage possesses a constellation of traits, enabling it to succeed in a particular habitat and disturbance regime, through a particular life history and physiology. The lineage will therefore leave most descendants in similar niches. This niche conservatism in turn will tend to sustain a similar constellation of traits in descendants of the lineage" (Westoby *et al.*, 1995a).

Harvey *et al.* (1995) tried to provide an answer to this criticism by emphasising that the independent contrasts method does not remove phylogenetic effects, but produces plots in which all the variation of the data set in one variable is graphed against all the variation in the other variable. In this way, phylogenetic niche conservatism means that adaptations to different components of the niche will be correlated (Harvey *et al.*, 1995). In any event, perhaps the best resolution of this debate would be to use methods that allow both the variance due to phylogenetic influences and that due to present ecological causes to be evaluated; we now turn to one such method.

9. PHYLOGENETICALLY STRUCTURED ENVIRONMENTAL VARIATION

Desdevises *et al.* (2001 (unpublished) and 2002a) have re-examined the controversy initiated by Westoby *et al.* (1995a) (but see also Ackerly and Donoghue, 1995; Fitter, 1995; Harvey *et al.*, 1995; Rees, 1995; Westoby et al., 1995b, c). Westoby et al. (1995a) pointed out that the phylogenetic portion of the total variance may contain a phylogenetic effect related to ecology, which Harvey and Pagel (1991) called "phylogenetic niche conservatism". This includes the shared attributes that related species have acquired because they tend to occupy similar niches during their evolutionary history. Westoby et al. (1995) proposed to partition the variance of the data set in three components (Figure 1 from Westoby et al., 1995a; here Figure 7, equivalent to fraction [a], [b] and [c]): a part strictly due to ecology ([a]), a part strictly due to phylogeny ([c]), and a part due to the common influence of these two factors ([b]), which was called "phylogenetically-structured environmental variation", an expression equivalent to "phylogenetic niche conservatism" as used by Desdevises (2001) and Desdevises et al. (2002a).

Desdevises et al. (2002a) proposed a method for variation partitioning in a phylogenetic context, which was already proposed by Borcard et al. (1992) and Borcard and Legendre (1994) for the analysis of spatially structured ecological data sets.

In a phylogenetic context, we express phylogeny as a distance matrix following PER previously discussed. The phylogeny is expressed in the form of principal coordinates computed from a patristic distance matrix, following Diniz-Filho et al. (1998).

The method used to partition the variance is as follows. Let Y be the dependent variable (i.e. parasite species richness or PSR), X_E the ecological explanatory variable(s) (i.e. BMR, host weight, host longevity), and PCs the principal coordinates representing the phylogeny.

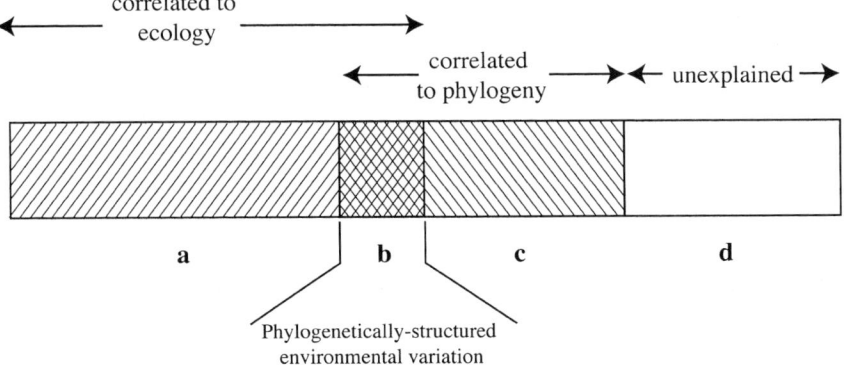

Figure 7 Partitioning the variance of the data, with a part strictly due to ecology ([a]), a part strictly due to phylogeny ([c]), a part due to the common influence of these two factors ([b]), and the unexplained variance (d) (after Desdevises et al., 2002).

Table 1 Some software packages available for comparative and statistical analyses.

Software	Website
ACAP	http://www.stanford.edu/~dackerly/ACAP.html
ANCML	http://www.zoology.ubc.ca/~schluter/ancml.html
CAIC	http://www.bio.ic.ac.uk/evolve/software/caic/
MacroCAIC	http://www.bio.ic.ac.uk/evolve/software/macrocaic/
COMPARE	http://compare.bio.indiana.edu/
DISCRETE	http://www.ams.rdg.ac.uk/zoology/pagel/mppubs.html
PDAP	http://cnas.ucr.edu/~bio/faculty/Garland/PDAP.html
TFSI	http://life.bio.sunysb.edu/ee/ehab/
R, Permute!, ParaFit	http://www.fas.umontreal.ca/biol/casgrain/en/siteOutline.html

We first compute a regression of Y on X_E. The coefficient of multiple determination of the regression, R^2, is equal to the fraction [a + b] of the variance decomposition (a stepwise procedure can select a subset of explanatory variables). Second, we compute a multiple regression of Y on PCs. R^2 in this case is equal to [b + c]. Third, we compute a multiple regression of Y on both X_E and PCs. R^2 is now equal to [a + b + c].

We can find [d] = 1 − [a + b + c], which is the unexplained variance (note [a] can be found from [a + b + c] − [b + c]).

This method is here illustrated with an application to the same data set on mammal hosts and the species richness of their helminth parasites that we used earlier.

We find [a + b] = 0.581 ($p < 0.0001$) with host BMR, weight and longevity as explanatory variables of PSR.

We find [b + c] = 0.169 ($p = 0.008$) and [a + b + c] = 0.711 ($p < 0.0001$)

This allows us to estimate the phylogenetically structured environmental variation [b] = 0.039, i.e., about 4% of the variance in parasite species richness among mammal species is due to phylogenetically structured environmental variation. The phylogenetic inertia is given by [c] = 0.13, i.e., 13% of the variance in parasite species richness among mammal species.

This result can mean that hosts with comparable PSR tend to occupy the same kind of ecological niche, but this trend is very weak because the phylogenetic inertia [b] is low; also, because of the very low value of phylogenetically structured environmental variation [c], phylogenetic and ecological influences on PSR are almost independent.

This method of partitioning the variance between phylogenetic and ecological causes not only serves to silence the critics of phylogenetically based comparative approaches, but it allows one to quantify precisely the importance of phylogeny for any character.

10. CONCLUSIONS

Data on a wide range of biological features of parasites have been accumulating steadily over the past several years. This information may contain the key to understanding how parasites evolve in response to selective pressures from their hosts and the external environment. Modern comparative methods that incorporate phylogenetic information allow us to make full use of these data. They allow several questions to be answered. For instance, how are two traits correlated with each other independently of the phylogenetic relationships of their bearer? What portion of the variation in one trait is due to ecological influences as opposed to historical or phylogenetic effects? Here, we briefly reviewed some of the current comparative methods available to address these and related questions, providing references where readers can find out more about these methods. As pointed out recently (Poulin, 1998), the comparative approach has not been used much by parasitologists. Our hope is that this essay will stimulate more researchers to adopt these methods and apply them to the study of parasite evolutionary ecology.

ACKNOWLEDGEMENTS

We thank two anonymous referees for their helpful comments and Yves Desdevises for stimulating discussions.

REFERENCES

Abouheif, E. (1998). Random trees and the comparative method: a cautionary tale. *Evolution* **52**, 1197–1204.
Ackerly, D.D. and Donoghue, M.J. (1995). Phylogeny and ecology reconsidered. *Journal of Ecology* **83**, 730–732.
Agapow, P.-M. and Isaac, N.J.B. (2002). MacroCAIC: revealing correlates of species richness by comparative analysis. *Diversity and Distributions* **8**, 41–43.
Anderson, R.C. (2000). *Nematode parasites of vertebrates: Their Development and Transmission.* Wallingford: CABI Publishing.

Barraclough, T.G., Hogan, J. and Vogler, A.P. (1999). Testing whether ecological factors promote cladogenesis in a group of tiger beetles (Coleoptera: Cicindelidae). *Proceedings Royal Society London B* **266**, 1061–1067.

Blaxter, M.L., De Ley, P., Garey, J.R., Liu, L.X., Scheldeman, P., Vierstraete, A., Vanfleteren, J.R., Mackey, L.Y., Doriis, M., Frisse, L.M., Vida, J.T. and Thomas, W.K. (1998). A molecular evolutionary framework for the phylum Nematoda. *Nature* **392**, 71–75.

Björklund, M. (1997). Are 'comparative methods' always necessary? *Oikos* **80**, 607–612.

Borcard, D. and Legendre, P. (1994). Environmental control and spatial structure in ecological communities: an example using oribatid mites (Acari, Oribatei). *Environmental and Ecological Statistics* **1**, 37–53.

Borcard, D., Legendre, P. and Drapeau, P. (1992). Partialling out the spatial component of ecological variation. *Ecology* **73**, 1045–1055.

Brooks, D.R. and McLennan, D.A. (1991). *Phylogeny, Ecology, and Behavior. A Research Program in Comparative Biology*. Chicago, The University of Chicago Press.

Cooper, A. and Fortey, R. (1998). Evolutionary explosions and the phylogenetic fuse. *Trends in Ecology and Evolution* **13**, 151–156.

Derrickson, E.M. and Ricklefs, R.E. (1988). Taxon-dependent diversification of life history traits and the perception of phylogenetic constraints. *Functional Ecology* **2**, 417–423.

Desdevises, Y. (2001). Recherche des déterminants de la spécificité parasitaire dans le modèle *Lamellodiscus* (Diplectanidae, Monogenea)-Sparidae (Teleostei) en Méditerranée. Unpublished PhD thesis, Université de Perpignan, 317 pp.

Desdevises Y., Morand, S. and Oliver, G. (2001). Linking specialisation to diversification in the Diplectanidae Bychowsky 1957 (Monogenea, Platyhelminthes). *Parasitology Research* **87**, 223–230.

Desdevises, Y., Legendre, P. and Morand, S. (2002a). Evolution and determinants of host specificity in the genus *Lamellodiscus* (Monogenea). *Biological Journal of the Linnean Society* **77**, 431–443.

Desdevises, Y., Morand, S., Jousson, O. and Legendre, P. (2002b). Coevolution between *Lamellodiscus* (Monogenea: Diplectanidae) and Sparidae (Teleostei): the study of a complex host-parasite system. *Evolution* **56**, 2459–2471.

Diaz-Uriarte, R. and Garland, T. Jr. (1996). Testing hypotheses of correlated evolution using phylogenetically independent contrasts: sensitivity to deviations from brownian motion. *Systematic Biology* **45**, 27–47.

Diniz-Filho, J.A.F., de Sant'Ana, C.E.R. and Bini, L.M. (1998). An eigenvector method for estimating phylogenetic inertia. *Evolution* **52**, 1247–1262.

Felsenstein, J. (1985). Phylogenies and the comparative method. *American Naturalist* **125**, 1–15.

Fitter, A.H. (1995). Interpreting quantitative and qualitative characteristics in comparative analyses. *Journal of Ecology* **83**, 730.

Garland, T. Jr., Harvey, P.H. and Ives, A.R. (1992). Procedures for the analysis of comparative data using phylogenetically independent contrasts. *American Naturalist* **41**, 18–32.

Gittleman, J.L. and Luh, H.K. (1992). On comparing comparative methods. *Annual Review of Ecology and Sytematics* **23**, 383–404.

Gould, S.J. and Vrba, E.S. (1982). Exaptation: a missing term in the science of form. *Paleobiology* **8**, 4–15.

Gower, J.C. and Legendre, P. (1986). Metric and Euclidean properties of dissimilarity coefficients. *Journal of Classification* **3**, 5–48.

Grafen, A. (1989). The phylogenetic regression. *Philosophical Transactions of the Royal Society of London B* **326**, 119–157.

Hafner, M.S. and Page, R.D.M. (1995). Molecular phylogenies and host-parasite cospeciation. *Philosophical Transactions of the Royal Society of London B Biological Sciences* **349**, 129–143.

Hafner, M.S., Sudman, P.D., Villablanca, F.X., Spradling, T.A., Demastes, J.W. and Nadler, P.A. (1994). Disparate rates of molecular evolution in cospeciating hosts and their parasites. *Science* **265**, 1087–1090.

Harvey, P.H. and Keymer, A.E. (1991). Comparing life histories using phylogenies. *Philosophical Transactions of the Royal Society of London B Biological Sciences* **332**, 31–39.

Harvey, P.H. and Pagel, M. (1991). *The Comparative Method in Evolutionary Biology*. Oxford: Oxford University Press.

Harvey, P.H., Read, A.F. and Nee, S. (1995). Why ecologists need to be phylogenetically challenged. *Journal of Ecology* **85**, 535–536.

Hugot, J.-P. (1999). Primates and their pinworm parasites: the Cameron hypothesis revisited. *Systematic Biology* **49**, 523–546.

Legendre, P. and Legendre, L. (1998). *Numerical Ecology*. Elsevier, Amsterdam.

Leroi, A.M., Rose, M.R. and Lauder, G.V. (1994). What does the comparative method reveal about adaptation? *American Naturalist* **143**, 381–402.

Littlewood, D.J.T. and Bray, R.A. (2001). *Interrelationships of the Platyhelminthes*. London, Taylor and Francis.

Losos, J.B. (1994). An approach for the analysis of comparative data when a phylogeny is unavailable or incomplete. *Systematic Biology* **43**, 117–123.

Martins, E.P. (1996). Conducting phylogenetic comparative studies when phylogeny is not known. *Evolution* **50**, 12–22.

Martins, E.P. and Garland, T. Jr. (1991). Phylogenetic analyses of the correlated evolution of continuous characters: a simulation study. *Evolution* **45**, 534–557.

McKitrick, M.C. (1993). Phylogenetic constraint in evolutionary theory: has it any explanatory power? *Annual Review of Ecology and Systematics* **24**, 307–330.

Morand, S. (1996). Life-history traits in parasitic nematodes: a comparative approach for the search of invariants. *Functional Ecology* **10**, 210–218.

Morand, S. (2000). Wormy world: comparative tests of theoretical hypotheses on parasite species richness. In: *Evolutionary Biology of Host-Parasite Relationships: Theory Meets Reality* (R. Poulin, S. Morand and A. Skorping, eds), pp. 63–79. Amsterdam: Elsevier.

Morand, S. and Harvey, P.H. (2000). Mammalian metabolism, longevity and parasite species richness. *Proceedings Royal Society London B* **267**, 1999–2003.

Morand, S. and Poulin, R. (1998). Density, body mass and parasite species richness of terrestrial mammals. *Evolutionary Ecology* **12**, 717–727.

Morand, S., Legendre, P., Gardner, S.L. and Hugot, J.-P. (1996). Body size evolution of oxyurid (Nematoda) parasites: the role of hosts. *Oecologia* **107**, 274–282.

Morand, S., Page, R.D.M., Hafner, M.S. and Reed, D.L. (2000). Comparative evidence of host-parasite coadaptation: body-size relationships in pocket gophers and their chewing lice. *Biological Journal of the Linnean Society* **70**, 239–249.

Murphy, W.J., Eizirik, E., Johnson, W.E., Ping Zhang, Y., Ryder, O.A. and O'Brien, S.J. (2001). Molecular phylogenetics and the origins of placental mammals. *Nature* **409**, 614–616.

Orme, C.D.L., Isaac, N.J.B. and Purvis, A. (2002a). Are most species small? Not within species-level phylogenies. *Proceedings of the Royal Society of London B* **269**, 1279–1287.

Orme, C.D.L., Quicke, D.L.J., Cook, J.M. and Purvis, A. (2002b). Body size does not predict species richness among the metazoan phyla. *Journal of Evolutionary Biology* **15**, 235–247.

Poulin, R. (1995). Phylogeny, ecology, and the richness of parasite communities in vertebrates. *Ecological Monographs* **65**, 283–302.

Poulin, R. (1998). *Evolutionary Ecology of Parasites: from Individuals to Communities.* London, Chapman & Hall.

Poulin, R. (2002). The evolution of monogenean diversity. *International Journal for Parasitology* **32**, 245–254.

Poulin, R. and Morand, S. (2000). The diversity of parasites. *Quarterly Review of Biology* **75**, 277–293.

Price, T. (1997). Correlated evolution and independent contrasts. *Philosophical Transactions of the Royal Society of London B Biological Sciences* **352**, 519–529.

Purvis, A. and Garland, T. Jr. (1993). Polytomies in comparative analyses of continuous characters. *Systematic Biology* **42**, 569–575.

Purvis, A. and Rambaut, A. (1995). Comparative analysis by independent contrasts (CAIC): an Apple Macintosh application for analysing comparative data. *Computer Applications for the Biosciences* **11**, 247–251.

Purvis, A., Gittleman, J.L. and Luh, H.-K. (1994). Truth or consequences: effects of phylogenetic accuracy on two comparative methods. *Journal of Theoretical Biology* **167**, 293–300.

Rees, M. (1995). EC-PC comparative analyses. *Journal of Ecology* **83**, 891–893.

Ricklefs, R.R. and Starck, J.M. (1996). Applications of phylogenetically independent contrasts: a mixed progress report. *Oikos* **77**, 167–172.

Simková, A., Desdevises, Y., Gelnar, M. and Morand, S. (2000). Coexistence of gill ectoparasites (*Dactylogyrus*: Monogenea) parasitising the roach (*Rutilus rutilus* L.): history and present ecology. *International Journal for Parasitology* **30**, 1077–1088.

Skorping, A., Read, A.F. and Keymer, A.E. (1991). Life history covariation in intestinal nematodes of mammals. *Oikos* **60**, 365–372.

Sorci, G., Morand, S. and Hugot, J.-P. (1997). Host-parasite coevolution: comparative evidence for covariation of life-history traits in primates and oxyurid parasites. *Proceedings of the Royal Society of London B* **264**, 285–289.

Symonds, M.R.E. (2002). The effects of topological inaccurrency in evolutionary trees on the phylogenetic comparative method for independent contrasts. *Systematic Biology* **51**, 541–543.

Westoby, M., Leishman, M.R. and Lord, J.M. (1995a). On misinterprinting the 'phylogenetic correction'. *Journal of Ecology* **83**, 531–534.

Westoby, M., Leishman, M.R. and Lord, J.M. (1995b). Further remarks on 'phylogenetic correction'. *Journal of Ecology* **83**: 727–730.

Westoby, M., Leishman, M.R. and Lord, J.M. (1995c). Issues of interpretation after relating comparative datasets to phylogeny. *Journal of Ecology* **83**, 892–893.

Recent Results in Cophylogeny Mapping

Michael A. Charleston

Royal Society University Research Fellow,
Department of Zoology, University of Oxford,
South Parks Road, Oxford OX1 3PS, UK

Abstract	303
1. Introduction	304
2. Cophylogenetic Events	306
3. Cophylogeny Mapping	308
3.1. Interpretability	308
3.2. Optimality	309
3.3. Jungles	310
3.4. Optimality and Event Costs	312
4. Complexity	312
5. Modelling Cophylogeny	316
5.1. Phase 1, Codivergence: Location in host tree: node x	319
5.2. Phase 2, Duplication, Extinction and Host Switch: Location in host tree: branch a	319
6. Tests of Significance	322
7. Confounding Cophylogeny	324
8. Discussion	326
8.1. The Future	327
Acknowledgements	328
References	328

ABSTRACT

Virtually every problem in biology benefits from consideration within an evolutionary frame work. Parasitism has been part of life ever since one organism was able to provide an environment for another: questions of parasitology naturally lend themselves to consideration of the shared ancient history of parasites and hosts. The derivation of that shared history is therefore an area of great interest to theoreticians and practitioners alike. The most intuitive approach to this is by cophylogeny mapping. Mathematically the problem is that of optimally mapping the dependent tree into the independent one, e.g., parasite into host or gene tree into organismal phylogeny. This article describes some of the recent advances in cophylogenetic simulation, significance testing, and theoretical properties of maps. In simulation the author shows that the number of ways of mapping the parasite phylogeny into that of the hosts does indeed grow exponentially quickly in most cases and shows no close correlation with the similarity between the phylogenies, and that under a simple coevolutionary model, the range of behaviours of simulated parasite phylogenies is extremely broad and would appear to confound efforts to infer model parameters from observed cases. In the area of significance testing the author demonstrates that the maximal number of inferred codivergence events is not necessarily the best statistic for measuring cophylogenetic agreement, and that significance testing by randomisation which does not alter the parasite tree substantially biases results, and provides a new test to determine whether phylogenetic similarity is consistent with preferential host switching.

1. INTRODUCTION

The recovery of phylogenetic history is a steadily advancing science. The 'phylogeny problem', that of inferring a phylogenetic tree based on the limited information presented by extant organisms, is seeing significant improvements, not least of which being the recent inclusion of Bayesian and Markov chain Monte Carlo methods into the systematist's toolbox. Such developments mean that not only are phylogenies becoming more frequently available, but that they are becoming more reliable. The phylogeny is not the end of the story however: for the most part we find it in order to answer some other question, such as how a particular morphological character may have evolved or the origin of a certain pathogen (Hahn *et al.*, 2000).

Cophylogeny mapping is a relatively recent development in the study of the relationships among and between ecologically related organisms. In this discipline, a dependent phylogeny, such as that of a taxonomic group of parasites, is mapped into an independent one, such as that of their hosts. The favourite example is of pocket gophers and their chewing lice (Hafner and Nadler, 1988; Huelsenbeck *et al.*, 1997). Parasites and their hosts, genes and the organisms which house them, geographical areas and their endemic species, retrotransposons and their host genomes (Martin *et al.*, 1999, 2002), even words and the people that utter them (Gray and Jordan, 2000): all these are very closely related problems, and are amenable to solution in much the same way. The theory involved is intricate and subtle, and therefore appealing to (bio)mathematicians, but far greater is its appeal to practising biologists interested in the associations of parasites with their hosts. At the smallest scale coevolutionary study revolves around the microscopic and even molecular: incremental changes in both parasite and host resulting in an arms race through concomitant evolution. At the phylogenetic level, such coevolution can lead to similarity between the phylogenies of hosts and parasites, that of the parasites coming to 'mirror' that of their hosts (Fahrenholz, 1913; Eichler, 1948). Unfortunately, there are not only numerous processes which can lead to disparity between the phylogenies, but even some which can suggest cophylogenetic evolution (in fact, codivergence) when there is none.

In cophylogeny mapping, the most intuitive method of discovering these processes, a dependent phylogeny is mapped into an independent one – in this case, that of a group of parasites into that of their hosts. The map is constructed so as to provide the best possible explanation of why the phylogenies are as they are: whether the parasites have consistently codiverged with their hosts, or whether they have undergone host switching, duplication by independent speciation or extinction events. The number of maps is extremely large: it grows approximately exponentially with the number of host and of parasite taxa, and so exhaustive listing of all solutions rapidly becomes prohibitive.

Another method of uncovering ancient associations between dependent and independent taxonomic groups arose in the subject of biogeography. In that case the 'hosts' are geographical regions and their 'parasites' are the species endemic to those regions, undergoing codivergence in vicariant speciation, loss by extinction and 'missing the boat' *sensu* Paterson *et al.* (1993), independent divergence (sympatric speciation), and horizontal transfer by migration. In a similar way we may consider genes (or other genomic material) as parasitising their host organisms, with horizontal transfer taking the place of host switching, multigene families arising from multiple duplications, and genetic loss (Page and Charleston, 1997). The parallels are easy to see, and are listed in Table 1. The biogeographical method mentioned above, known as Brooks' Parsimony Analysis (Brooks, 1981, 1990) re-codes the parasite tree as a set of binary characters and then fits the resulting 'data matrix' to the host tree in a parsimonious fashion, that is, minimizing the number of changes of character 'state' on the host tree to arrive at the pattern of character states at the tips of the host tree. It was indeed a 'valiant effort', but is now widely recognised as inadequate since it fails to deal correctly with host switching (Page, 1990) and at any rate it is not clear what exactly a change of character state *means* in this situation.

Table 1 Parallel problems in cophylogeny.

General problem	Codivergence	Duplication	Loss	Host switch
Host/parasite	codivergence/ cospeciation	independent speciation/ divergence	"missing the boat" and extinction	host switch
Biogeography	vicariant speciation	sympatric speciation	"missing the boat" and extinction	migration
Genes/organisms	codivergence	duplication	loss	horizontal transfer

There are various 'fixes' to the approach (Brooks, 2001) but its fundamental flaw remains: the characters are derived from the non-independent branches of the parasite tree, and then treated independently. As such it cannot be considered as a valid solution to the cophylogeny problem and will not be considered further here, even though the problem to which it was applied is mathematically identical to that presented by host–parasite systems.

All the empirical results presented here were gained using the Macintosh® programs TREEMAP2 (Charleston and Page, 2002) and CoSPEC 0.51 (Charleston, 2002b). TREEMAP2 is available from the evolution web site at http://evolve.zoo.ox.ac.uk/software, and CoSPEC is still in development prior to public release.

There has been some interesting new work in a maximum likelihood approach to solving the cophylogeny mapping problem (Huelsenbeck *et al.*, 1997, 2000), but the models are simplistic at present and much remains to be done. A difficulty in such methods is lack of confidence: since even a relatively restrictive stochastic model of cophylogenetic evolution can give rise to a huge variance in observed behaviours, the converse problem of estimating a set of parameters from a single instance of the cophylogeny problem will have huge confidence limits (Figure 9). The approach taken here is a necessary compromise, since at some level we must 'put our necks out' and say that *if* these are the correct phylogenies, then *these* are the sets of ancient associations which we must consider as the best explanations of what took place to give rise to their (dis)similarities.

2. COPHYLOGENETIC EVENTS

Essentially, there are just four kinds of events which we can hope to recover from an analysis of pairs of phylogenetic trees. These are *codivergence, duplication, loss* and *host switching*, represented in Figure 1(a)–(d) respectively.

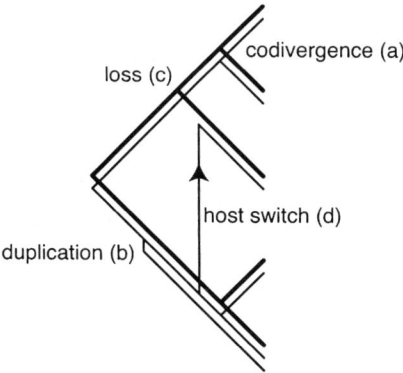

Figure 1 Recoverable cophylogenetic events. The codivergence event (a) is the most obvious one to recover and is the only kind present in a history of strict codivergence. We have to posit other events if the trees disagree either in their branching pattern or their divergence times, to include duplication (b) and loss (c), and possibly host switching (d).

Codivergence, also called cospeciation but more general than that term, is the concurrent divergence of host and parasite lineages into two new lineages each, as in Figure 1(a). Concurrence is not assumed to be at the exact same instant in time of course: speciation takes a finite period, and there is bound to be some slight variation in divergence times of host and parasite, but codivergence must be inferred when we cannot resolve a significant time difference between the two. Codivergence is a more general term than cospeciation since it is concerned only with lineages in a phylogeny, and avoids the whole species concept altogether. Cospeciation is codivergence, but codivergence may be concurrent divergence of populations, viral strains, genera, species and their areas of endemism, etc.

Duplication (b) is the divergence of the parasite lineage into two new lineages, independently of the host. The term comes from, and is easiest to think about in terms of, gene duplication in an organism. The organism does not speciate, but suddenly there become two new copies of a gene coexisting in that organism, for instance in the α- and β-haemoglobin genes in primates (Goodman *et al.*, 1979).

Loss (c) encompasses three kinds of events, but we cannot distinguish them from each other just from the phylogenies (Paterson *et al.*, 2002). These are 'missing the boat' *sensu* Paterson *et al.* (1993) in which a newly divergent host lineage is simply not occupied by the parasite lineage, and so the parasite misses out on occupying that host; next is extinction, in which the parasite just goes extinct on its host; the last is sampling error, in which we do not find that parasite occupying that host. It is easy to see why we cannot distinguish these events solely by using phylogenetic information of extant taxa. (Page uses 'sorting event' to include both missing the boat and extinction.)

Host switching (also known as *horizontal transfer*) Figure 1(d), is the bane of all cophylogeny analysis. If there is no host switching present, or indeed permitted in the analysis, then the problem is remarkably easy and simple methods can be used to find guaranteed optimal solutions to the problem, such as Page's 'reconciled tree analysis' (Page and Charleston, 1998). However the presence of host switching, in which a newly diverged parasite lineage invades another host lineage, dramatically increases the number of possible solutions to the problem.

3. COPHYLOGENY MAPPING

Informally, the cophylogeny mapping problem can be phrased as follows: Given a host tree H and parasite tree P, and known associations expressed as a mapping φ from the tips of P to the tips of H – this (H, P, φ) triplet collectively known as a *tanglegram* as in Figure 3(a), we must find the set of maps of P into H which optimise some cost function and which are interpretable biologically.

3.1. Interpretability

Beyond the cost function, we must be able to ensure of our maps that they have explanatory power: they must be biologically reasonable, interpretable, and applicable in principle to any appropriate P, H, and φ, regardless of their size and phylogenetic agreement. A *feasible* map will be one that makes biological sense. In this case maps which require time-travelling parasites are not feasible, and those with more than one cophylogenetic event occurring contemporaneously are likewise infeasible. For instance we shall discount all solutions with combinations of events like those shown in Figure 2, which can either be interpreted in several different ways (with equal lack of confidence), or which require comparatively rare events to take place in the same instant, which must be deemed impossibly unlikely.

Such uninterpretable and untraceable events have to be discounted from our analyses simply because we can never recover them. Their inclusion would only add confusion and increase the complexity of this already difficult problem.

COPHYLOGENY MAPPING

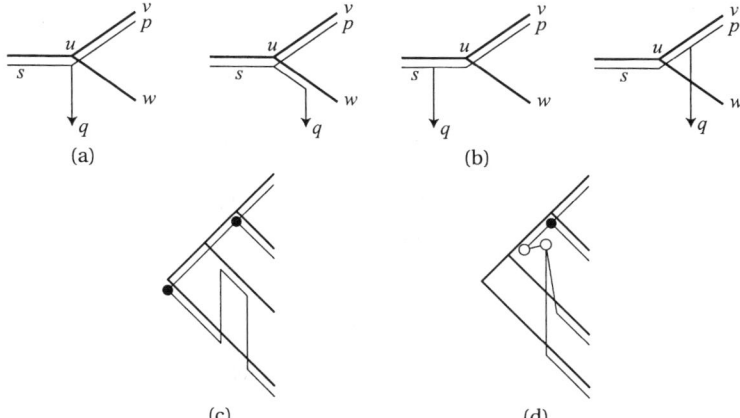

Figure 2 Uninterpretable (a, b) and untraceable (c, d) cophylogenetic events. In (a) both the host and parasite diverge, and one of the new parasite lineages switches host at the same moment. (b) shows some possible resolutions of these two events which cannot realistically be taken to occur in the same instant. In (c) a parasite lineage indulges in a "short stay", which is untraceable from the phylogenies, as is the switching of both descendent parasite lineages in (d).

3.2. Optimality

The simplest cost function is to use the total number of codivergence events (CEs), and that has been used successfully in the past in numerous studies (Hafner and Nadler, 1988). Alternatively we might count the total number of *non*-codivergence events (NCEs) and minimise that. Both counts will give the correct optimal result if the trees match perfectly, but they differ if the match is imperfect. It makes sense to use the number of NCEs required in addition to the number of CEs possible as a measure of agreement, particularly when determining the significance of fit of P into H, since its distribution is wider. Because there is a greater range of categories available in terms of NCEs, we can get a more accurate measure of the significance than if we just used CEs on its own (see Tests of Significance, Section 6).

A slightly more general scheme gives a positive cost to each of the non-codivergence events which are recoverable from a given tanglegram: duplication, host switch, and loss. (We cannot distinguish extinction, 'missing the boat' or simple taxon sampling from each other given only the tanglegram.) The cost of a map is then the sum of the costs of all the events which the map posits. This is the most general model of its kind, since no more events can be recovered from a mapping of P into H.

The next step might be to allow more complex cost functions for the various events, such as an affine cost function for host switches, which increased the event cost proportionally to the phylogenetic distance between source and target host. Such a measure would be appropriate when preferential host switching is believed to be an important factor in the coevolution of host and parasite (Charleston and Robertson, 2002).

A convenient interpretation of event costs is as a measure of their 'unlikelihood' – say as their negative log-likelihoods (Huelsenbeck *et al.*, 2000). However all biological systems are different and it cannot ever be expected that global cost measures will be found that are appropriate for all host–parasite systems, let alone gene/species or biogeographical ones. Our best bet at the moment is to accommodate all maps which could be optimal under some cost scheme and, once we have found them, use other external biological information to deduce which provides the best explanation of the similarities and differences between P and H.

3.3. Jungles

Given all the constraints which a cophylogeny map must satisfy in order to be feasible, the next problem is to find one, in fact, to find all those which might be feasible or optimal. To that end, the *jungle* was developed (Charleston, 1998).

A jungle is a *graph* whose *vertices*, called *j*-vertices, are associations of elements of the associate phylogeny P with locations in the host phylogeny H. A location in H can be either a vertex (elsewhere called a node) or an *arc* (elsewhere, a branch or edge). I use the graph theoretic terminology here for convenience in later discussions.

One of the central assumptions in all cophylogeny methods to date, either explicit or implicit, has been that once associate lineages diverge they can be treated independently. If we adopt this assumption one more time, we can then treat each component of a mapping of P into H as essentially independent, possibly to be repeated in many other maps. Thus, we can construct a graph which contains all the components of all the solutions which we care to consider, from the extreme of all feasible solutions right down to only those which have the maximal number of codivergence events or minimal number of non-codivergence events.

In Figure 3 there is one of the simplest possible tanglegrams, on just three taxa for H and P, and the jungle containing all the maps of P into H. A detailed description of jungle construction is not appropriate here: the interested reader is directed to Charleston (1998). The root of P is labelled with a square □ and the other internal vertex with a circle ○. Two of the

COPHYLOGENY MAPPING

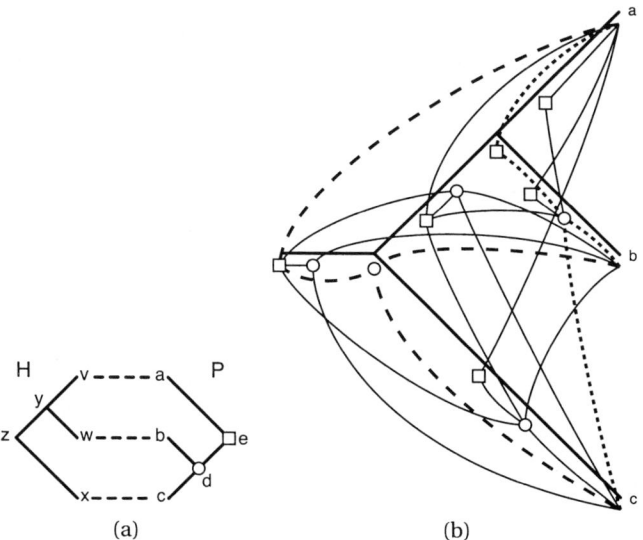

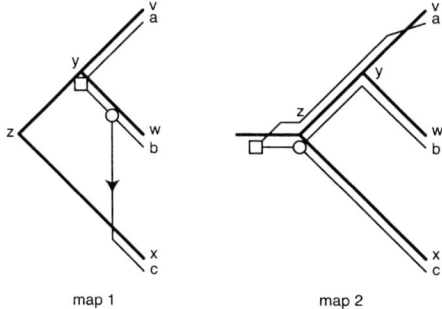

Figure 3 A simple tanglegram and jungle. In the above figure we see a simple tanglegram with host (H) and parasite (P) trees on three taxa each (a), and the jungle containing all the maps of P into H. Two of the maps are selected with dashed lines (see Figure 4).

Figure 4 Maps 1 and 2 from Figure 3.

maps have been selected using dotted and dashed lines, maps 1 and 2 respectively, and are shown in Figure 4. The arcs in the jungle correspond to arcs in the parasite tree, so the history of the parasite lineages can be read directly from the maps embedded within the jungle.

Map 1 in Figure 4 requires a host switch and no losses, whereas map 2 needs no host switch but requires 3 losses. Both solutions have one codivergence event, but without assigning event costs it is impossible to say which one is to be preferred. If having one host switch is more costly (i.e., less likely) than having three losses then we would infer that map 1 is more likely to be correct; else if three losses are less more likely than 1 host switch we would infer that map 2 was to be preferred.

3.4. Optimality and Event Costs

Clearly an issue central to the determination of which solution(s) is (are) optimal is that of the event costs. The assignment of general event costs is likely to remain very difficult since every biological system is different, and empirical testing of the likelihood of the various events is difficult and expensive (Clayton, 1990, 1991; Johnson and Clayton, 2002). It certainly will never be possible to measure the likelihood of ancient host switches between species long gone. It is fortunate then that in many cases the exact value of event costs does not change significantly which solutions are optimal (Charleston and Perkins, 2002). This lack of change in the optimal solutions in the face of changing event costs can be viewed as a kind of sensitivity analysis: those solutions which remain optimal over a wide range of event costs could be thought of as being more robustly supported than those which are only optimal for a smaller range. There is a caveat to this kind of thinking however, since in general we do not have a good idea in advance as to the probability distribution of those event costs: our prior information is not so much limited as virtually non-existent. Therefore we cannot justify saying that a particular solution is 90% supported just because it is optimal in 90% of the set of event costs which we consider, nor can we say that a particular codivergence (or any other) event is 90% supported just because it appears in 90% of the potentially optimal solutions.

Statistical inference in cophylogeny mapping is, and will remain, a very difficult process, until we have a better understanding of the distribution of likely event costs which we should consider. For that reason we must retain *all* those solutions which could be optimal for some combination of event costs, and judge them on their own merits using external information where available.

4. COMPLEXITY

Crucially important to any complex optimisation problem, such as this one, is an understanding of computational requirements involved. Too often one hears 'why not just get a faster computer?', in connection with the solution of larger and larger combinatorial problems, like the phylogeny problem for instance. The phylogeny problem is, in most formulations, NP-complete, that is, no polynomial time method exists for its solution, and the chance that one exists is becoming more and more remote (Garey and Johnson, 1979). The exponential growth in the number of solutions possible means that even though it may be quick to find the 'goodness' of a solution (e.g., its likelihood value) there are simply too many feasible solutions to guarantee ever finding the optimal one. Most of the interesting optimisation problems in systematics are NP-complete, or worse.

It appears that the cophylogeny mapping problem is one of these characteristically 'interesting' combinatorial problems also. In order to determine just when heuristics must be used, and how we may find faster methods, we must therefore find out how the number of feasible solutions grows with the size of the problem input, i.e., the number of taxa involved.

Empirical complexity testing of cophylogeny mapping is slow almost by definition since a thumbnail calculation shows the number of feasible cophylogeny maps to increase approximately exponentially with the number of host and associate taxa. For example, consider the case where both H and P are the completely unbalanced trees with n leaves and a 1-1 relationship between the tips of each tree. Provided we preserve the partial ordering of the $(n-1)$ internal nodes in P when mapped into the $(2n-1)$ branches in H (including one prior to the root), there are $\binom{3n-1}{n-1} = (3n-1)!/((2n!)(n-1)!)$ maps with no codivergences at all. This number, which is fewer than the total number of maps, grows exponentially in n, so the total number of cophylogeny maps for such tree shapes will also grow at least exponentially. Unfortunately, a complete combinatorial assessment of the number of feasible maps is beyond easy computation since the number of maps is dependent on the shape of each tree, and on the given relationships between their tips. Therefore it is necessary, in order to get a global impression of the complexity of this problem, to perform an empirical test and estimate the growth in number of maps for some small values of n, by elucidating all the maps and counting them. Since by our thumbnail calculation this number will grow approximately exponentially, we will therefore be limited in the number of taxa which we can use. In this case we begin with $n=2$ and increase to $n=7$ and construct all the feasible maps from P into H for several random instances of the cophylogeny mapping problem.

The procedure then amounts to a repetition of the following steps: For each value of n,

1. create random host and parasite trees H and P respectively, according to the standard Markov model of tree growth;
2. randomly associate each tip of P with a unique tip of H;
3. construct all the feasible maps (using jungles in TREEMAP2);
4. repeat.

Due to the computational constraints mentioned above the sample sizes are limited to 1000 instances of the cophylogeny problem for $n = 2,3,4,5$ and 6, and 250 instances for $n = 7$. Results are displayed in Figure 5.

Our thumbnail calculation of the complexity of cophylogenetic mapping is thankfully not too far off: Figure 5 illustrates the growth in the number of maps with n, the number of taxa in both host and parasite trees. We find that the best (least squares) estimate of the exponential growth function is (#Number of feasible maps) $\sim 0.2116e^{1.2501n}$. This complexity is 'worst case' in a sense because this is for random instances of the problem, not necessarily representative (indeed we hope not) of the kinds of problems which we expect to work on. However, it is evident for these random instances that the number of feasible maps does not appear to be correlated at all with the degree of fit of the potentially optimal maps: there the correlation is non-significant and the relationship appears random (Figure 6).

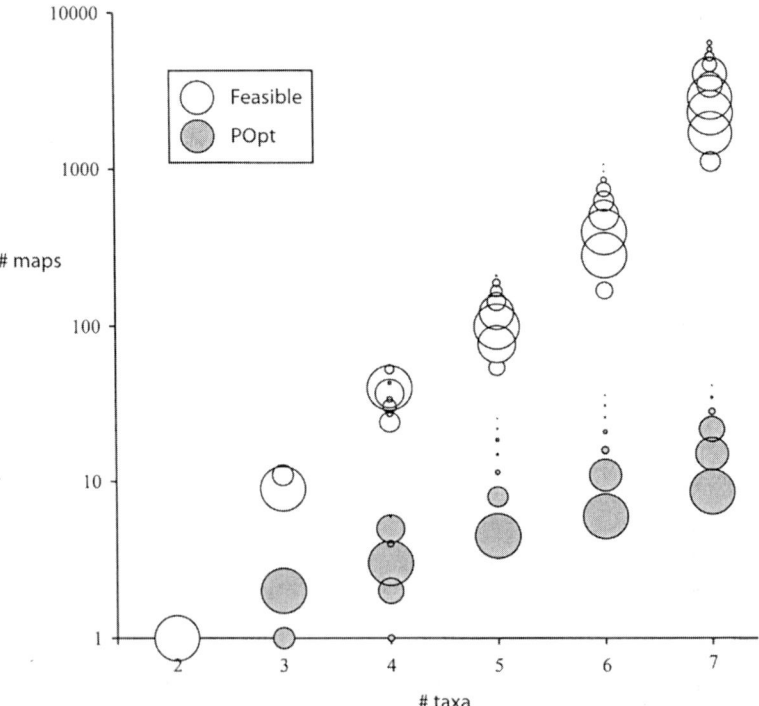

Figure 5 Numbers of cophylogeny maps vs. number of taxa. The number of feasible cophylogeny maps (white circles) increases exponentially with the number of taxa involved, but the number of potentially optimal maps (grey circles) does not increase so rapidly.

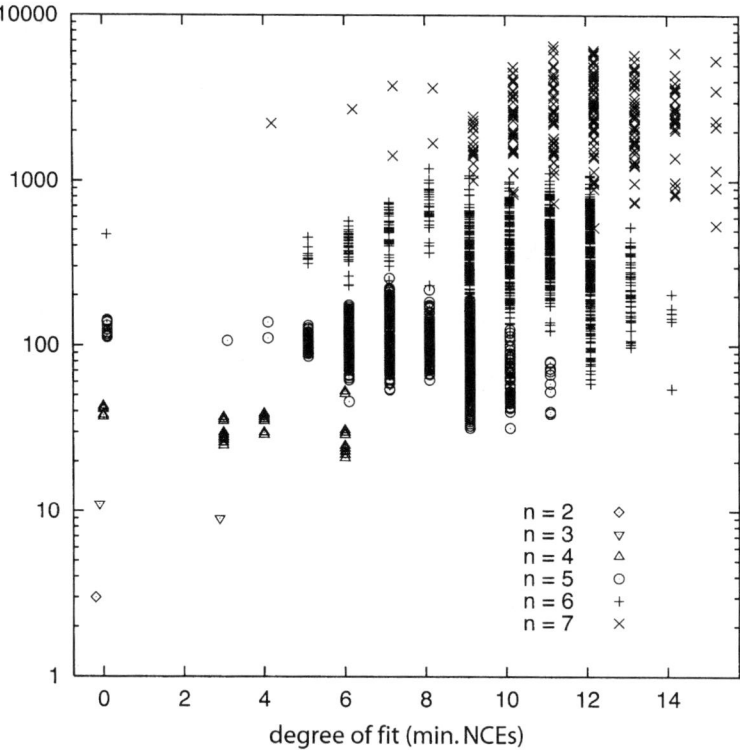

Figure 6 The number of maps plotted against the degree of fit for different numbers of taxa. The plot above shows no strong relationship between the number of feasible maps and the optimal degree of fit, measured in minimal number of non-codivergence events, of those maps, for different numbers *n* of taxa, though the number of feasible maps increases with *n*; the data points are slightly offset for clarity.

This is a slightly irritating, if unsurprising, result in that if we are to seek all the feasible maps of a given *P* into a given *H*, then we will be bounded in the number of taxa we can consider to relatively small numbers, since the number of solutions will increase approximately exponentially no matter how well matched our trees are. But there is some good news: the number of potentially optimal maps does not increase anything like as much as the number of feasible ones (Figure 5), so we can reduce the number of maps which we need to consider enormously, by only accepting those which could be optimal under some scheme of event costs.

While the set of feasible maps of parasite into host phylogeny increases exponentially with the number of host and of parasite taxa, it is therefore

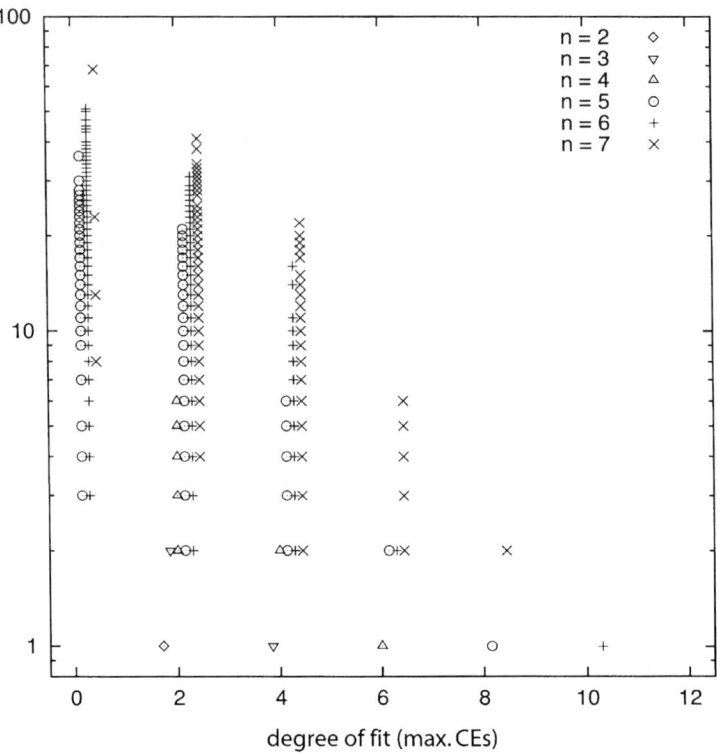

Figure 7 Number of potentially optimal maps vs. max. codivergences. The plot above shows the negative relationship between the size of POpt and the degree of agreement as measured by maximum number of codivergence events, split by the number of taxa; again, the data points are slightly offset for clarity.

comforting to discover that the rate of growth in the number of potentially optimal maps is much lower (Figure 5). In fact the size of POpt is negatively correlated with the degree of fit when measured in terms of number of codivergence events, as Figure 7 shows.

This means that while the number of all feasible maps remains large, we can narrow down the potentially optimal ones to manageable numbers, provided we can fit enough codivergences between P and H.

5. MODELLING COPHYLOGENY

I have described above those cophylogenetic events which we can recover. These are closely linked to, but not precisely the same as, those events we must include in any stochastic model of cophylogenetic evolution. It seems clear that such a model should be based on growing an associate tree on an existing host tree with given branching order and branch lengths, and include

(i) a probability of codivergence q_c at the divergence of a host lineage;
(ii) an instantaneous rate of duplication r_d;
(iii) a probability of host switching of a newly diverged associate lineage r_x;
(iv) an instantaneous rate of extinction of associates, r_x;
(v) a probability of sampling each associate on its host(s) q_s.

Recently it has also become apparent that a further event is required: the *failure to diverge* must be included to take into account those cases in which the parasite continues to exist, undifferentiated, on two or more diverged host lineages. Clonal parasites for instance cannot realistically be judged to be different taxonomic units simply because they have different hosts. Thus we might also aim to have in our model

(vi) a probability of failure to diverge.

The inclusion of a parameter in a model does not mean that it may be recovered from data generated by that model. In the above, we must realistically include parameters for missing the boat, extinction, and non-sampling, but it is impossible to determine from a map of P into H which of these events led to any losses observed. This shows part of the difficulty in inferring model parameters such as the probability of codivergence in a single host–parasite system, from a single instance of cophylogeny.

A further difficulty lies in the huge variability in the behaviours which can arise from a single set of input parameters in such a model. Huelsenbeck and Rannala have made some progress with modelling cophylogenetic processes (Huelsenbeck and Rannala, 1997; Rannala and Michalakis, 2002) but much remains to be done: as yet the models are very simple, and cannot account for all the complexities which are commonly apparent in host–parasite, gene–organism, and species–area systems. One of the main aims in a likelihood analysis involving a stochastic model is the estimation of parameters of that model. For instance we include a Ts/Tv (transition/transversion) ratio as a parameter in many models of molecular evolution, and such programmes as PAUP* (Swofford, 2002) can be used to estimate the Ts/Tv ratio which maximises the likelihood that we observe the data which we do. In cophylogeny we would like to be able to estimate rates of duplication and extinction in the parasite tree, probabilities of host switching and of codivergence, and more complex parameters when we graduate to more sophisticated models.

Perhaps the most immediately interesting parameter is the probability of host switching, since that will provide insight into the risk of new human pathogens emerging through zoonosis. Some work has already been done on this as mentioned previously (Huelsenbeck et al., 2000), but we must be cautious: we do not yet know how phylogenies evolving under even the simplest cophylogenetic models are expected to behave. It is not clear how much variation we should expect if we were to 'rewind the clock' and start over with the same host tree (which we consider to be an independent variable here) and regrow the parasite tree on it. Hence it is crucial to find out that variation, in order to assess how confident we may be in the values of parameters which we hope to estimate from our H, P and φ.

Unfortunately we must treat each instance of a cophylogeny problem as a single data point in such an estimation, and in order to assess how accurate that will be, we need to find out the converse relation, of how much variation we expect to see in the behaviour of an evolved parasite tree P given a single parameter set in the model.

The program CoSpec (version 0.51) (Charleston, 2002b) was used in this part of the study, with prescribed numbers of host taxa, but with the number of parasite taxa surviving to the present being a consequence of the stochastic processes involved in the model.

The parameters of the model are quite simple: they are

- r_d, the rate of duplication (independent speciation) of the parasite tree,
- r_x, the rate of extinction in the parasite tree,
- q_c, the probability that, given that a parasite is currently occupying a host lineage which undergoes a divergence (speciation) event, the parasite does too,
- q_{hs}, the probability that, given that a parasite duplicates, one lineage switches to another randomly chosen contemporaneous host, rather than remain on the original one.

The host tree is created in the same way as are random host and parasite trees in the parts of the study using TreeMap, that is, according to a Yule–Markov model (Yule, 1924) in which each tip is considered equally likely to undergo a divergence event at any instant, that rate remaining constant over time. The extinction rate for the host tree was set to equal zero for computational feasibility; this will be allowed to vary positively in later studies. Once H is constructed with the correct number n of tips and divergence times consistent with the Yule model, a single 'parasite' lineage is placed at the root of H. The growth of P continues stepwise through H in different phases: the first phase concerns the nodes of H, and the second the cophylogenetic events which take place between those nodes. The growth proceeds from the root of H to its tips.

5.1. Phase 1, Codivergence: Location in host tree: node x

At each internal node of H, if there are any parasites currently occupying it, there is a finite probability that they will codiverge with the host lineage. Hence for each parasite p occupying x we randomly determine whether it codiverges with x with probability q_c, or whether the parasite misses the boat and continues down just one of the descendent host lineages, chosen at random.

5.2. Phase 2, Duplication, Extinction and Host Switch: Location in host tree: branch a

After Phase 1 we consider the events which may befall any parasite on any branch at a time t which is not at a divergence event in H. (There will be more than one such branch.) For each extant tip of P we first determine the time of the next duplication or extinction event according to an exponential distribution based on the instantaneous rates of each, and then compare that with the time of the next relevant event on the host tree, i.e., the divergence of that host lineage or the arrival of the host lineage at the present time. If either event occurs before the next relevant host event, then we modify the parasite tree accordingly: if the event is an extinction then we simply remove that parasite from its extant status (though it is left in the tree for the purpose of counting events later), whereas if the event is a duplication, that parasite lineage diverges. After a duplication event we next determine whether one of the two descendent parasite lineages also undergoes a host switch, with probability q_{hs}, to a uniform randomly chosen target host lineage.

The simulation is performed many times (typically 1000) for each host tree in order to assess the variance in the numbers of events of each kind. Figure 8 shows the distribution of numbers of parasite lineages surviving to the present, and Figure 9 shows the variance in the number of 'real' events which occur under the model described above, for the parameter sets listed in Table 2.

In Figures 8 and 9 we can see the wide range of possible outcomes for quite 'sensible' sets of parameters in this simple model of cophylogenetic evolution. In Figure 8 we see the distributions of numbers of parasite lineages which survive to the present using parameter sets 1–4 in Table 2.

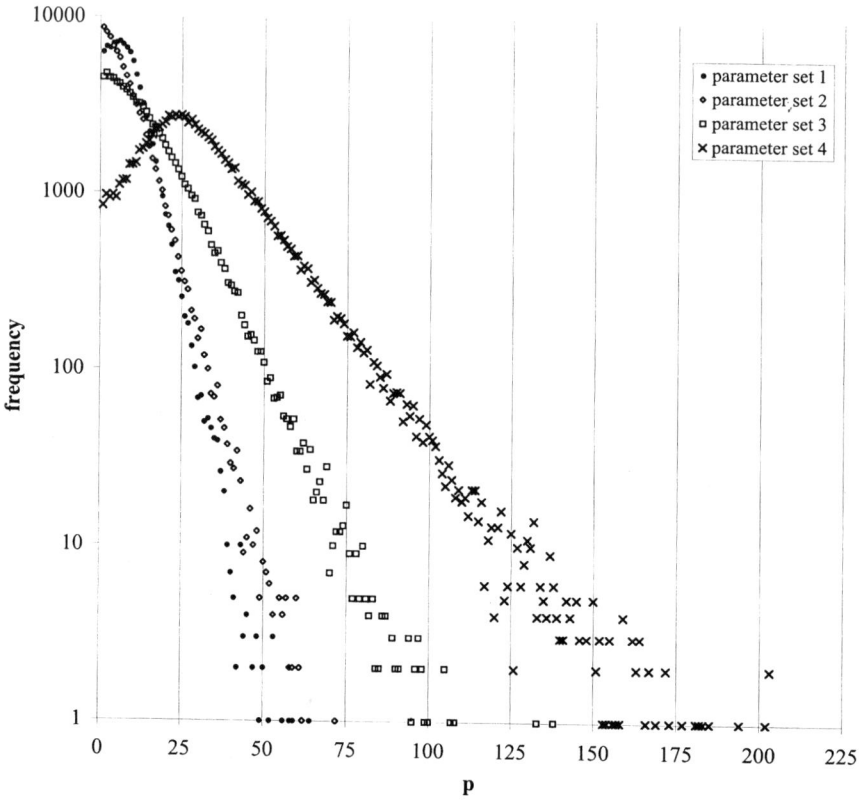

Figure 8 Number of parasite lineages surviving. The above plots show the distribution of the number of parasite lineages which survived in 10^5 simulations for each parameter set.

Given that we observe, say, 10 parasite taxa surviving to the present on H, we would be hard pressed to distinguish which parameter set were acting in our model, even from the small set available in Table 2, let alone the problems we'd have if we had to choose from the entire continuum of possible model parameters. The plots for parameter sets 1 and 2 are very similar over a large range. Note that all the plots show a clear exponential decay to the right hand of the frequency distribution (to the right-hand side of the peak, $R^2 \approx 0.95$ is typical). This is to be expected under the Yule model of tree growth of H (modulated by the codivergence probability) and of P apart from codivergence events.

Figure 9 shows plots of the number of real cophylogenetic events which occur in the coevolution of P and H (study is underway to show how the

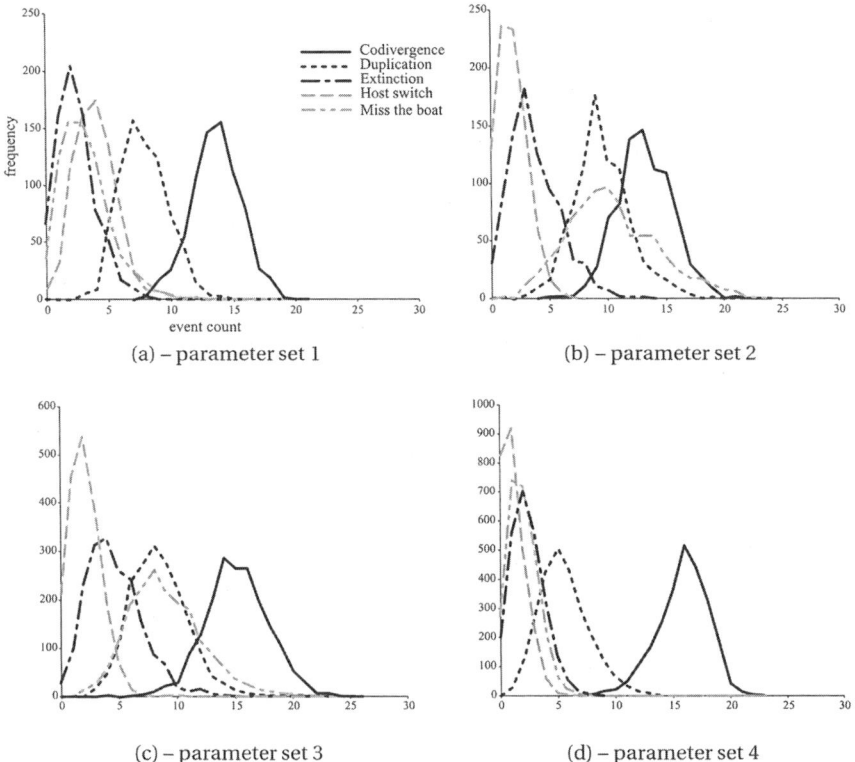

Figure 9 Distributions of numbers of cophylogenetic events.

Table 2 Parameters used in cophylogenetic simulation with the CoSpec program. q_c is the probability of a codivergence in the parasite lineage given that its host has diverged; r_d is the instantaneous rate of duplication of the parasites, quoted as 1/ (mean time between events), as is r_x, the instantaneous rate of extinction of the parasites; q_{hs} is the probability that, given a duplication has occurred, one of the two child lineages switches to a randomly chosen contemporaneous host. The number of trials is quoted as (number of host trees × number of parasite trees per host tree). The expected length of each branch in the host tree was 1.0.

Parameter set	Num. hosts	q_c	r_d	q_{hs}	r_x	Num. samples
1	10	0.75	1/5	0.5	1/10	100 × 1000
2	20	0.5	1/5	0.2	1/10	100 × 1000
3	30	0.6	1/5	0.25	1/10	100 × 1000
4	20	0.9	1/5	0.5	1/20	100 × 1000

number of *inferred* cophylogenetic events relates to this). These plots were constructed from those simulations generating the data in Figure 8 in which 10 parasite lineages survived to the present.

The plots in Figure 9 show the number of cophylogenetic events which occurred on a set of simulated host trees – 100 host trees and 1000 parasite trees 'grown' on each, for each parameter set. All the plots show quite similar distributions of numbers of codivergences: if we were considering just the number of inferred codivergence events (even assuming we recovered them correctly) we would have little chance of distinguishing between the parameter sets. Of course there is more than one kind of event; perhaps in concert we might use the number of reconstructed codivergences, of host switches, and of losses (extinctions + missings of the boat). Again, this is not likely to prove fruitful: supposing we reconstructed 12 codivergences, 8 host switches and a total of 8 loss events. Each of these values is near the middle of its distribution both in parameter set 2 and in set 3, and not far off (in host switches) for parameter set 1 either. It would be difficult to choose between just one of these four possibilities, let alone finding the maximum likelihood estimation from the complete range of parameters available.

It does appear that the recovery of model parameters from a single instance of the cophylogeny problem is going to remain a difficult statistical issue, to be plagued by a great lack of precision, in addition to the methodological problems in reconstructing most likely scenarios of coevolutionary history between H and P.

6. TESTS OF SIGNIFICANCE

It is not enough to provide a degree of fit of P to H without being able to quantify it. Asserting that we can fit P into H with 12 codivergence events is meaningless unless we know how surprising that is: is 12 codivergences *a lot*?

A standard statistical method in biology uses randomisation to create a null distribution, with which a test statistic can be compared. The standard test in cophylogeny mapping uses randomisation of the associate tree P, to create instances of the cophylogeny mapping problem (H, P', φ), and then solve them to minimize the number of NCEs or maximise the number of CEs. The former is to be preferred since it has a wider distribution and hence provides a more accurate estimation of the real significance of the level of agreement between P and H. Note that randomising P effectively also randomises φ.

Again, with significance testing it is not possible to calculate the distribution of goodness of fit which we would expect from real biological data; we must estimate it from random instances of the cophylogeny problem and use that as a first guess. In most cases in practice, we will not be testing the significance of agreement between trees which are unrelated to each other, rather, we are more likely to be investigating the relationship between the trees of taxonomic units which we believe to be ecologically linked, and therefore have phylogenies which are non-randomly related to each other. The method employed to estimate the distribution of significance in cophylogeny mapping is slightly different from that used to measure complexity. In testing significance we are only concerned with whether a given triplet of (H, P, φ) has a solution with a certain degree of fit, and if it does not, we may discard it completely. Hence this part of testing can use the bounding method built in to jungle construction in TREEMAP2, as follows: For each n,

1. create random instances of H and P and the tip associations φ;
2. choose a measure of degree of fit, e.g., 5 or more codivergence events;
3. construct a jungle stepwise and halt if no map can exist with this degree of fit;
4. repeat many times and keep count of the number of acceptable maps.

At the highest values of degree of fit (whether they be the maximum number of codivergence events possible or the minimum number of non-codivergence events required), there will be very few random instances of P and H (and therefore φ) which will permit such maps. Since jungle construction can be often halted before the jungle is complete if no acceptable solution is possible, many more random instances can be used at the highly significant tail of the distribution than at less significant values. Hence a more precise estimate of the distribution is possible at this tail. Test trees with 8 taxa were used, and at least 1000 random instances were created for each significance category (10 000 replicates were used for the most significant categories for both measures) in order to increase precision.

Results are shown in Figure 10. In the figure we see the marked difference between using maximum number of codivergence events (CEs) and the minimum number of non-codivergence events (NCEs) as a measure of goodness of fit between P and H, for the case where $n = 8$. Results are plotted on the same figure for easy comparison; hence the standard method is given as minimum required duplications, rather than maximum possible codivergences. There are just two categories which are significant in terms of numbers of codivergence (duplication) events, while there are seven categories of significance of numbers of NCEs required to reconcile the trees. This trend is more marked as we increase the number of taxa, as the disparity in numbers of possible classes increases (data not shown).

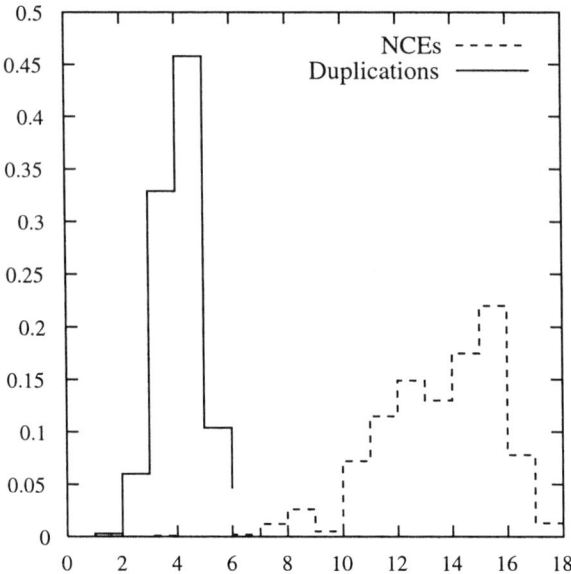

Figure 10 Distribution of significance of CEs and NCEs with 8 taxa. Plotted are the frequency distributions of minimum required non-codivergence events (NCEs) and of duplication events estimated from random instances of the cophylogeny problem with 8 taxa on each tree.

It is clear that using NCEs as a measure is therefore to be preferred since the distribution has more classes of significance possible, therefore will provide a more accurate measure of the significance of cophylogenetic agreement between P and H.

There may well be more biologically appropriate measures which could be used, but as yet they are not apparent. For the moment the statistical argument must carry sway, that it makes sense to use the distribution with the greater spread. When we learn more about cophylogenetic processes we will probably find measures more appropriate to specific systems.

7. CONFOUNDING COPHYLOGENY

Coevolution is not the same thing as codivergence.

A number of cases have arisen recently in which codivergence is taken as the logical consequence of 'host-dependent evolution' (Beer *et al.*, 1999), which is making a rather unjustified leap of logic. Just because host and parasite are *coevolving*, that is not the same thing as codivergence, even if the pattern of divergence of the parasite is dependent on that of the host (Charleston and Robertson, 2002).

It was shown in that study that much of the significance of match between accepted primate and primate lentivirus phylogenies could be attained through a model involving no codivergence at all: merely the presence of 'preferential host switching', in which parasites are more likely to successfully switch to hosts which are more closely related to their current host than they are to more distant ones. Quite a lengthy simulation was used to give this result. It would be convenient if there were to be a simple test of such behaviour, and one has recently been added to TREEMAP.

This 'cherry-picking test' (Charleston and Robertson, in prep.) makes use of the following observation: if preferential host switching is causing much of the apparent cophylogenetic agreement between P and H, then this will be reflected in inflated numbers of apparent codivergence events near the tips of both trees (as in Figure 1, at (a), but absent in the tanglegram in Figure 3(a)). A cherry is the term given to a pair of extant sister taxa (McKenzie and Steel, 2000).

In Figure 11 we see the tanglegram of a pair of phylogenies, of primates (left) and lentiviruses (right), with their known associations (compiled from various sources). One of each pair in a matched cherry is removed and the significance of the best possible match is measured, both using maximum codivergences and minimum non-codivergence events as the test statistic. From a significance of $p = 0.05 \pm 0.02$ (based on a sample size of 100 and

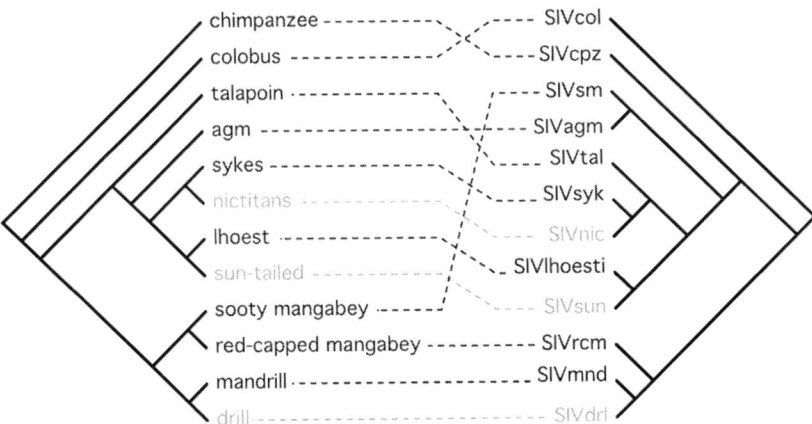

Figure 11 Primate/lentivirus tanglegram. The tanglegram above shows an estimate of the phylogenies of a group of primates and their lentiviruses. The significance of the match is 0.05 ± 0.02 (based on 100 replicates) with all taxa included, and 0.28 ± 0.04 without the greyed-out taxa.

using the normal approximation to the binomial distribution), the significance drops to $p = 0.28 \pm 0.04$ – a huge change. But that does not tell us whether that change is itself significant: we must determine whether randomly removing the same number of associated tips from both trees would engender the same drop. Thus we must compare the significance drop in many replicates (a lengthy process) and count how many replicates have an equal or greater drop in significance (or, equivalently, how many have a lower significance than 0.28). If we find that 95% of tanglegrams with an equal number of tips removed have $p < 0.28$, then the 'pruning' of the matched cherries had a significant effect, and we may take this as an indication (though not proof) that preferential host switching, possibly in addition to codivergence, may be playing a large part in the coevolution of these two groups of organisms. This is the subject of current study.

8. DISCUSSION

The discipline of cophylogeny mapping is a new and growing one. There are few working in the field, but that number is gradually increasing, and the number of studies published which include a cophylogenetic study is also on the increase. Statistical, computational and theoretical methods are however in their infancy, and much needs to be done to extend the bounds of what information we can recover from related phylogenies.

In the above sections I have introduced the basic concepts of cophylogeny mapping and described some recent results in that field.

The *complexity* of the cophylogeny mapping problem has been explored empirically, and some interesting results unearthed: though it is unsurprising that the number of cophylogeny maps increases exponentially with the number of taxa, it is interesting to note also that the number of potentially optimal maps (those which are optimal for some scheme of event costs) and the optimal number of codivergences is negatively correlated, that is, the better the fit when measured in CEs, the fewer the potentially optimal solutions. This is a useful result because we can therefore constrain the map construction (using jungles) to maximise the fit of parasite into host tree (since it is that maximal degree of fit will be tested for significance): jungle construction is much faster when the constraints are tighter, and thus cophylogenetic analysis can be made more speedy when we do have closely congruent host and parasite phylogenies.

The downside of this analysis is that there is no apparent correlation between the number of feasible maps and the maximum number of CEs which can be posited to reconcile P with H, nor does there seem to be one between the number of potentially optimal maps and the degree of fit when measured in terms of NCEs, which the later section on *statistical testing* concluded was the better statistical test to use. This argument was based on the more spread out distribution in minimum number of non-codivergence events required to reconcile P with H, leading to a more precise measure of significance.

Another challenge is presented by the variability in cophylogenetic behaviour for particular sets of model parameters, described in the section on *modelling cophylogeny*, above. It has become clear that similar distributions of numbers of codivergence, duplication, host switching, extinction and 'missing the boat' events, can occur with quite different model parameters. This means that the converse problem of estimating model parameters from a single instance of the cophylogeny problem is fraught with difficulty, and the confidence intervals for these parameters are likely to be very broad. It is not clear what to do about this problem: since every biological system is different, we cannot expect to arrive at global parameters which can be agreed upon to apply to many systems, but perhaps within a range of similar systems some progress can be made. Again the cry must go up for more data, else we shall not be able to recover the global patterns of change which relate the large-scale evolution of host and parasite.

8.1. The Future

There is much to do in order to advance further the study of how phylogenies relate to each other.

The sampling problem is clearly an important issue which must be resolved – to date there has been little work on the effect of which hosts we choose, or which parasites we find on those hosts, or how we deal with widespread parasites (those whose life cycles are closely linked with more than just the definitive host).

A recent theoretical advance in cophylogeny mapping is the ability to map non-treelike phylogenies – either or both of the host and parasite phylogenies can be networks. The method, using 'lianas' (Charleston, 2003), makes use of the fact that none of the several qualities which we require of cophylogeny maps require either phylogeny to be a tree. We can let either (or both) be the more general structures of directed acyclic graphs, i.e., each must have a root, and be directed away from the root (towards the present). It is therefore possible to perform cophylogeny mapping with recombinant viruses as parasites, or hybridised plant phylogenies, or non-treelike histories of geographical areas. This generalisation represents a substantial advance in the practice of cophylogeny mapping, but still is of little use to the practising parasitologist or systematist until there is software to implement the method.

Another complication arises naturally when we start to consider the relationships among more than two phylogenies: unearthing the dependencies among figs, their pollinating and parasitoid wasps, and *their* parasites, is computationally intractable and must eventually be solved with heuristics (noting again that methods that take non-independent characters from phylogenies and then treat them independently cannot deal correctly with multiple phylogenies, despite the attractively intuitive optimisation of such multiple-derived 'data' sets on the single host tree). No methodology exists for such situations, though it would surely be desirable, if only to avoid the statistical problem of multiple testing relationships between the same phylogenies. Perhaps there is room for a 'hyperjungle' – though the computational complexity of such a structure promises to be prohibitively high, even with the possibility of added temporal data to further constrain the solutions.

ACKNOWLEDGEMENTS

The author is grateful to Andrew Jackson, Rod Page and Susan Perkins for useful comments, and to the Royal Society for continued support. This manuscript was prepared using LaTeX; graphs were created using gnuplot.

REFERENCES

Beer, B.E., Bailes, E., Goeken, R., Dapolito, G., Norley, S.G., Kurth, R., Gautier, J.P., Gautier-Hion, A., Vallet, D., Sharp, P.M. and Hirsch, V.M. (1999). Simian immunodeficiency virus (SIV) from sun-tailed monkeys (*Cercopithecus solatus*): evidence for host-dependent evolution of SIV within the *C. lhoesti* superspecies. *Journal of Virology* **73**, 7734–7744.

Brooks, D.R. (1981). Hennig's parasitological method: a proposed solution. *Systematic Zoology* **30**, 229–249.

Brooks, D.R. (1990). Parsimony analysis in historical biogeography and coevolution: methodological and theoretical update. *Systematic Zoology* **39**, 14–30.

Brooks, D.R. (2001). How to do BPA, really. *Journal of Biogeography* **28**, 345–358.

Charleston, M.A. (1998). Jungles: A new solution to the host/parasite phylogeny reconciliation problem. *Mathematical Biosciences* **149**, 191–223.

Charleston, M.A. (2003). Cophylogenetic maps generalize to reticulate phylogenies. *Journal of Mathematical Biology* (in prep.).

Charleston, M.A. (2002b). CoSpec: a Macintosh® program for modelling cospeciation.

Charleston, M.A. and Page, R.D.M. (2002). TREEMAP program. Version 2.0β. Platform: Macintosh®., http://evolve.zoo.ox.ac.uk/software/treemap/.
Charleston, M.A. and Perkins, S.L. (2002). Lizards, Malaria, and Jungles in the Caribbean. In: *Tangled Trees: Phylogeny, Cospeciation, and Coevolution* (R.D.M. Page, ed), pp. 65–92. Chicago: University of Chicago Press.
Charleston, M.A. and Robertson, D.L. (2002). Host switching in lentiviruses can account for phylogenetic similarity with the primate phylogeny. *Systematic Biology* **51**, 528–535.
Clayton, D.H. (1990). Host specificity of strigiphilus owl lice (Ischnocera: Philopteridae), with descriptions of new species and host associations. *Journal of Medical Entomology* **27**, 257–265.
Clayton, Dale, H. (1991). Coevolution of avian grooming and ectoparasite avoidance. In: *Ecology, Evolution and Behaviour* (J.E. Loye and M. Zuk, eds), pp. 258–289. Oxford: Oxford University Press.
Eichler, W. (1948). Some rules in ectoparasitism. *Annals and Magazine of Natural History (Series 12)* **1**, 588–598.
Fahrenholz, H. (1913). Ectoparasiten und abstammungslehre. *Zoologischer Anzeiger* **41**, 371–374.
Garey, M.R. and Johnson, D.S. (1979). *Computers and Intractability: A Guide to the Theory of NP-Completeness*. Series of books in the Mathematical Sciences. New York: W.H. Freeman & Co.
Goodman, M., Czelusniak, J., Moore, G.W., Romero-Herrera, A.E. and Matsuda, G. (1979). Fitting the gene lineage into its species lineage: a parsimony strategy illustrated by cladograms constructed from globin sequences. *Systematic Zoology* **28**, 132–168.
Gray, R.D. and Jordan, F.M. (2000). Language trees support the express-train sequence of austronesian expansion. *Nature* **405**, 1052–1055.
Hafner, M.S. and Nadler, S.A. (1988). Phylogenetic trees support the coevolution of parasites and their hosts. *Nature* **332**, 258–259.
Hahn, B.H., Shaw, G.M., De Cock, K.M. and Sharp, P.M. (2000). Aids as a zoonosis: scientific and public health implications. *Science* **287**, 607–614.
Huelsenbeck, J.P. and Rannala, B. (1997). Phylogenetic methods come of age: testing hypotheses in an evolutionary context. *Science* **276**, 227–232.
Huelsenbeck, J.P., Rannala, B. and Larget, B. (2000). A Bayesian framework for the analysis of cospeciation. *Evolution* **54**, 352–364.
Huelsenbeck, J.P., Rannala, B. and Yang, Z. (1997). Statistical tests of host-parasite cospeciation. *Evolution* **51**, 410–419.
Johnson, K.P. and Clayton, D.H. (2002). Coevolutionary history of ecological replicates: Comparing phylogenies of wing and body lice to Columbiform hosts. In: *Tangled Trees: Phylogeny, Cospeciation, and Coevolution* (R.D.M. Page, ed), pp. 262–286. Chicago: Chicago University Press.
Martin, J., Herniou, E., Cook, J., Waugh O'Neill, R. and Tristem, M. (1999). Interclass transmission and phyloetic host tracking in Murin Leukemia virus-related retroviruses. *Journal of Virology* **73**, 2442–2449.
Martin, J., Kabat, P. and Tristrem, M. (2002). Cospeciation and horizontal transmission rates in the murin leukaemia-related viruses. In: *Tangled trees: Phylogeny, Cospeciation, and Coevolution* (R.D.M. Page, ed), pp. 174–194. Chicago: Chicago University Press.
McKenzie, A. and Steel, M. (2000). Distributions of cherries for two models of trees. *Mathematical Biosciences* **164**, 81–92.

Page, R.D.M. (1990). Component analysis: a valiant failure? *Cladistics* **6**, 119–136.
Page, R.D.M. and Charleston, M.A. (1997). From gene to organismal phylogeny: Reconciled trees and the gene tree/species tree problem. *Molecular Phylogenetics and Evolution* **7**, 231–240.
Page, R.D.M. and Charleston, M.A. (1998). Trees within trees: phylogeny and historical associations. *Trends in Ecology and Evolution* **13**, 356–359.
Paterson, A.M., Gray, R.D. and Wallis, G.P. (1993). Parasites, petrels and penguins: does louse presence reflect seabird phylogeny? *International Journal for Parasitology* **23**, 515–526.
Paterson, Adrian, M., Palma, R.L. and Gray, R.D. (2002). Drowning on arrival, missing the boat and x-events: how likely are sorting events? In: *Tangled Trees: Phylogeny, Cospeciation and Coevolution* (R.D.M. Page, ed), pp. 287–309. Chicago: University of Chicago Press.
Rannala, B. and Michalakis, Y. (2002). Population genetics and cospeciation: From process tp pattern. In: *Tangled Trees: Phylogeny, Cospeciation, and Coevolution* (R.D.M. Page, ed), pp. 120–143. Chicago: University of Chicago Press.
Swofford, D.L. (2002). PAUP*: Phylogenetic Analysis Using Parsimony, version 4β. Computer program.
Yule, G.U. (1924). A mathematical theory of evolution based upon the conclusions of Dr J. C. Willis, FRS. *Phil. Trans. R. Soc. London (B)* **213**, 21–87.

Inference of Viral Evolutionary Rates from Molecular Sequences

Alexei Drummond[1,2], Oliver G. Pybus[1] and Andrew Rambaut[1]

[1]*Department of Zoology, University of Oxford, South Parks Road, Oxford, OX1 3PS, UK*
[2]*Department of Statistics, University of Oxford, South Parks Road, Oxford, OX1 3TG, UK*

Abstract	332
1. Introduction	332
1.1. Ancestral Diversity and Evolutionary Non-independence	334
2. General Linear Regression and Other Distance-Based Methods	337
2.1. Root-to-tip Linear Regression	337
2.2. Pairwise Distance Linear Regression	338
2.3. Generalised Least-squares on a Tree	339
2.4. Hypothesis Testing and Estimation of Errors	339
2.5. Examples of Linear Regression Methods	341
3. Maximum Likelihood Estimation	343
3.1. Hypothesis Testing and the Likelihood Ratio Test	344
3.2. Rate Variation Through Time	344
3.3. Examples of Maximum Likelihood Methods	345
3.4. Shortcomings of Current Maximum Likelihood Implementations	346
4. Bayesian Inference of Evolutionary Rates	347
4.1. Estimation of Errors Using MCMC	348
4.2. Examples of MCMC Estimation Methods	349
5. Discussion	350
5.1. Estimation of Divergence Times	351
5.2. The Neutral Theory of Molecular Evolution and the Molecular Clock	352
5.3. Estimating Generation Length	353
5.4. Conclusion	354
Acknowledgements	354
References	355

ABSTRACT

The processes of mutation and nucleotide substitution contribute to the observed variability in virulence, transmission and persistence of viral pathogens. Since most viruses evolve many times faster than their human hosts, we are in the unusual position of being able to measure these processes directly by comparing viral genes that have been isolated and sequenced at different points in time. The analysis of such data requires the use of specific statistical methods that take into account the shared ancestry of the sequences and the randomness inherent in the process of nucleotide substitution. In this paper we describe the various statistical methods for estimating evolutionary rates, which can be classified into three general approaches: linear regression, maximum likelihood, and Bayesian inference. We discuss the advantages and shortcomings of each approach and illustrate their use through the analysis of two example viruses; human immunodeficiency virus type 1 and dengue virus serotype 4. Reliable estimates of viral substitution rates have many important applications in population genetics and phylogenetics, including dating evolutionary events and divergence times, estimating demographic parameters such as population size and generation time, and investigating the effect of natural selection on molecular evolution.

1. INTRODUCTION

As a general rule, parasites have faster rates of mutation than their hosts. Parasites tend to be smaller in size, with shorter generation times, and therefore undergo more rounds of reproduction per unit time. This difference in mutation rates is particularly clear for viruses and their human hosts, not only because viral generation times are often very short, but also because replication of their genetic material is commonly many times more error-prone than in humans. That said, viruses do vary widely in mutation rate as a result of differences in their life cycles and mode of replication (e.g., Holland *et al.*, 1982; Smith and Inglis, 1987; Jenkins *et al.*, 2002).

The most significant consequence of the high mutation rate of viruses is their ability to quickly adapt to their environment, as illustrated by the rapid evolution of human immunodeficiency virus (HIV) strains that are resistant to anti-viral drugs or are capable of evading the hosts' immune response (e.g., Nijhuis *et al.*, 1997; Goulder *et al.*, 2001). In other circumstances, mutation may allow a virus to productively infect new cell types or new host species. As the rate of mutation contributes to the adaptive potential of a virus it is obviously important to accurately measure this value. In addition, the mutation rate is a key parameter in population genetic and phylogenetic analyses of viral populations and is therefore necessary to understand both the pattern of viral genetic diversity observed today and the timescale of past evolutionary and epidemiological events.

In this article we describe the various methods by which mutation rates can be estimated from molecular sequence data, and discuss the advantages and disadvantages of each. We illustrate these methods by applying them to two human viruses that cause worldwide morbidity and mortality; human immunodeficiency virus type 1 (HIV-1) and dengue virus serotype 4 (DEN-4). Both are RNA viruses whose large genetic diversity and high mutation rates directly contribute to their virulence and pathogenicity. Our choice of data sets also illustrates the range of evolutionary timescales across which the methods we describe can be applied, as the HIV-1 data are taken from a study of viral evolution within an individual infected patient (Shankarappa *et al.*, 1999), whereas the DEN-4 data have been sampled from infected individuals across several decades in many different countries (Lanciotti *et al.*, 1997). Both these data sets are characterized by the fact that the sequences were sampled at different points in time (commonly referred to as temporally spaced, or serially sampled sequences).

Although we only consider viruses here, the methods described are equally applicable to any population from which gene sequences sampled at different points in time show a statistically significant number of genetic differences. Populations from which estimates of mutation rates can be readily obtained are characterised by some combination of the following properties: (i) a high mutation rate, (ii) long periods of time between samples, as is the case for "ancient DNA", and (iii) long stretches of sampled sequence data.

At this point we must introduce the distinction between mutation rates and substitution rates, although the two terms are sometimes confused in the literature. The former is the rate at which mutational errors are incorporated into a genome during replication, and can be expressed as the number of mutations per nucleotide site per replication event. This rate is largely determined by the particular viral or host polymerase used and the presence or absence of post-replicative repair systems. RNA viruses and small DNA viruses tend to lack such repair systems and thus have higher mutation rates. Molecular biology techniques can be used to estimate the mutation rate of viruses *in vitro* and *in vivo* (e.g., Mansky and Temin, 1995).

In contrast, the substitution rate of a virus depends on many factors and is a property of the viral population as a whole. It is the rate at which new mutations spread and become fixed in the population as a result of natural selection or random genetic drift, and is expressed as the number of substitutions per nucleotide site per unit time (days, years or generations). The substitution rate depends on the complex interaction between the effective size of the population and the distribution of mutational selection coefficients, that is, the relative proportion of mutations that are advantageous, neutral or disadvantageous. Some of these interactions can be unravelled by comparing the substitution rates separately at synonymous and non-synonymous nucleotide sites – mutations at synonymous sites do not change the encoded amino acid and can therefore be considered as having little or no selective effect (for example, Kimura, 1977). One of the most important theoretical results in molecular evolution is that if all mutations are selectively neutral then the substitution rate is equal to the mutation rate (Kimura and Ohta, 1971; Kimura, 1987). Although useful, the argument that synonymous sites are neutral should be applied carefully as it assumes the absence of several factors, some of which are common in viruses, namely (i) fitness differences arising from the use of alternative codons, (ii) secondary RNA or DNA structure in coding and non-coding regions, and (iii) overlapping reading frames. In general, if selection is acting at the nucleotide level as well as the encoded protein absolute statements about selection at the protein level can be difficult, although relative statements can often still be made.

All the procedures outlined below estimate the substitution rate, not the mutation rate. Although the substitution rate depends on several parameters and may sometimes be difficult to interpret, it is important precisely because it does contain information about many fundamental evolutionary processes. For example, if two genes in the same viral genome have unequal substitution rates then we might conclude that different selection pressures have been acting on them, since the underlying mutation rate is unlikely to differ. Furthermore, comparison of substitution rates among different virus species and strains may shed light on the varied roles that mutation and genetic diversity play in maintenance of viral infection and transmission.

1.1. Ancestral Diversity and Evolutionary Non-independence

Intuitively, one might expect to be able to estimate substitution rates using a simple argument along the lines of "distance equals rate multiplied by time". If two gene sequences differ at d nucleotide sites and were sampled at different points in time, t_1 and t_2, then this rationale suggests that the substitution rate μ equals $d/(t_2-t_1)$ (Figure 1a). However, this will only be true if one sequence is a direct ancestor of the other. In practice, as illustrated in Figure 1b, this method will overestimate μ when the time of most recent common ancestor (t_{root}) of the two sequences exists prior to t_1. This is known as the problem of "ancestral diversity" and occurs because the population at time t_1 contains some genetic variation. In most cases t_{root} will be unknown so the time over which genetic distance d has accumulated will also be unknown. Ancestral diversity can be taken into account by adding a third outgroup sequence. The difference between the genetic distances (d_1 and d_2) of the two sampled sequences to this outgroup thus reflects the difference between their sampling times (Figure 1c; Li et al., 1988). This argument holds even if individual nucleotide sites have undergone multiple substitutions, provided that an accurate probabilistic model of nucleotide substitution is used to estimate d (see Swofford et al., 1996). In common with many other methods this approach assumes that the substitution rate remains constant through time, an assumption known as the molecular clock.

Extending this methodology to multiple sequences appears straightforward; for example, if y_i is the genetic distance from sequence i to the most recent common ancestor of the sampled sequences (measured off a phylogenetic tree) and t_i is the sampling time of sequence i, then the

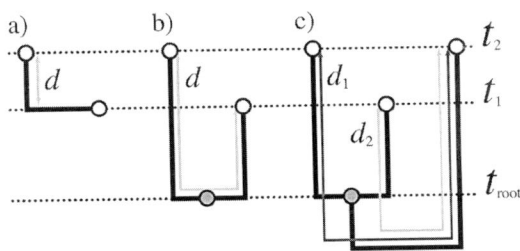

Figure 1 The problem of ancestral diversity. Gene sequences (open circles) have been sampled at two time points, t_1 (earlier) and t_2 (later). The vertical dimension represents genetic distance. (a) There is zero genetic diversity at the earlier time point (t_1) so the genetic distance d between the sequences reflects the difference in sampling times. (b) Due to genetic diversity at the earlier time point, the common ancestor of the sampled sequences (t_{root}) exists prior to t_1. Hence the genetic distance d is erroneously large. (c) The problem of ancestral diversity can be avoided by using an outgroup. The difference between the distances d_1 and d_2 correctly reflects the difference in sampling times.

gradient of a linear regression of y_i against t_i should provide an estimate of the substitution rate μ (Figure 2). We call this method the root-to-tip linear regression method and it has often been used to estimate substitution rate, but unfortunately it has serious shortcomings. It assumes that each y_i is statistically independent, whereas the sampled sequences are linked by a common evolutionary history and are thus not independent. The non-independence arises from the internal branches of the phylogeny that describes this shared history, as these branches contribute to multiple pairs of y_i and t_i values. This problem of non-independence arises in many other evolutionary problems (e.g., Harvey and Pagel, 1991) and can be solved by developing methods that explicitly incorporate the phylogenetic structure implicit in sampled sequence data. The use of models that do not incorporate this information will produce unpredictable biases in inference and hypothesis-testing procedures.

While many computational methods have been developed to estimate the phylogenetic structure of sequence data under the assumption of a molecular clock, few allow for temporally spaced sequences. Methods such as UPGMA (Sokal and Michener, 1958), likelihood ratio tests of the molecular clock (Felsenstein, 1981) and coalescent methods in population genetics (Kingman, 1982; Hudson, 1990; Kuhner et al., 1995) all assume that there are no significant differences between the sampling times of the individual sequences. Rather than being a potential problem, we show here that the unique structure of temporally spaced sequence data is an asset that can be exploited to a number of novel ends, including the accurate estimation of substitution rates.

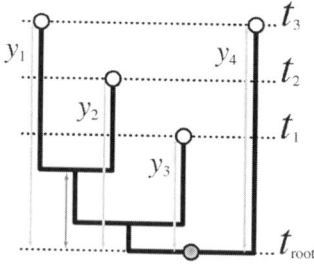

Figure 2 Root-to-tip distances measured on a phylogeny. Four gene sequences (open circles) have been sampled at three different time points (t_1, t_2, t_3). The y values represent the genetic distances from each sequence to the root (filled circle). Many of the y values are non-independent because of shared ancestry of the sequences. For example, the dark arrow denotes the shared part of distances y_1 and y_2.

The remainder of this article is organised with each section focusing on a different approach for estimating molecular evolutionary rates. We start with the simplest methods and progress to the more sophisticated. In each section we demonstrate the relevant methods on HIV-1 and DEN-4 data sets, discussing the advantages and shortcomings of each.

2. GENERAL LINEAR REGRESSION AND OTHER DISTANCE-BASED METHODS

Some of the first estimates of substitution rates using temporally spaced sequences were obtained by comparing sequences of human influenza A strains (Krystal *et al.*, 1983; Martinez *et al.*, 1983; Hayashida *et al.*, 1985). All of this early work involved direct comparison of the genetic distance between two sequences with the interval separating their isolation times (Figure 1a). As explained in the previous section, this method is only accurate if the genetic diversity of the population at the time of sampling is negligible, such that sequences isolated at different times differ only by substitutions accumulated during the time interval. If this condition is not met then this method has an upward bias and can provide an upper limit for the estimate of substitution rate. Consequently, the outgroup method described in Figure 1c was introduced in the context of estimating the rate of evolution of HIV-1 (Li *et al.*, 1988).

From the mid 1980s to the present, a series of distance-based regression methods were employed by various researchers to remove the "ancestral diversity" bias generated by the substantial population polymorphism that exists in most viral populations (for example, Buonagurio *et al.*, 1986; Saitou and Nei, 1986; Gojobori *et al.*, 1990; Fitch *et al.*, 1991; Leitner and Albert, 1999; Pagel, 1999; Shankarappa *et al.*, 1999; Drummond and Rodrigo, 2000; Korber *et al.*, 2000). With some exceptions, these methods were largely introduced in an informal manner, and often not linked to past research. In the next sections we will describe the three major classes of regression methods that this body of research fall into.

2.1. Root-to-tip Linear Regression

The "root-to-tip" linear regression method, described briefly above, has been a common choice for the estimation of substitution rate. This method proceeds by first estimating a rooted phylogeny of the sequences under analysis and then performing a linear regression between the time of sampling of each tip and the genetic distance (sum of reconstructed branch lengths) from the root to each tip. The linear model is thus:

$$E[d_{\text{root},i}] = \mu(t_i - t_{\text{root}}) = \mu t_i - \mu t_{\text{root}}$$

where t_i is the time of tip i, μ is the unknown substitution rate and t_{root} is the unknown time of the root (in population genealogies this is equal to the time of the most recent ancestor). Under this model, the gradient of a linear regression of $d_{\text{root},i}$ against t_i provides an estimate of the substitution rate and the y-intercept is equal to $-\mu t_{\text{root}}$. By definition, the x-intercept is equal to t_{root}. Shankarappa et al. (1999) used this linear regression technique to study the long-term intra-host rate of HIV-1 evolution in nine infected patients, and Korber et al. (2000) used it to date the origin of HIV-1 group M.

2.2. Pairwise Distance Linear Regression

The second regression method was introduced by Leitner and Albert (1999) and formally described and extended upon in Drummond and Rodrigo (2000). This method relies on a result from the population genetics literature; that the expected distance between two random sequences sampled at the same time in a haploid population is equal to $\Theta = 2N_e\mu_g$, where N_e is the effective population size and μ_g is the mutation rate per site per generation. Extending this model to pairs of sequences i and j with times t_i and t_j, results in the following linear model:

$$E[d_{i,j}] = \mu|t_i - t_j| + \Theta$$

Following this model, the gradient of a linear regression of $d_{i,j}$ against $\triangle t_{ij} = |t_i - t_j|$ is an estimate of the substitution rate and the y-intercept is an estimate of the population genetic parameter Θ. Because μ_g is measured in mutations per generation, while the slope of the regression is typically in units of substitutions per year or per day, it is only possible to interpret the x-intercept of this regression as the product of $2N_e$ and generation length in the time units used. Unlike the root-to-tip method, this method does not require an estimate of the tree, and has been shown to be an unbiased estimator (Drummond and Rodrigo, 2000). However, because it does not use information about the correlation of the sequences due to shared ancestry the power of this method is significantly reduced, typically leading to very large confidence intervals. Leitner and

Albert (1999) used a parametric bootstrapping technique to demonstrate that the distribution of pairwise distances in the HIV-1 transmission history they analysed was not over-dispersed. This suggests that a simple Poisson process could not be rejected as an adequate description of molecular evolution of HIV-1 in the transmission history they studied. However, this result could also be due to the low power of this method.

2.3. Generalised Least-squares on a Tree

An interesting approach to substitution rate estimation arises from literature concerning the use of the comparative method in evolutionary biology (Harvey and Pagel, 1991; Pagel, 1999). The comparative method is a general framework for estimating the covariance of phenotypic traits among species, which correctly accounts for the non-independence arising from shared ancestry. These methods can be regarded as a class of generalised least squares (GLS) approaches. Within this framework the correlation between a continuous trait and the branch lengths of the tree itself can also be examined. If we regard time itself as a continuous trait, then we can consider the correlation of time with the branch lengths of the tree. If the branch lengths of the tree are in units of substitutions per site then we can obtain an estimate of the rate of substitution per site per unit time (see Pagel, 1999). Because this method explicitly considers the correlation structure of the tree, the non-independence of observed sequences is correctly accounted for. However, the interpretation of this method is difficult, as it assumes that time is a random variable, when clearly the substitutions themselves are the source of stochasticity, not time. Nevertheless, this method represents an interesting intermediate between the distance methods outlined above and the fully probabilistic methodologies outlined in later sections.

2.4. Hypothesis Testing and Estimation of Errors

Point estimates are meaningless without a measure of the error associated with the estimate. This is particularly important when the aim of estimation is hypothesis testing. Unfortunately we cannot make use of standard linear regression tests and statistics because these assume that the data are independent, whereas the sequences are not independent because they share evolutionary history. For example, the use of confidence intervals about a regression line (e.g., Tanaka *et al.*, 2002) can produce an underestimate of the true error about substitution rate.

However, the linear regression procedures are amenable to general statistical techniques for error estimation such as bootstrapping (parametric and non-parametric; Efron and Tibshirani, 1993) and jackknifing (Wu, 1986). These methods involve the construction of pseudo-replicate data sets to estimate the stochastic error in the original data. In the case of bootstrapping, this can be done by randomly sampling the original data with replacement (non-parametric bootstrapping), or by simulating data using an assumed or inferred model of evolution (parametric bootstrapping). However, care must be taken because of the non-independence of the genetic distance data used. For example, Korber *et al.* (2000) used the root-to-tip regression method to estimate the evolutionary rate and age of the root (t_{root}) of HIV-1 (group M) viruses. Korber *et al.* attempted to estimate confidence intervals around their point estimate of t_{root} by re-sampling with replacement (non-parametric bootstrapping) the linear regression data points. Their analysis gave a tight confidence interval around their estimate of t_{root} and thus enabled them to reject a recent hypothesis for the origin of HIV-1. However, in this setting the bootstrap procedure they used will underestimate the true confidence intervals as it does not take into account the correlation of bootstrap replicates due to shared ancestry (i.e., it treats each root-to-tip distance as an independent piece of information about substitution rate). A statistically rigorous approach would have been to bootstrap the nucleotide sites in the sequence alignment, rather than the root-to-tip distances, as each site in an alignment represents an independent realization of the substitution process. In the next section we verify this theoretical result and show that bootstrapping the nucleotide sites rather than the regression points produces significantly larger and more realistic confidence intervals for a set of DEN-4 virus sequences.

Non-parametric bootstrapping of sequence data could also be used to estimate the confidence intervals of the GLS method. However, non-parametric bootstrapping of sequences is not sufficient for error estimation in the pairwise distance regression. The pairwise distance regression has two sources of statistical error: (i) error due to the substitution process, and (ii) error due to the coalescent process. Thus, a full parametric bootstrap must be employed to correctly assess the error associated with the pairwise distance method (Drummond and Rodrigo, 2000).

In all three cases the statistical error due to phylogenetic reconstruction is difficult to accommodate. However, these problems can be circumvented by the use of full likelihood or Bayesian methods, which we describe in Sections 3 and 4, below.

2.5. Examples of Linear Regression Methods

In this section we compare the root-to-tip and pairwise distance linear regression methods on two example datasets (DEN-4 and HIV-1) in order to illustrate their relative performance. For both methods, a matrix of pairwise genetic distances was calculated using an empirically derived F84 model of substitution (described in Swofford et al., 1996). For the root-to-tip regression method, this matrix was used to estimate a neighbour-joining tree (Saitou and Nei, 1987), and the root of the tree was picked so as to maximise the R^2 value of the regression. In a real-life application, some form of model selection process should be used to choose the substitution model that best describes the data.

Figures 3 and 4 display the results of both root-to-tip and pairwise distance regression analyses on the DEN-4 and HIV-1 datasets, respectively. The root-to-tip estimate of substitution rate for the DEN-4 dataset was 8.14×10^{-4} substitutions per site per year with a confidence interval of $[4.69 \times 10^{-4}, 14.1 \times 10^{-4}]$ estimated from 5000 bootstrap replicates of the sequence data. The corresponding estimate of the date of the root is 1928 with a confidence interval of [1901, 1946]. The estimated rate compares with 9.10×10^{-4} $[0, 33.7 \times 10^{-4}]$ for the pairwise distance method. These confidence intervals are very large and include a rate of zero. The two

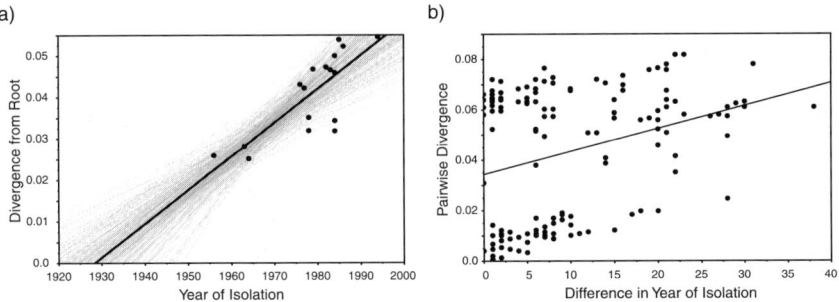

Figure 3 Application of the linear regression methods to 17 Dengue virus serotype 4 (DEN-4) sequences, isolated from different patients across five decades. (a) Root-to-tip linear regression. The gradient of the regression (bold line) is an estimate of the substitution rate (μ) and the x-axis intercept is an estimate of the time of the most recent common ancestor (t_{root}). The pale lines represent 500 bootstrap replicates of the original sequence alignment and provide an estimate of the statistical error in these estimates. (b) Pairwise distance linear regression. The gradient of the regression (bold line) is an estimate of the substitution rate (μ) and the y-axis intercept is an estimate of Θ (see text for details).

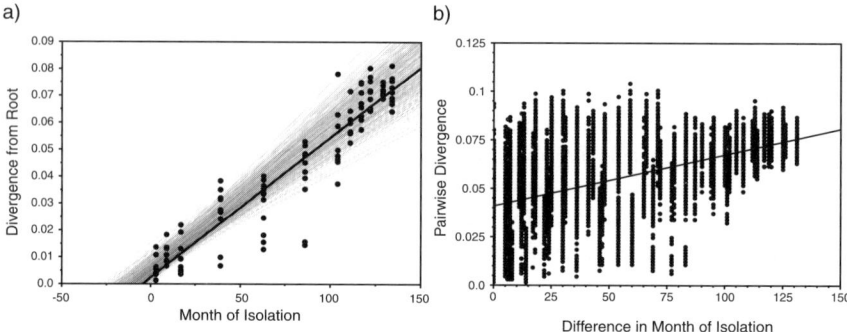

Figure 4 Application of the linear regression methods to 117 HIV-1 sequences, isolated over 134 months from a single infected patient. (a) Root-to-tip linear regression. (b) Pairwise distance linear regression. See Figure 3 legend for more details.

methods are congruent, and as we will see in later sections they agree well with the results of more sophisticated methods, but suffer from a lack of power, and less flexibility in model specification.

Note that if we had followed Korber *et al.* (2000) and bootstrapped the root-to-tip distances instead of the sequence data we would have calculated a confidence interval for substitution rate of only [5.83×10^{-4}, 11.1×10^{-4}], and a confidence interval for t_{root} of [1912, 1942]. These intervals are 56% and 67% of the intervals produced by the correct bootstrapping method. Discrepancies of this magnitude can easily result in incorrect conclusions when testing specific hypotheses.

Figure 4 shows the results of the distance-based analyses of the HIV-1 dataset. The pairwise regression method gave an estimate rate of 3.17×10^{-3} [0.26×10^{-3}, 8.61×10^{-3}] substitutions per site per year, whereas the root-to-tip regression method gave an estimate of 6.24×10^{-3} [4.64×10^{-3}, 7.89×10^{-3}], with confidence intervals that exclude the pairwise estimate. A possible reason for this discrepancy is model misspecification in one or both of the methods. However because of the very wide confidence intervals on the pairwise method, the discrepancy between these results may not be very important.

In general, linear regression procedures are fast and useful for visualising new data sets. They can assist in model selection, by suggesting whether a uniform or variable model of substitution rate through time is necessary to explain the temporal structure in a given data set (for example, Shankarappa *et al.*, 1999). However, they make several limiting assumptions and we do not recommend that they provide the final result of an analysis of temporally spaced viral data. Newer methods, such as maximum likelihood and Bayesian statistical inference, can utilise more information from the sequences and can potentially allow much more complex models of molecular evolution and demography to be investigated. We will describe these methods in the next two sections.

3. MAXIMUM LIKELIHOOD ESTIMATION

Since the initial attempts to estimate rates using distance-based methods, a number of researchers have developed and tested ML methods that accommodate the time structure of temporally spaced sequences (Rambaut, 2000; Drummond et al., 2001; Seo et al., 2002b). Each tip of the tree has a known time (the isolation date of the sequence). The times of the internal nodes of the tree are initially unknown and are given arbitrary starting times consistent with their order in the tree. An additional parameter, the substitution rate, is then used to scale these times into units of expected number of substitutions per site. Given the tree and the expected number of substitutions per site for each branch of that tree, the likelihood of the model can be calculated (Felsenstein, 1981). The vector of internal node times, $(t_0, t_1, \ldots, t_n)$ along with the substitution rate (μ) and any parameters of the substitution model (such as the transition–transversion ratio) are then put into a standard multi-dimensional optimisation procedure to find the values that provide the maximum likelihood. This model has been labelled as the 'single rate dated tips' (SRDT) model (Figure 5; Rambaut, 2000).

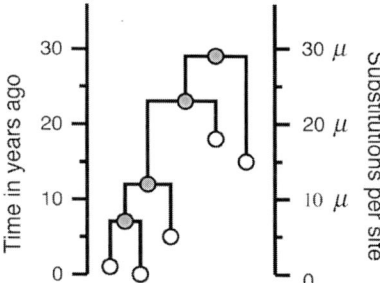

Figure 5 Maximum likelihood estimation of substitution rate using the single rate dated tips (SRDT) model. Ancestral divergence times (filled circles) are unknown and free to vary. The isolation times of the sampled sequences (open circles) are known and fixed. The substitution rate (μ) is used to convert the isolation times into genetic distances (measured in units of substitutions per site). The ancestral divergence times and μ are then found by maximum likelihood.

These methods are more sensitive and accurate than distance-based methods, as can be seen by the reduced confidence intervals reported in Section 3.3.

3.1. Hypothesis Testing and the Likelihood Ratio Test

One of the strengths of the ML inference framework is that it provides powerful tools for hypothesis testing and model comparison through tests such as the likelihood ratio test (LRT). This test uses the difference in log likelihood between two hypotheses to assess whether one provides a significantly better fit to the data than the other. The LRT requires that the hypotheses are nested, that is, one or more parameters of the more general hypothesis are constrained to particular values in order to obtain the more specific hypothesis. For example, the substitution rate parameter could be constrained to a particular *a priori* value (perhaps a previously inferred value estimated from different data). The likelihood ratio would then be used to test whether this substitution rate was a significantly worse fit to the data than the maximum likelihood estimate of rate (the more general hypothesis). For such cases, an approximate null distribution of the likelihood ratio statistic has been described (Wilks, 1938) or it can be generated using simulation (e.g., Huelsenbeck and Rannala, 1997). One of the first descriptions of such a test in phylogenetics was the test of the molecular clock by Felsenstein (1981) but this test assumes that all sequences were sampled contemporaneously. The application of the likelihood ratio test to temporally sampled data is described in more detail by Rambaut (2000) and Drummond *et al.* (2001).

3.2. Rate Variation Through Time

There are a number of reasons why evolutionary rates may vary through time across an entire viral population. For example it has been suggested that HIV-1 viruses in a host exhibit a slowdown in substitution rate at the end of the asymptomatic period (Shankarappa *et al.*, 1999). These patterns of concerted population-wide changes in rate through time have also been observed due to external changes in environment such as application of anti-retroviral drugs (Drummond *et al.*, 2001). It is relatively straightforward to design an LRT that will test the hypothesis of concerted rate variation through time, and this has already been described for the case of stepwise changes in substitution rate (Drummond *et al.*, 2001). Models of this variety may be described as multiple rates dated tips (MRDT) models.

Further advances in this direction will assist in the rigorous and detailed dissection of the molecular evolutionary process and its variation through time.

3.3. Examples of Maximum Likelihood Methods

Under a maximum likelihood framework we have greater flexibility in model selection. Using the program TipDate (Rambaut, 2000) we estimated the substitution rate of the DEN-4 dataset using the HKY model of substitution with a different rate at each codon position. The input tree topology was estimated under the different rates (DR) model and the same substitution model in PAUP* (Swofford, 1998). The maximum likelihood estimate of the mean substitution rate was 7.91×10^{-4} [6.07×10^{-4}, 9.86×10^{-4}] substitutions per site per year and the estimated age of the root is 1922 [1900, 1936]. Figure 6 shows the maximum likelihood trees for DEN-4 under the SRDT model of evolution.

Using the program PAML (Yang, 1997), which also allows estimation of rate under the SRDT model but is better able to handle large data sets, we

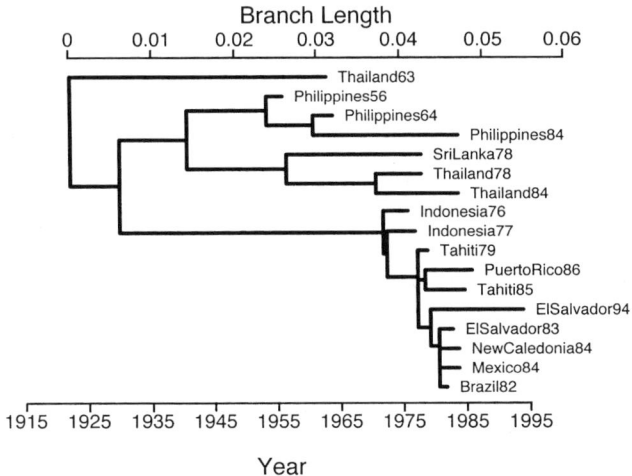

Figure 6 Application of the maximum likelihood approach to the DEN-4 data set. The phylogeny shown was estimated under the SRDT model, so that each sequence is positioned correctly with respect to its sampling date. The tips are labelled with the year of sampling and the top scale gives the genetic distance from the root. At the bottom is the timescale in years estimated using maximum likelihood.

estimated the rate of evolution of the HIV-1 sequences. Again, the input tree topology was estimated under the different rates (DR) model and under the HKY model of substitution in the program PAUP*. The maximum likelihood estimate of the mean substitution rate was 4.24×10^{-3} [3.26×10^{-3}, 5.36×10^{-3}] substitutions per site per year. The associated confidence intervals do not contain either the pairwise or the root-to-tip regression point estimates. This inconsistency in error estimates between different methods arises from the implicit assumptions of each method, such as the assumption of perfect knowledge of the tree topology, which we discuss in the next section.

3.4. Shortcomings of Current Maximum Likelihood Implementations

One limitation of current ML implementations, such as PAML and TipDate, is that only a single tree is considered. This can be a problem for two reasons: Firstly, there is usually considerable uncertainty in our estimation of the true tree, so that it becomes important to reflect this uncertainty in the confidence interval associated with the estimated evolutionary rate. Secondly, the maximum likelihood tree topology under the single rate dated tips (SRDT) model can be different from the maximum likelihood tree topology under the DR model (Drummond et al., 2001), so using an ML topology from PAUP* can bias substitution rate estimation using TipDate. As with the root-to-tip regression analysis, the use of a single tree introduces the potential for bias.

One could attempt to simultaneously find the maximum likelihood tree and the evolutionary rate using heuristic optimisation. This would involve progressively making changes to the tree accepting those that improve the likelihood (the hill-climbing approach). Such techniques are used to estimate the maximum likelihood tree in phylogenetics packages such as PAUP* and PHYLIP, although not for the case of temporally sampled sequences. Whilst such methods are feasible for small numbers of sequences, the number of possible trees increases explosively as the number of sequences increases (Schröder, 1870). This, in addition to the complex nature of the constraints of the SRDT model, would make hill-climbing extremely susceptible to producing sub-optimal solutions. On the other hand, if we were not directly interested in the ancestral tree itself, it would be preferable to have a method that took into account the shared ancestry of the data without basing inference on a single estimation of ancestral genealogy. Markov chain Monte Carlo (MCMC) methods provide exactly this opportunity.

4. BAYESIAN INFERENCE OF EVOLUTIONARY RATES

Markov chain Monte Carlo (MCMC) integration is often used in statistical inference to summarise high-dimensional probability densities where analytical solutions are difficult or impossible to calculate. MCMC works by sampling the probability density function of interest, so as to provide a representative sample of parameter values of the chosen model, given the data. To estimate substitution rates, the chosen model generally includes the tree topology, the times of ancestral nodes in the tree, the substitution rate, and substitution parameters such as the transition/transversion ratio (Drummond *et al.*, 2002).

In phylogenetics and population genetics we often want to estimate parameters, such as substitution rate, despite not knowing the true ancestral history of the sequences. A good solution to this problem would be to estimate the substitution rate from each of a large set of different ancestral histories and then combine these individual estimates such that each rate is weighted proportional to the likelihood of the corresponding tree. Trees that make the data highly probable contribute most to the overall estimate. By making the tree a nuisance parameter of the model, it becomes possible to sample all plausible trees in an MCMC analysis in order to find the range of plausible substitution rates. Unlike ML, which typically employs some kind of hill-climbing procedure, MCMC is a stochastic algorithm and is thus able to avoid getting stuck in local sub-optimal solutions because it samples the whole distribution of interest. At each step in the algorithm, MCMC proceeds by proposing a new set of parameter values (of which the tree topology is one) and then either accepting or rejecting the newly proposed state based on the Metropolitan-Hastings criterion (Metropolis *et al.*, 1953; Hastings, 1970). In essence, if the proposed state is better than the previous state, it is accepted. However, if the proposed state is, say, 10 times worse than the current state, it is accepted with a probability of $p = 1/10 = 0.1$. If the proposed state is rejected then the MCMC retains the current state and the process is repeated. Using this acceptance criterion, the proportion of times the MCMC algorithm visits a particular tree is an estimate of its relative probability given the data.

Sample-based inference using MCMC readily lends itself to Bayesian inference, in which prior information can be incorporated into the analysis. Probability theory tells us that *Posterior Probability* $\propto$ *Likelihood* $\times$ *Prior Probability*. One natural approach to assigning prior probabilities to genealogies that represent large populations is the coalescent process (Kingman, 1982). In addition, parameters such as t_{root} and effective population size (N_e) can be given prior distributions that reflect information from independent sources, or are simply used in an exploratory manner to investigate different *a priori* assumptions and hypotheses. For example, in the case of a set of viruses sampled from a single infected host, a potential prior distribution on t_{root} is the age of the host. This prior represents the assumption that the initial infection was from a single viral particle or a small homogeneous population with no double-infection.

The historical population processes that shape the genetic diversity of a population can be illuminated by genealogical methods such as the coalescent (Kingman, 1982). The coalescent is the most appropriate framework for studying the evolutionary genetics of a large population from which a sample of sequences is drawn, and provides a number of opportunities for inference in viral populations (e.g., Pybus *et al.*, 2000, 2001). A description of the coalescent for serially sampled sequences has recently been given (Rodrigo and Felsenstein, 1999). This formulation of the coalescent has been used to develop methods that estimate population sizes and substitution rates from serially sampled sequences whilst taking into account the uncertainty of the tree topology using MCMC (Drummond *et al.*, 2002). Others have used the coalescent to describe a pseudo-maximum likelihood method of estimating population size or substitution rates when the tree is known (Seo *et al.*, 2002a). However, it should be noted that currently implemented coalescent methods make a number of limiting assumptions, specifically, no population subdivision, no recombination within the genome region under investigation, and no selection.

Theoretical developments in the future will enable these assumptions to be relaxed, but for the time being our understanding of the molecular biology and life cycle of the virus concerned should be used to carefully interpret the results of each analysis. For example, strong natural selection acts on many viruses, but often acts unequally at different levels: within an infected individual, HIV is constantly adapting in response to the host's cellular and humoral immune responses and selection is obviously strong. However, successful transmission to a new host almost always leads to the establishment of a persistent infection, so the number of "offspring" infections generated by one infected host is primarily determined by that host's behaviour, rather than by particular mutations in the viral genome. Thus the reproductive success of an HIV infection at the epidemiological level has a low heritability, and consequently selection at this level will be weak and slow-acting.

4.1. Estimation of Errors Using MCMC

Highest posterior density (HPD) intervals and central posterior density (CPD) intervals are Bayesian analogues of the confidence interval. HPD and

CPD intervals of a parameter of interest, such as substitution rate, can be obtained empirically from the frequency distribution of the parameter's values sampled by the MCMC algorithm. This is valid because after the MCMC algorithm has had an appropriate time to converge (referred to as the burn-in period) it will begin to sample values of a parameter at a frequency proportional to their (posterior) probability density. The resulting frequency distribution of a parameter of interest is thus an empirical estimate of the marginal posterior probability density of the parameter. These marginal densities can be used to reject specific *a priori* hypotheses; for example, the substitution rates of two genes are the same.

4.2. Examples of MCMC Estimation Methods

MCMC was used to estimate the substitution rate of DEN-4 without assuming exact knowledge of the tree topology. In addition, a coalescent prior on node heights was introduced to investigate its effect on rate estimation. Figure 7 shows the marginal probability distributions for each of the three codon positions as well as the mean substitution rate across all nucleotide positions. The posterior estimate of mean substitution rate in DEN-4 was 8.29×10^{-4} [6.33×10^{-4}, 10.4×10^{-4}] substitutions per site per year. Notice that this HPD interval is slightly larger than the ML confidence interval (Figure 9). This reflects the increased uncertainty in the rate due to the uncertainty in the exact genealogical relationships of the sequences. By assuming a single tree topology the ML analysis gave artificially tight confidence intervals. Interestingly, the confidence intervals for the age of the root are smaller in the MCMC analysis. This probably reflects the effect of the coalescent prior on the tree topology, as assumptions about population processes will tend to reduce the variance in estimates of node times.

Figure 8 shows the resulting probability densities of substitution rate for two independent MCMC analyses of HIV-1 dataset. The only difference between the models used was that the first analysis assumed a constant population size whereas the second allowed an exponentially expanding population (with the growth rate included as a parameter of the model). The two distributions are remarkably similar demonstrating a robustness of the estimate of rate to the exact choice of prior on the distribution of internal node ages. The posterior estimates and HPD intervals of substitution rate for the constant size and exponentially expanding population models were 6.19×10^{-3} [5.32×10^{-3}, 7.07×10^{-3}] and 6.11×10^{-3} [5.33×10^{-3}, 6.88×10^{-3}] substitutions per site per year, respectively. With these examples, we have tried to demonstrate that properties such as (i) low variance, (ii) flexibility of modelling and (iii) accurate assessment of

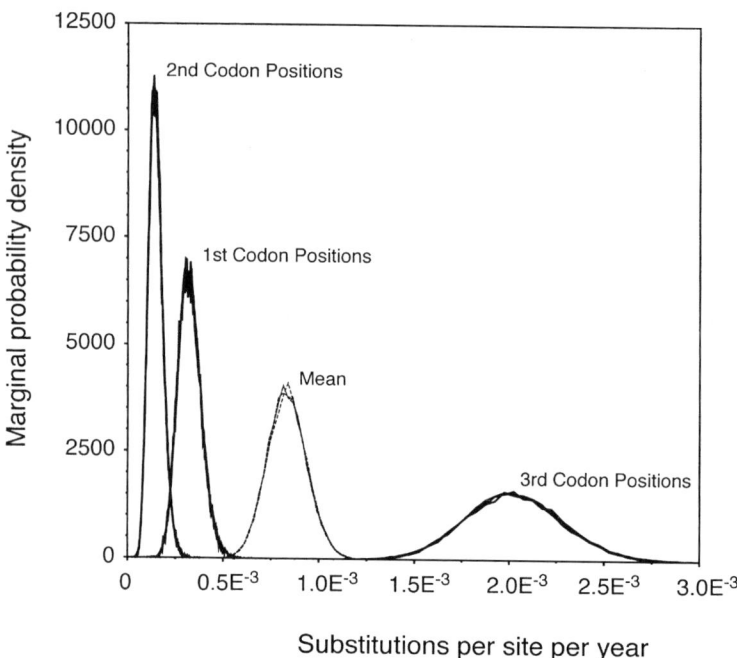

Figure 7 Application of the Bayesian inference approach to the DEN-4 data set. The figure shows the estimated posterior distributions of substitution rate for each codon position. In addition, the estimated posterior distribution of the mean rate is shown. Interestingly, the mean rate distribution does not overlap with any of the codon position distributions. The figure overlays results from four separate runs of the MCMC algorithm on the same data each with different random starting topologies. The similarity of the four distributions indicates that the algorithm has converged to the correct posterior distribution and thus has sampled the parameter space of the model adequately.

statistical errors, make MCMC an attractive and practical option for the estimation of evolutionary parameters such as substitution rate.

5. DISCUSSION

Temporally spaced data from rapidly evolving viruses provide an opportunity to ask questions about population dynamics and molecular evolution that are not possible with slow-evolving organisms or contemporaneous sequence data. Although we have concentrated on the estimation of molecular evolutionary rates there are a number of closely

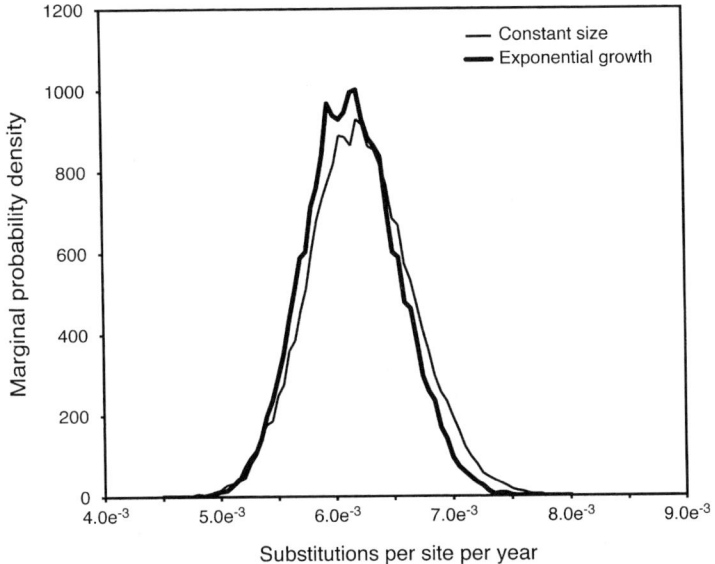

Figure 8 Application of the Bayesian inference approach to the HIV-1 data set. The figure shows results from two separate runs of the MCMC algorithm, the first assuming a constant population size for the coalescent model (thin line) and the second allowing an exponentially expanding population (bold line).

related problems that can be tackled using the methods above. We outline a few of them below.

5.1. Estimation of Divergence Times

Temporally spaced sequence data allows for the independent estimation of divergence times in viral phylogenies and genealogies. Traditionally, in the wider field of phylogenetic inference, independent calibration information has been used to determine the divergence time of an anchor node and then, assuming a molecular clock, used to estimate the ages of other divergences in the tree (for example, Shields and Wilson, 1987). However internal-node calibration methods suffer difficulties when there are few calibration points and when the substitution rates over long timescales are used to calibrate divergences over short timescales. It is also generally unlikely that internal-node calibrations will be available for viral sequence data, although a few examples do exist (e.g., Leitner and Albert, 1999; Pybus *et al.*, 2001; Van Dooren *et al.*, 2001).

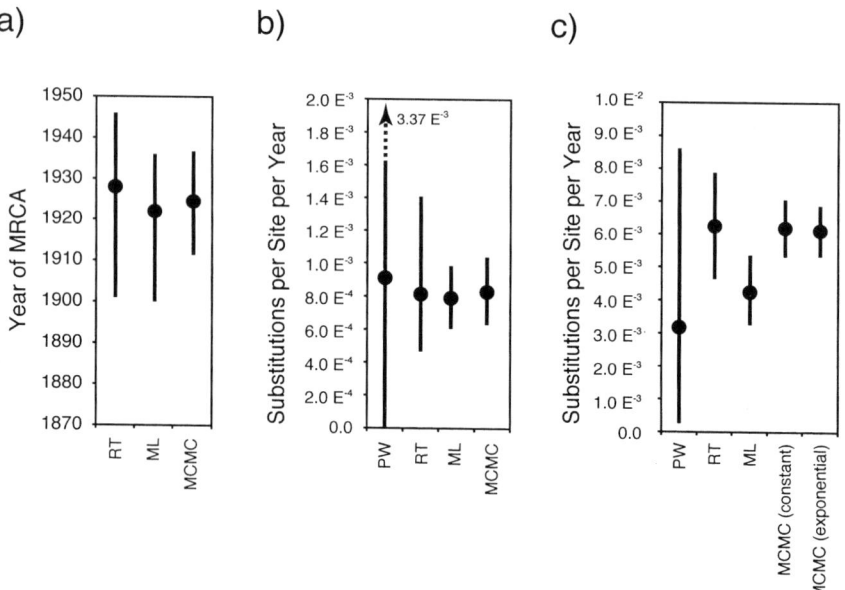

Figure 9 A comparison of the parameter estimates produced by the different methods discussed in this paper; root-to-tip linear regression (RT), pairwise linear regression (PW), maximum likelihood (ML) and Bayesian Markov chain Monte Carlo inference (MCMC). (a) Comparison of the results obtained for the time of the most recent common ancestor for the DEN-4 data set. (b) Comparison of the results obtained for the rate of substitution for the DEN-4 data set. (c) Comparison of the results obtained for the rate of substitution for the HIV-1 data set. For this data set, the MCMC estimates under both the constant population size and the exponential growth models are shown.

5.2. The Neutral Theory of Molecular Evolution and the Molecular Clock

The clock-like nature of many rapidly evolving viruses has been used to support both the molecular clock hypothesis (e.g., Leitner and Albert, 1999) and Kimura's neutral theory of evolution (e.g., Gojobori et al., 1990). Although there is now fairly strong evidence of positive selection in HIV-1 (Nielsen and Yang, 1998), it still appears to be a relatively minor contribution to the molecular evolution of the HIV-1 genome as a whole. In fact, recent preliminary evidence of a negative correlation between population size and mutation rate suggests that negative selection imposed by functional constraints is more important and ubiquitous in HIV-1 evolution than positive selection (Seo *et al.*, 2002a). This observation can be explained by either the nearly neutral theory or the slightly deleterious model of molecular evolution (Ohta and Kimura, 1971; Ohta, 1987; Tachida, 1991; discussed in Gillespie, 1995). The difficulty is that the strict molecular clock hypothesis does not appear to survive careful scrutiny. For example, only 7 out of 50 RNA viruses fit a strict molecular clock when tested in one recent comprehensive study (Jenkins *et al.*, 2002). However, the authors of that study went on to show by simulation that even for the viruses that did not obey a strict molecular clock, the substitution rates estimated could still be regarded as an accurate reflection of the average substitution rate. A more satisfactory solution to this problem is the recent development of 'relaxed clock' models of substitution (Thorne *et al.*, 1998; Huelsenbeck *et al.*, 2000). These models allow molecular evolutionary rates to vary over time and across lineages. In the future, incorporation of these methods into the analysis of temporally spaced sequence data should allow both estimation of average evolutionary rate and the extent of rate variation across lineages. This has already begun, with a recent description of an MCMC method (though without considering phylogenetic uncertainty) that allows for dated-tips and lineage-specific rate variation (Thorne and Kishino, 2002). The chief concern in the further development of tests of the molecular clock will be in assessing the relative merits of rate-per-lineage models and MRDT models in uncovering the trends in the variation of evolutionary rate.

5.3. Estimating Generation Length

If the ages of sequences are known in calendar units (for example, days or years) then it is possible to estimate the substitution rate per site per calendar unit. However, population genetic theory tells us that in a haploid population the expected genetic diversity, Θ, is two times the product of population size and mutation rate per *generation*. Hence in order to estimate population size we need to know the conversion factor τ, the number of calendar units per generation (i.e. the generation length). This problem can be turned on its head if the mutation rate is already known from some external source. In this case, one can estimate the generation length from serially sampled genetic data, given the mutation rate. A number of methods have been described to do this for HIV-1 (Rodrigo *et al.*, 1999; Fu, 2001; Seo *et al.*, 2002a) and all agree closely with methods based on viral load dynamics. This congruence between genetic methods and viral load dynamics is encouraging because it occurs despite completely different sources of data. The most recent of these methods, a pseudo-likelihood method (Seo et al., 2002a), was used to estimate the generation length of nine intra-patient data sets. Assuming a single underlying mutation rate, they estimated that generation length in HIV-1 varied from 0.73 to 2.43 days among the nine patients, again showing close congruence with early work.

5.4. Conclusion

Recent maximum likelihood and Bayesian methods of analysis have filled an important gap in the study of viral evolution. These methods both provide a wealth of options for hypothesis testing and model comparison by providing a solid statistical basis for genealogy-based inference of molecular rates, based on coalescent theory and likelihood models of molecular evolution. However, as mentioned above, the methods described here are still limited by a number of simplifying assumptions. Substantial population subdivision, recombination or selection may adversely affect analysis of temporally spaced viral sequences. Most of the methods described here assume single panmictic populations, free of recombination and selection. Therefore, extensions of the Bayesian inference framework described here to take into account migration between subpopulations, substantial recombination and selection effects are needed. Most of these processes fall squarely within the purview of population genetics and are already understood in the context of contemporaneous samples of sequences. We expect that in the near future methods that allow incorporation of all of these effects will exist for analysis of rapidly evolving viruses. In fact, very early on it was predicted that temporally spaced data would provide the opportunity to shed new light on these forces:

> "To sum up, selective trends will be detectable only if data from the past are available." (Cavalli-Sforza and Edwards, 1967)

The use of the methods outlined in this article, and their derivatives, will assist in answering fundamental questions about the tempo and mode of viral molecular evolution.

Software packages for performing some of these analyses and links to other resources are available from http://evolve.zoo.ox.ac.uk/VirusRates/.

ACKNOWLEDGEMENTS

This work was funded by EPSRC and MRC (AJD), The Wellcome Trust (OGP) and The Royal Society (AER). Thanks to Ziheng Yang and an anonymous referee for helpful comments.

REFERENCES

Buonagurio, D.A., Nakada, S., Parvin, J.D., Krystal, M., Palese, P. and Fitch, W.M. (1986). Evolution of human influenza A viruses over 50 years: rapid, uniform rate of change in NS gene. *Science* **232**, 980–982.

Cavalli-Sforza, L.L. and Edwards, A.W.F. (1967). Phylogenetic analysis: models and estimation procedures. *American Journal of Human Genetics* **19**, 233–257.

Drummond, A. and Rodrigo, A.G. (2000). Reconstructing genealogies of serial samples under the assumption of a molecular clock using serial-sample UPGMA. *Molecular Biology and Evolution* **17**, 1807–1815.

Drummond, A., Forsberg, R. and Rodrigo, A.G. (2001). The inference of stepwise changes in substitution rates using serial sequence samples. *Molecular Biology and Evolution* **18**, 1365–1371.

Drummond, A.J., Nicholls, G.K., Rodrigo, A.G. and Solomon, W. (2002). Estimating mutation parameters, population history and genealogy simultaneously from temporally spaced sequence data. *Genetics* **161**, 1307–1320.

Efron, B. and Tibshirani, R. (1993). An introduction to the bootstrap. London: Chapman and Hall.

Felsenstein, J. (1981). Evolutionary trees from DNA sequences: a maximum likelihood approach. *Journal of Molecular Evolution* **17**, 368–376.

Fitch, W.M., Leiter, J.M., Li, X.Q. and Palese, P. (1991). Positive Darwinian evolution in human influenza A viruses. *Proceedings of the National Academy of Sciences of the United States of America* **88**, 4270–4274.

Fu, Y.X. (2001). Estimating mutation rate and generation time from longitudinal samples of DNA sequences. *Molecular Biology and Evolution* **18**, 620–626.

Gillespie, J.H. (1995). On Ohta's Hypothesis: most amino acid subsitutions are deleterious. *Journal of Molecular Evolution* **40**, 64–69.

Gojobori, T., Moriyama, E.N. and Kimura, M. (1990). Molecular clock of viral evolution, and the neutral theory. *Proceedings of the National Academy of Sciences of the United States of America* **87**, 10015–10018.

Goulder, P.J.R., Brander, C., Tang, Y.H., Tremblay, C., Colbert, R.A., Addo, M.M., Rosenberg, E.S., Nguyen, T., Allen, R., Trocha, A., Altfeld, M., He, S.Q., Bunce, M., Funkhouser, R., Pelton, S.I., Burchett, S.K., McIntosh, K., Korber, B.T.M. and Walker, B.D. (2001). Evolution and transmission of stable CTL escape mutations in HIV infection. *Nature* **412**, 334–338.

Harvey, P.H. and Pagel, M.D. (1991). The comparative method in evolutionary biology. In: *Oxford Studies in Ecology and Evolution* (R.M. May and P.H. Harvey, eds), Oxford: Oxford University Press.

Hastings, W.K. (1970). Monte Carlo sampling methods using Markov chains and their applications. *Biometrika* **57**, 97–109.

Hayashida, H., Toh, H., Kikuno, R. and Miyata, T. (1985). Evolution of influenza virus genes. *Molecular Biology and Evolution* **2**, 289–303.

Holland, J., Spindler, K., Horodyski, F., Grabau, E., Nichol, S. and Vandepol, S. (1982). Rapid evolution of RNA genomes. *Science* **215**, 1577–1585.

Hudson, R.R. (1990). Gene genealogies and the coalescent process. *Oxford Surveys in Evolutionary Biology* **7**, 1–44.

Huelsenbeck, J.P. and Rannala, B. (1997). Phylogenetic methods come of age: testing hypotheses in an evolutionary context. *Science* **276**, 227–232.

Huelsenbeck, J.P., Larget, B. and Swofford, D. (2000). A compound poisson process for relaxing the molecular clock. *Genetics* **154**, 1879–1892.

Jenkins, G.M., Rambaut, A., Pybus, O.G. and Holmes, E.C. (2002). Rates of molecular evolution in RNA viruses: a quantitative phylogenetic analysis. *Journal of Molecular Evolution* **54**, 156–165.

Kimura, M. (1977). Preponderance of synonymous changes as evidence for the neutral theory of molecular evolution. *Nature* **267**, 275–276.

Kimura, M. (1987). Molecular evolutionary clock and the neutral theory. *Journal of Molecular Biology* **26**, 24–33.

Kimura, M. and Ohta, T. (1971). Protein polymorphism as a phase of molecular evolution. *Nature* **229**, 467–469.

Kingman, J.F.C. (1982). The coalescent. *Stochastic Processes and their Applications* **13**, 235–248.

Korber, B., Muldoon, M., Theiler, J., Gao, F., Gupta, R., Lapedes, A., Hahn, B.H., Wolinsky, S. and Bhattacharya, T. (2000). Timing the ancestor of the HIV-1 pandemic strains. *Science* **288**, 1789–1796.

Krystal, M., Buonagurio, D., Young, J.F. and Palese, P. (1983). Sequential mutations in the NS genes of influenza virus field strains. *Journal of Virology* **45**, 547–554.

Kuhner, M.K., Yamato, J. and Felsenstein, J. (1995). Estimating effective population size and mutation rate from sequence data using Metropolis-Hastings sampling. **140**, 1421–1430.

Lanciotti, R.S., Gubler, D.J. and Trent, D.W. (1997). Molecular evolution and phylogeny of dengue-4 viruses. *Journal of General Virology* **78**, 2279–2286.

Leitner, T. and Albert, J. (1999). The molecular clock of HIV-1 unveiled through analysis of a known transmission history. *Proceedings of the National Academy of Sciences, USA* **96**, 10752–10757.

Li, W.-H., Tanimura, M. and Sharp, P.M. (1988). Rates and dates of divergence between AIDS virus nucleotide sequences. *Molecular Biology and Evolution* **5**, 313–330.

Mansky, L.M. and Temin, H.M. (1995). Lower in-vivo mutation-rate of human-immunodeficiency-virus type-1 than that predicted from the fidelity of purified reverse-transcriptase. *J. Virol.* **69**, 5087–5094.

Martinez, C., del Rio, L., Portela, A., Domingo, E. and Ortin, J. (1983). Evolution of the influenza virus neuraminidase gene during drift of the N2 subtype. *Virology* **130**, 539–545.

Metropolis, N., Rosenbluth, A., Rosenbluth, M., Teller, A. and Teller, E. (1953). Equations of state calculations by fast computing machines. *Journal of Chemical Physics* **21**, 1087–1091.

Nielsen, R. and Yang, Z. (1998). Likelihood models for detecting positively selected amino acid sites and applications to the HIV-1 envelope gene. *Genetics* **148**, 929–936.

Nijhuis, M., Schuurman, R., de Jong, D., van Leeuwen, R., Lange, J., Danner, S., Keulen, W., de Groot, T. and Boucher, C.A.B. (1997). Lamivudine-resistant human immunodeficiency virus type 1 variants (184V) require multiple amino acid changes to become co-resistant to zidovudine in vivo. *Journal of Infectious Diseases* **176**, 398–405.

Ohta, T. (1987). Very slightly deleterious mutations and the molecular clock. *Journal of Molecular Evolution* **26**, 1–6.

Ohta, T. and Kimura, M. (1971). On the constancy of the evolutionary rate of cistrons. *Journal of Molecular Evolution* **1**, 18–25.

Pagel, M. (1999). Inferring the historical patterns of biological evolution. *Nature* **401**, 877–884.
Pybus, O.G., Rambaut, A. and Harvey, P.H. (2000). An integrated framework for the inference of viral population history from reconstructed genealogies. *Genetics* **155**, 1429–1437.
Pybus, O.G., Charleston, M.A., Gupta, S., Rambaut, A., Holmes, E.C. and Harvey, P.H. (2001). The epidemic behavior of the hepatitis C virus. *Science* **292**, 2323–2325.
Rambaut, A. (2000). Estimating the rate of molecular evolution: Incorporating non-contemporaneous sequences into maximum likelihood phylogenies. *Bioinformatics* **16**, 395–399.
Rodrigo, A.G. and Felsenstein, J. (1999). Coalescent approaches to HIV population genetics. *In*: Molecular evolution of HIV (K. Crandall, ed), Baltimore, MD: Johns Hopkins University Press.
Rodrigo, A.G., Shpaer, E.G., Delwart, E.L., Iversen, A.K., Gallo, M.V., Brojatsch, J., Hirsch, M.S., Walker, B.D. and Mullins, J.I. (1999). Coalescent estimates of HIV-1 generation time in vivo. *Proceedings of the National Academy of Sciences of USA* **96**, 2187–2191.
Saitou, N. and Nei, M. (1986). Polymorphism and evolution of influenza A virus genes. *Molecular Biology and Evolution* **3**, 57–74.
Saitou, N. and Nei, M. (1987). The neighbor-joining method: a new method for reconstructing phylogenetic trees. *Molecular Biology and Evolution* **4**, 159–166.
Schröder, E. (1870). Vier Combinatorische Probleme. *Zeitschriften für Mathematik und Physik* **15**, 361–376.
Seo, T.K., Thorne, J.L., Hasegawa, M. and Kishino, H. (2002a). Estimation of effective population size of HIV-1 within a host. A pseudomaximum-likelihood approach. *Genetics* **160**, 1283–1293.
Seo, T.K., Thorne, J.L., Hasegawa, M. and Kishino, H. (2002b). A viral sampling design for testing the molecular clock and for estimating evolutionary rates and divergence times. *Bioinformatics* **18**, 115–123.
Shankarappa, R., Margolick, J.B., Gange, S.J., Rodrigo, A.G., Upchurch, D., Farzadegan, H., Gupta, P., Rinaldo, C.R., Learn, G.H., He, X., Huang, X.-L. and Mullins, J.I. (1999). Consistent viral evolutionary changes associated with the progression of human immunodeficiency virus type 1 infection. *Journal of Virology* **73**, 10489–10502.
Shields, G.F. and Wilson, A.C. (1987). Calibration of mitochondrial DNA evolution in geese. *Journal of Molecular Evolution* **24**, 212–217.
Smith, D.B. and Inglis, S.C. (1987). The mutation-rate and variability of eukaryotic viruses - an analytical review. *Journal of General Virology* **68**, 2729–2740.
Sokal, R.R. and Michener, C.D. (1958). A statistical method for evaluating systematic relationships. *University of Kansas Science Bulletin* **38**, 1409–1438.
Swofford, D.L. (1998). PAUP 4.0: Phylogenetic analysis using parsimony (and other methods). 4.0b10 ed. Sunderland, MA: Sinauer Associates, Inc.
Swofford, D.L., Olsen, G.J., Waddell, P.J. and Hillis, D.M. (1996). Phylogenetic inference. 2nd ed. *In*: Molecular Systematics (D.M. Hillis, C. Moritz and B.K. Mable, eds), pp. 407–514. Sunderland: Sinauer Associates, Inc.
Tachida, H. (1991). A study on a nearly neutral mutation model in finite populations. *Genetics* **128**, 183–192.
Tanaka, Y., Hanada, K., Mizokami, M., Yeo, A.E.T., Shih, J.W.-K., Gojobori, T. and Alter, H.J. (2002). A comparison of the molecular clock of hepatitis C virus in the United States and Japan predicts that hepatocellular carcinoma incidence

in the United States will increase over the next two decades. *Proceedings of the National Academy of Sciences of the United States of America* **99**, 15584–15589.

Thorne, J.L. and Kishino, H. (2002). Divergence time and evolutionary rate estimation with multilocus data. *Systematic Biology* **51**, 689–702.

Thorne, J.L., Kishino, H. and Painter, I.S. (1998). Estimating the rate of evolution of the rate of molecular evolution. *Molecular Biology and Evolution* **15**, 1647–1657.

Van Dooren, S., Salemi, M. and Vandamme, A.-M. (2001). Dating the origin of the African human T-cell lymphotropic virus Type-I (HTLV-I) subtypes. *Molecular Biology and Evolution* **18**, 661–671.

Wilks, S.S. (1938). The large-sample distribution of the likelihood ratio for testing composite hypotheses. *Annals of Mathematical Statistics* **9**, 60–62.

Wu, C.F.J. (1986). Jacknife, bootstrap and other resampling methods in regression analysis. *The Annals of Statistics* **14**, 1261–1295.

Yang, Z. (1997). PAML: a program package for phylogenetic analysis by maximum likelihood. *Computer Applications in Biosciences* **13**, 555–556.

Detecting Adaptive Molecular Evolution: Additional Tools for the Parasitologist

James O. McInerney[1], D. Timothy J. Littlewood[2] and Christopher J. Creevey[1]

[1] *Bioinformatics and Pharmacogenomics Laboratory, Department of Biology, National University of Ireland, Maynooth, Co. Kildare, Ireland*
[2] *Parasitic Worms Division, Department of Zoology, The Natural History Museum, Cromwell Road, London SW7 5BD, England, UK*

Abstract	360
1. What is Adaptive Molecular Evolution?	360
2. Methodological Advances	362
2.1. Pair-wise Comparisons	362
2.2. Phylogeny-Based Methods	363
3. Example of Adaptive Evolution in the Malaria Rifin Proteins	366
3.1. Methods	367
3.2. Results	368
3.2.1. Pair-wise Distance Method	368
3.2.2. Messier and Stewart Method	368
3.2.3. Relative Rate Ratio Test (RRRT)	370
3.2.4. Likelihood Ratio Test (LRT) of Positive Selection ..	370
3.2.5. Bayesian Inference of Codons under Positive Selection ...	373
4. Prospects	373
Acknowledgements	376
References	377

ABSTRACT

It is likely that infectious diseases have shaped the evolution of many vertebrates, including humans. The etiological agents of disease continuously strive to evade the immune response and the immune response, in turn, seeks to change in order to keep pace with the invaders. This 'arms race' may be characterized by the selection for new variant hosts and new variant parasites. Here we discuss the utility of phylogenetics in detecting adaptive evolution at the molecular level and, for illustration, we concentrate on a family of surface-exposed proteins (the rifins) found in the recently sequenced genome of *Plasmodium falciparum*. We employed phylogeny-based methods in order to characterize adaptive evolution in these proteins. We found evidence for adaptive evolution in many of the amino acid residues in at least one lineage. These results indicate that there has been selection for those strains of *P. falciparum* that contain the new genotypes. These proteins are likely to be of great importance for the survival of the parasite. Studies of the interaction of these proteins with the antigen-presenting cells of the immune system should lead to a better understanding of malarial infection.

1. WHAT IS ADAPTIVE MOLECULAR EVOLUTION?

Proteins are exquisitely designed molecular machines that have been honed and perfected by a system of mutation and selection over time. Mutation relentlessly increases variation which selection edits, either removing (as in negative selection) or accepting it (as in positive directional or non-directional selection). For that reason, selection is likely to have been a powerful agent of evolution, largely dictating the rate and pattern of protein change. Usually, we think of the main role of selection in terms of rejection of new variants that are considered deleterious. However, it is becoming more obvious that adaptive change (positive selection of new variants) is a frequent and essential component of molecular evolution (McDonald and Kreitman, 1991; Swanson and Vacquier, 1995; Messier and Stewart, 1997; Nielsen and Yang, 1998; Zanotto et al., 1999; Bielawski and Yang, 2001).

If a new mutation occurs, a number of factors will influence whether or not this mutation remains polymorphic in the population for a long period. If the new mutation is neither significantly advantageous nor disadvantageous, then the length of time it remains in the population as a polymorphism along with the wild type is dependent on the effective population size of the species. If the new mutation affects the phenotype then it either may be removed or become fixed in the population in a manner that is almost independent of the population size depending whether the mutation is under positive or negative selection (Hughes, 1999).

For the most part, the effects of selection are analysed in protein-coding regions. Protein-coding genes are a mosaic of nucleotide positions and mutations within them can be classified as either synonymous or nonsynonymous, depending on whether or not the mutation at that codon site changes the encoded amino acid. The most commonly used method of detecting adaptive evolution at the protein level usually has involved comparing rates of synonymous changes per synonymous site (dS) versus nonsynonymous changes per nonsynonymous site (dN). The synonymous rate is used to calibrate the properties of neutrally evolving DNA so that when it is compared against the nonsynonymous rate the action of negative or positive selection may be detected. Negative selection, being the retention of status quo, is characterized by a synonymous rate that is much greater than the nonsynonymous rate. Positive selection, being the advocate of change, is characterized by a nonsynonymous rate that is much greater than the synonymous rate.

It is usual to think of adaptive evolution as an ephemeral event that is associated with the development of a new function or the modification of an existing property of a protein. A period of positive selection might be supplanted by a period of purifying selection once this novel property has been established. The transient nature of this phenomenon, allied to its enormous potential for affecting phenotypic diversity, makes it important to understand its nature and extent.

For example, it has been reported that periods of adaptive evolution have characterized the development of new functions in the lysozyme proteins of foregut fermenting primates (Yang, 1998; Creevey and McInerney, 2002). In this situation, a lysozyme protein that is active in low pH conditions and is unusually resistant to cleavage by pepsin is found in the fermenting foregut of herbivorous primates. The unusual properties of lysozyme allow the digestion of plant material that would otherwise have remained refractory. In some marine invertebrates, positive selection for new variant sperm proteins has also been reported and implicated to be responsible for the origin of new species (Yang *et al.*, 2000a). The biological 'arms race' that characterizes the interaction between pathogenic micro-organisms and their vertebrate hosts has also been documented in terms of adaptive evolution of important components of both systems (Hughes, 1992; Hughes and Hughes, 1995).

In a seemingly contradictory finding, the importance of any individual gene for the survival of a species may be relatively small. In yeast, it has been shown that deleting a gene has little phenotypic effect owing to the existence of compensatory mechanisms (Winzeler et al., 2000). This could indicate that individual residues within genes are even more dispensable (and presumably not under a great deal of selective pressure). However, running contrary to this expectation are the observations of selective pressures on most genes (Creighton and Darby, 1989; Hughes, 1992; McInerney, 1998; Bielawski and Yang, 2001).

2. METHODOLOGICAL ADVANCES

Although it is becoming increasingly obvious that the evolutionary histories of many genes have been characterized by intermittent periods of adaptive evolution, it is still a controversial and difficult task to correctly identify those periods and to identify the amino acids that were influenced by positive selection (Yang and Bielawski, 2000). Initially, the most commonly used methods relied on simple pair-wise comparisons of protein-coding sequences (Li et al., 1985; Nei and Gojobori, 1986; Li, 1993; Ina, 1995). More recently, however, phylogenetic trees have been employed in order to pinpoint adaptive evolutionary events with greater accuracy (Messier and Stewart, 1997; Yang, 2000; Yang and Bielawski, 2000; Creevey and McInerney, 2002; Yang and Nielsen, 2002).

2.1. Pair-wise Comparisons

One of the first pair-wise distance methods was devised by Li et al. (1985). This method was later revised (Li, 1993) and other modifications of this principle have also been described (Ina, 1995). According to Nei and Gojobori (1986), for each pair of protein-coding sequences in an alignment, each codon position is classified according to whether or not there has been a change since the two sequences last shared a common ancestor. These variable positions are classified as synonymous (silent) or nonsynonymous (replacement) substitutions. Two pair-wise 'distances' are calculated – the proportion of synonymous substitutions per synonymous site (pS) and the proportion of nonsynonymous substitutions per nonsynonymous site (pN):

$$pS = Sd/S \qquad (2)$$
$$pN = Nd/N \qquad (3)$$

where Sd is the number of synonymous differences and S is the number of synonymously variable sites and Nd is the number of nonsynonymous differences and N is the number of nonsynonymous sites. A log-normal correction (Jukes and Cantor, 1996) for superimposed substitutions modifies the observed distance and produces an estimate of dN and dS that is larger than the observed distance:

$$dS = -3\log_e(1-4pS/3)/4 \qquad (4)$$
$$dN = -3\log_e(1-4pN/3)/4 \qquad (5)$$

These corrected distances are compared and, in those situations where dN is significantly greater than dS, a period of adaptive evolution is invoked as the reason for this occurrence. There are many variations of methods for calculating these distances (Li *et al.*, 1985; Li, 1993; Ina, 1995), but the principle remains the same.

This kind of approach works very well for closely related sequences. The distances can be compared statistically as the variance can be computed and differences that may be due to estimation error can be distinguished from differences that are due to selection.

There are, however, a number of problems with pair-wise distance analyses. First, the rate of change at silent sites is usually quite high. Since substitutions at these sites are either not deleterious or only very slightly deleterious (there may be small selective pressures for optimal codon usage; McInerney, 1998), the silent substitution rate is very close to the mutation rate (Kimura, 1968). Mutation rates have been measured to be of the order of 10^{-9} substitutions per site per year (Gaut and Clegg, 1991, 1993). This means that, while amino acid conservation may be high in some proteins, a considerable amount of change may have occurred at silent positions. This results in a tendency to underestimate the value of dS. The value of dN is not masked to the same extent, although superimposed substitutions also occur at amino acid replacement sites. Additionally, the overall pair-wise distance is an average of the rate of change at all positions. Therefore, if selective pressures are heterogeneous across the protein, this variability is masked. Even in those situations where pair-wise sequence analysis indicates that positive selection has occurred, it is not possible to say on which lineage it has occurred.

2.2. Phylogeny-Based Methods

New methods for detection of adaptive evolution are based on phylogenetic trees. This has been the case since the seminal contribution by Messier and Stewart (1997). Their innovation involved the use of a phylogenetic tree to try to pinpoint the time when an adaptive evolutionary event took place.

Using a dataset of primate lysozyme sequences, Messier and Stewart reconstructed the hypothetical ancestral sequences at each of the internal nodes of the phylogenetic tree. They then performed all possible pair-wise comparisons between all sequences, both hypothetical and contemporary, and found that this provided a greater power when trying to pinpoint the time of adaptive evolutionary events.

There are a number of advantages to this approach. The first stems from the use of both a phylogenetic tree and reconstructed ancestral sequences. Given a phylogenetic tree, it is possible to use either maximum parsimony (MP) or maximum likelihood (ML) methods (Felsenstein, 1981, 1996; Yang et al., 1995) to reconstruct the hypothetical sequences at internal nodes. This means that a particular dataset has $2n-2$ taxa: n terminal lineages and $n-2$ internal nodes for a bifurcating unrooted phylogeny. These hypothetical ancestral sequences are distributed throughout the evolutionary history of the analysed sequences, reducing the gap between sampling times. This means that if selective pressures differ across the tree, the increased sampling should allow a greater power in pinpointing when the period of adaptive evolution occurred.

According to the neutral theory, the rate of synonymous to nonsynonymous mutations should be the same for intra- and interspecific comparisons of sequences. Initially proposed by Maynard-Smith (1970), this was implemented by McDonald and Kreitman (1991) to detect adaptive evolutionary events in the alcohol dehydrogenase locus of *Drosophila* species. An increase in the number of intraspecific nonsyonymous mutations compared to interspecific nonsynonymous mutations indicates that the intraspecific substitutions were favoured by selection within that species. This approach also implicitly uses phylogenetic trees, albeit simple ones with a single branch separating two closely related species.

Creevey and McInerney (2002) suggested a method that is loosely based on the McDonald and Kreitman test. In this case, a phylogenetic tree is inferred and presumed to be correct. The ancestral sequences at each internal node on this tree are reconstructed. From each internal node, all the substitutions in the descendent clade are characterized and their numbers are counted. Substitutions are classified into replacement-invariable (RI), replacement-variable (RV), silent-invariable (SI) and silent-variable (SV) sites. Replacement-invariable sites are those where a replacement substitution has occurred at a particular codon position somewhere in the clade, but this new amino acid did not subsequently change. Replacement-variable sites are those where a replacement substitution occurred in the descendent clade and, subsequently, this amino acid position changed again. Similar statistics, SI and SV, are calculated for silent substitutions. As with the McDonald–Kreitman test, it is assumed that, in cases where the sequences are evolving neutrally, there should be no difference between the ratios RI:RV and SI:SV. Once again, the silent substitutions are considered to be neutrally evolving and are used to calibrate the properties of neutrally evolving DNA. In this case, the ratio of SI to SV substitutions in any clade should be determined by the size and shape of the clade and the mutation rate. If the internal branch that describes the clade is close to the tips of the tree, then SI sites will dominate (and indeed, so should RI sites if the sequences are evolving neutrally). If the internal branch is further from the tips, then the ratio of SI to SV will decrease. If positive selection has characterized the evolution of the sequences being examined, the number of either RI or RV substitutions will be significantly greater than is expected from neutrality. The former indicates the presence of positive directional selection, and the latter indicates positive nondirectional selection. This method has been shown to be very effective at identifying adaptive evolution in a number of instances (Creevey and McInerney, 2002).

Maximum likelihood (ML) and Bayesian methods have also been implemented for detection of adaptive evolution (Goldman and Yang, 1994; Yang, 1998; Yang and Nielsen, 2002). Maximum likelihood methods choose a hypothesis that maximizes the likelihood of observing the data. This hypothesis is usually composed of two parts – the tree and the evolutionary process, each with a certain set of parameters. More formally:

$$L = P(X|\tau, \upsilon, \theta) \tag{10}$$

where L is the likelihood and P is the probability of observing X, the alignment given the tree τ with branch lengths υ and the substitution process described by θ.

The substitution process can be described by a continuous time Markov process with transition/transversion rate bias and codon usage bias allowed to vary. In addition, the physicochemical distances between amino acids being is used to accommodate selective restraints at the protein level. Given that selective pressures are likely to vary across different sites in a protein sequence, models have been developed to incorporate heterogeneous selective pressures at different sites (Yang et al., 2000b). Variation in selective pressure is usually modelled according to a statistical distribution or a mixture of statistical distributions (Yang et al., 2000b). As a result, a site has a probability of belonging to a particular class. In addition, the process may vary over the tree. To account for this, mixed models have been developed that allow the dN:dS ratio to vary either among lineages or among sites. In this case, classes of sites may differ between 'foreground' lineages (pre-specified lineages of interest) and 'background' lineages (the other lineages in the tree).

The likelihood ratio test (LRT) is used to evaluate nested models of sequence evolution. Some models are more parameter-rich extensions of other models and, when this is the case, an LRT may be performed with twice the log-likelihood difference being compared with a χ^2 distribution with the degrees of freedom equal to the difference in the number of parameters between the two models (for a more complete description, see Yang et al., 2000b).

Finally, when after the ML optimization of the parameters is complete, an empirical Bayes approach is used to infer which class a site is most likely to come from (Nielsen and Yang, 1998). Those sites with a high probability of coming from a class of sites with a high dN : dS ratio are most likely to be under positive selection.

3. EXAMPLE OF ADAPTIVE EVOLUTION IN THE MALARIA RIFIN PROTEINS

We examine the RIFIN family of proteins in the newly sequenced genome of the malarial parasite *Plasmodium falciparum* to demonstrate some of the methods of detecting adaptive evolution mentioned previously. Malaria is a major contributor to mortality and morbidity in developing countries. Its economic impact alone is sufficient to cripple the economy of most countries in which it is endemic (World Health Organization, 2002). Globally, it is estimated that 300–500 million people are infected with *Plasmodium* annually, with an estimated 1.5 to 2.7 million people dying each year from the disease, mostly children under the age of five years (World Health Organization, 1997).

Immunity to malaria usually develops later in life, with individuals usually not acquiring the broad repertoire of antibodies necessary to suppress the disease until they are older (Abdel-Latif et al., 2002). The acquired immune system mounts a response to the polymorphic antigens that are expressed on the surface of infected erythrocytes during the blood stage of the *Plasmodium* life cycle (Fernandez et al., 1999; Good and Doolan, 1999). These antigens usually display both temporal (Kyes et al., 2000) and sequence (Forsyth et al., 1989) variation. The best-characterized *P. falciparum* surface antigen is the erythrocyte membrane protein 1 (PfEMP-1), which is encoded by about 50 *var* genes (Abdel-Latif et al., 2002). Malaria infection is linked to parasite-induced surface changes of infected red blood cells. These lead to sequestration of infected red blood cells in the microvasculature and rosetting of uninfected red blood cells, with consequent obstruction to microvascular blood flow, tissue damage and disease (Kyes et al., 1999). The PfEMP-1 family of proteins mediate adhesion to at least some host cell receptors, via CD36 with endothelial cells and via complement receptor 1 with uninfected red blood cells (Kyes et al., 1999). PfEMP-1 is, therefore, considered a major virulence factor, but it alone is not sufficient to explain all parasite-induced surface phenotype changes. Recent analysis of the P. falciparum genome identified other multicopy gene families unique to the genus, the largest of which belongs to the rif (repetitive interspersed family) gene family (Weber, 1988; Kyes et al., 1999). It is estimated that there are in excess of 200 rif genes per haploid genome making them at least four times as abundant as the var genes. Rif genes are located in close association with var genes in clusters within 50 kb of the telomeres, and the name rifins has been proposed for the putative protein products (Gardner et al., 1998).

Rif genes possess two exons, the first of which encodes a putative signal peptide and the second of which encodes an extracellular domain made up of a conserved and a variable region, followed by a transmembrane segment and a short intracellular portion (Abdel-Latif et al., 2002). Rifin proteins have been shown to express on the surface of infected red blood cells. While they have a function in rosette formation along with var genes, it is not thought to be their primary function, as some parasites that do not show appreciable levels of rosetting still express rifin proteins (Kyes et al., 1999). Unlike var genes, only one of which is expressed at any one time, several rif genes are believed to be concomitantly expressed on the surface of infected erythrocytes. The function of these clonally variant rifins on the surface of IE remains speculative but, because of their surface locality and sequence diversity, these proteins may play an essential role in the host–parasite interface during the asexual blood stage (Abdel-Latif et al., 2002). Supporting this hypothesis, Abdel-Latif et al. (2002) demonstrated a naturally acquired antibody response in a large number of hosts to a subset of erythrocyte surface-localized rifin proteins. However, to date the functional role of rifins as a multigene family of clonally variant surface proteins has not been elucidated, but it remains plausible that anti-rifin antibodies could interfere with some important aspect of the life or cell function of the parasite, such as cytoadherence or rosette formation.

3.1. Methods

A total of 51 sequences were extracted from the P. falciparum database available at http://www.plasmodb.org/. Members of the family were identified using a BLASTP search (Altschul et al., 1997) using various candidate members of the rifin family as query sequence against a protein-coding version of the annotated Plasmodium genome. The DNA version of all sequences was retrieved, conceptually translated into the corresponding protein, aligned using the ClustalW alignment software (Thompson et al., 1994) and the indel positions that were introduced by the alignment software were put into the DNA sequences according to where they were found in the protein alignment. Any amino acid position that could be aligned in more than one way was removed from the alignment. Also, some sequences were removed from the final alignment as a result of alignment difficulty. The result was a 573 bp alignment of protein-coding DNA sequences, where every position in the alignment was aligned with a reasonable degree of confidence.

Analyses of signatures of adaptive evolution were carried out using four different methods. The first was the pair-wise method comparing the dN and dS values (Li, 1993). The second method involved the modification of this approach as described by Messier and Stewart (1997), where all the hypothetical ancestral sequences are reconstructed at each internal node of the phylogeny and dN and dS values are calculated between each node and its descendants. The third method was the relative rate ratio method (Creevey and McInerney, 2002) and the fourth method used the ML approach (Goldman and Yang, 1994; Yang et al., 2000b; Yang and Swanson, 2002). The first three methods are implemented in the program CRANN (Creevey and McInerney, 2003) and the ML calculations were carried out using the program package PAML (Yang, 1997).

3.2. Results

3.2.1. Pair-wise Distance Method

A pair-wise distance analysis was carried out for all possible pairs of sequences. In each case the averaged dN and the dS values were calculated. For just 205 of the total of 1275 comparisons, the dN value was greater than the dS value. Among these 205 cases, 149 (73%) involved just three sequences: PFD0070, PFD0125 and PFD1220. An examination of the phylogenetic position of these three sequences (Figure 1) indicates that these sequences are closely related to one another. This result provides strong evidence that the evolution of these sequences has been influenced by positive selection.

3.2.2. Messier and Stewart Method

Using Messier and Stewart's method (Messier and Stewart, 1997), two analyses were carried out. Two phylogenetic trees were constructed, one

DETECTING ADAPTIVE MOLECULAR EVOLUTION 369

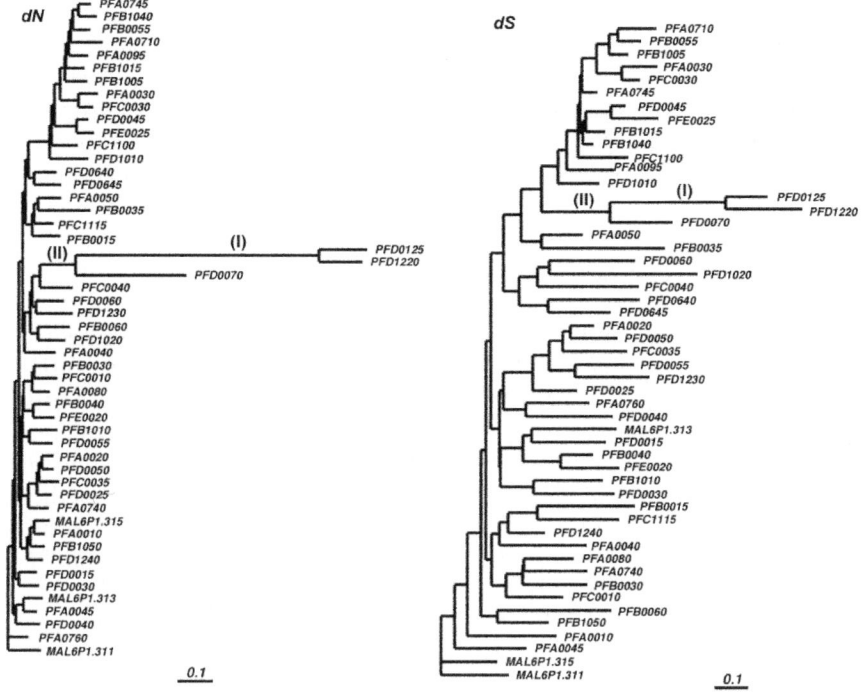

Figure 1 Phylogenetic trees of the rifin protein family from *P. falciparum*. The tree on the left has been inferred by the neighbor-joining method based upon dN distances calculated using the Li (1993) method. The tree on the right is based upon dS distances calculated using the same method. Scale bars represent 0.1 substitutions per site.

using the neighbor-joining method based on distances derived from synonymous sites (dS-distances) and one using the neighbor-joining method based on distances derived from nonsynonymous sites (dN-distances) calculated as described above. Both trees are shown in Figure 1. Although the trees differ slightly in their topology, there is broad agreement among the groupings on both trees. Of interest are the branches leading to the three sequences mentioned previously (PFD0070, PFD0125 and PFD1220). This clade appears to be rapidly evolving in both trees, however the branches in this clade are much longer on the tree based on dN distances, relative to the branch lengths in the rest of the tree. For each internal node on the tree, hypothetical ancestral sequences were inferred using parsimonious character reconstruction. Using the dN tree, the estimated dN : dS ratio comparing

PFD0070 with the hypothetical common ancestor of the clade was 6.5. The estimated $d\text{N}:d\text{S}$ ratio for the comparison of the common ancestor of PFD1220 and PFD0125 and the common ancestor of the entire clade was 4.3. The $d\text{N}:d\text{S}$ ratio of the common ancestor of PFD0125 and PFD1220 and the extant sequence of PFD0125 was 2.37, while the $d\text{N}:d\text{S}$ ratio for the common ancestor of these two sequences and PFD1220 was less than unity.

3.2.3. *Relative Rate Ratio Test (RRRT)*

The method proposed by Creevey and McInerney (2002) represents a different kind of approach to the problem of identifying adaptive evolution. In this approach, there is no attempt to estimate $d\text{N}$ and $d\text{S}$ ratios, rather the attempt is to detect clade-specific deviations from neutrality. An entire clade is chosen and the mutational process from the internal branch that describes the clade to the tips of the clade are evaluated. Using this method, and choosing a variety of outgroups, the clade that describes the sequences PFD1220, and PFD0125 was again consistently selected as one that has potentially undergone a period of adaptive evolution. Interestingly, the clade that included PFD0070 was not deemed significantly different from neutrality. This indicates that the most probable location of an adaptive evolutionary event is on the internal branch separating PFD1220 and PFD0125 from the rest, not on the internal branch that separates the clade that also includes PFD0070.

3.2.4. *Likelihood Ratio Test (LRT) of Positive Selection*

A variety of analyses was carried out on the alignment using different models of sequence evolution. The results of these analyses are summarized in Table 1. There were four different categories of models employed in these analyses. The first kind of model, designated M0 in Table 1, is the model of Goldman and Yang (1994). In this model a single $d\text{N}:d\text{S}$ ratio (designated ω) is assumed for all positions in the alignment across all lineages. Using this model, the ω value was estimated to be 0.1859, indicating strong purifying selection. However, this model is somewhat unrealistic and restrictive and represents an average across all positions in the alignment and all lineages, so a number of other models were also employed.

The next model was that of Yang (1998). This model allows the ω value to vary across the tree. In this model, we allow a single ω value for background lineages and an ω value for 'foreground' lineages. These ω values are iteratively optimized during the procedure. From our initial analyses

DETECTING ADAPTIVE MOLECULAR EVOLUTION

Table 1 Results of ML analysis of the rifin data using a variety of ML models. p_0, p_1, p_2 and p_3 refer to the proportion of sites in categories 0, 1, 2 and 3 respectively. ω refers to the $dN:dS$ ratio in these categories of sites. Descriptions of the various models are to be found in the text.

	p	ln L	Estimates of parameters
M0: one-ratio	1	−14735.843499	$\omega = 0.1859$
Branch-specific		**(PFD0125, PFD1220)**	
Two Ratios	2	−14734.339974	$\omega_0 = 0.1864$ $\omega_1 = 0.0097$
Branch-specific		**(PFD0070, (PFD0125, PFD1220))**	
	2	−14731.799842	$\omega_0 = 0.1868$ $\omega_1 = 0.0014$
Site-specific			
M1–Neutral	1	−15278.599347	$p_0 = 0.05446$ $p_1 = 0.94554$
M2–Selection	3	−14280.751708	$p_0 = 0.04307$ $p_1 = 0.28258$ $p_2 = 0.67435$ $\omega_2 = 0.13284$
M3 – $k = 2$	3	−14219.571671	$p_0 = 0.57091$ $\omega_0 = 0.06189$ $\omega_1 = 0.45720$
M3 – $k = 3$	5	−14134.152650	$p_0 = 0.46970$ $p_1 = 0.37161$ $p_2 = 0.15869$ $\omega_0 = 0.04819$
			$\omega_1 = 0.27168$ $\omega_2 = 0.80995$
M7 – beta	2	−14108.801679	$p = 0.56930$ $q = 1.55267$
M8 – beta&ω	4	−14095.766769	$p_0 = 0.90962$ $(p = 0.80147$ $q = 3.53496)$ $\omega = 1.02483$
Branch-site		**(PFD0125, PFD1220)**	
Model A	3	−15217.403731	$P_0 = 0.03549$ $p_1 = 0.43042$ $p_2 = 0.04068$ $p_3 = 0.49341$ $\omega_2 = 128.05992$
Model B	5	−14208.574035	$P_0 = 0.21796$ $p_1 = 0.16293$ $p_2 = 0.35428$ $p_3 = 0.26484$ $\omega_0 = 0.06207$
			$\omega_1 = 0.45667$ $\omega_2 = 32.01223$
Branch-site		**(PFD0070, (PFD0125, PFD1220))**	
Model A	3	−15247.783057	$p_0 = 0.04589$ $p_1 = 0.65750$ $p_2 = 0.01935$ $p_3 = 0.27725$ $\omega_2 = 561.08899$
Model B	5	−14209.806432	$P_0 = 0.47544$ $p_1 = 0.34650$ $p_2 = 0.10300$ $p_3 = 0.07507$ $\omega_0 = 0.06123$
			$\omega_1 = 0.46200$ $\omega_2 = 519.22796$

described above, we found that three sequences were potentially interesting, so we have chosen to perform different analyses of clades that contain these sequences. The first branch (branch I in Table 1 and Figure 1) was labelled as the foreground lineage and surprisingly, while the background lineages were estimated to have an ω value similar to M0 (0.1864), the foreground lineage was estimated to have an ω value of 0.0097, with a consequent, though not significant, increase in the log-likelihood. A similar situation was observed when the branch leading to the larger clade (branch II in Table 1 and Figure 1) was labeled.

The next series of models that were employed were those models that allow the ω value to vary across the sequence. These models allow for heterogeneity in selective pressures at different amino acid sites, however they do not account for rate variation across lineages. Model M1 is the least realistic of the models. It includes two classes of sites, one class where nonsynonymous substitutions are completely deleterious and removed $\omega = 0$ and one class that are neutral and have an ω value of 1. This particular model had the lowest log-likelihood score of all (-15278.59935). The next model (M2) includes an additional class of site with ω estimated from the data. The likelihood for this model is better than the simpler model M1 and the estimated ω for the additional class of sites is 0.13284, with an estimated proportion of 0.67435 sites in this category. This indicates that there is variation in selective pressure across these sites that is not accounted for by the neutral model.

We then used two models that use an unconstrained discrete distribution to model variability at different sites (M3). In the first of these models, we allow two separate discrete classes of sites ($k = 2$). This model fits the data much better that the one ratio model as evidenced by the increase of approximately 516 log-likelihood units. The M3 model with three site classes is an even better fit to the data by an additional 85 log-likelihood units. The model M7 does not allow for positively selected sites, but allows variation in ω across sites in the interval $(0, 1)$, however it again is a better fit to the data than any model so far. The model that appears to be the best fit is the model M8. This model can be compared with M7 using a χ^2 test with d.f. $= 2$ (i.e. $2\Delta l = [2 \times 13] = 26$, $P < 10^{-4}$). This is much greater than the χ^2 critical value. This model does not identify any amino acid positions that have a high probability of being under positive selection.

The conclusion so far is that models that allow variation in selective pressure across the alignment are preferable to those models that do not allow differences in ω at different sites.

The next category of models we have examined are those mixed models that allow ω to vary among sites and also allow ω to vary in different parts of the tree. These are designated 'Model A' and 'Model B' in Table 1. Given that we may select different lineages as the 'foreground' and 'background' lineages, we have decided to examine the same two internal branches that were examined in the previous branch-specific models.

Model A is a special case of the site-specific neutral model M1 and can be compared with this model using a χ^2 test with d.f. = 2. Model B is a special case of the discrete model M3 with $k = 2$ and can be compared with this model using a χ^2 test with d.f. = 2. In all cases, the branch-site models are significantly better than the respective null models with all p-values $< 10^{-4}$. This indicates that models that account for differences along these two lineages are a significantly better fit to the data than models that do not account for this variation. The best fitting branch-site model is model B. This model does not fix any category of site to any particular value. We note, however, that there are significant differences in the outcomes of labelling the different branches.

When the branch leading to the larger clade is labelled as the foreground branch, approximately 17% of sites are predicted to be under strong positive selection ($p_2 + p_3$). When the branch leading to the smaller clade is labelled, a much larger proportion of sites are predicted to be under positive selection. In fact, 61% of the sites are predicted to be under positive selection in this lineage. This result may help to explain the result from the RRRT and the unusual branch lengths for branch I on the dN tree in Figure 1.

3.2.5. Bayesian Inference of Codons under Positive Selection

The branch-site models in our analyses have proposed that a large proportion of sites are under the influence of positive selection. The sites with a high probability of being in the classes of site that are under positive selection are outlined in Table 2. Clearly, when branch I is labelled, a much higher number of sites have a significant probability of being in the class of site that is undergoing positive selection. A total of 28 sites have a probability of greater than 0.95 of being in the class of site that is undergoing positive selection in the internal branch labelled (I) in Figure 1. There are only two sites with this high a probability of being under positive selection on the branch labelled (II) in Figure 1.

4. PROSPECTS

Episodic adaptation to new ecological niches has been reported in a wide variety of proteins, ranging from primate stomach proteins (Messier and

Table 2 A list of the amino acid positions that have a probability greater than 0.95 of having been under positive selection along either of the two branches labelled in Figure 1. Branch I refers to the branch labelled (I) in Figure 1, branch II refers to the branch labelled (II) in Figure 1.

Branch	Branch II
7 N 0.9800*	119 S 0.9917**
10 L 0.9978**	174 R 0.9925**
19 L 0.9900**	
24 C 0.9635*	
31 P 0.9940**	
54 R 0.9784*	
76 D 0.9588*	
82 I 0.9818*	
86 I 0.9927**	
92 E 0.9783*	
93 K 0.9968**	
101 T 0.9960**	
102 L 0.9990**	
113 T 0.9977**	
114 C 0.9969**	
116 C 0.9705*	
118 K 0.9833*	
121 A 0.9978**	
131 C 0.9898*	
136 G 0.9958**	
153 S 0.9960**	
168 Y 0.9967**	
170 I 0.9938**	
172 R 0.9616*	
179 V 0.9872*	
181 K 0.9992**	
188 L b0.9815*	
191 E 0.9849*	

Stewart, 1997) to invertebrate sperm proteins (Swanson and Vacquier, 1995; Galindo *et al.*, 2003), and bacterial membrane proteins (Smith *et al.*, 1995). In all cases, these adaptive events have been associated with lifestyle changes in the organism in which these changes are found. In this chapter, we have analysed a set of 51 paralagous genes from *P. falciparum*. In three of these genes, we have found evidence that positive selection has occurred. We have also used LRTs in order to evaluate different models of sequence evolution. The preferred models directly point to adaptive evolution in the clade that defines these sequences. In addition, this finding is supported by pair-wise analyses of the sequences and a relative rate ratio test of deviation from neutrality. We have also identified amino acids within these proteins that are likely to be under positive selection for change.

It is interesting to note that only three proteins, or 6% of the members of this family, appear to be under positive selection for change. Overall, the proteins are not too dissimilar, with typical pair-wise amino acid distances being less than 0.3 substitutions per site. The rifin proteins generally elicit an antibody response from malaria-infected individuals and presumably, because they are recognized by the immune system, they are exposed to strong selective pressures. The reason for positive selection in just these three members of the family may therefore be their location on the cell surface, their level of expression, or perhaps their slightly different function. Based on sequence analysis alone, it is not possible to say with certainty why there are differences in selective pressure among different members of this family, however there is some strong evidence that different members interact differently with the human immune system.

In a study of individuals living in an area where malaria is endemic, Abdel-Latif *et al.* (2002) found that one particular rifin protein caused a greater induction of anti-rifin antibodies than other rifin proteins in the same study. This could suggest that this gene belongs to a subset of stably expressed dominant and commonly recognized *rif* proteins, which are present on the infected red blood cell surface. If this were the case, members of this subset would be continually under selective pressure for change, unlike the less commonly recognized proteins. The results described here of variable selective pressures in different members of the rifin gene family support this theory. This study has identified three rifin proteins that appear to be under positive selection for change. This contrasts with the rest of the members of the family. Positive selection will manifest itself only if the selective advantage offered by the mutation is sufficient to overcome random genetic drift. If a protein is only transiently expressed or if it does not elicit a strong antibody response, then it will not be subject to the same selective pressure as one that is expressed more regularly and/or is capable of inducing a strong adaptive immune response.

The analyses described in this paper have succeeded in finding a lineage within a multigene family where positive selection appears to characterize its evolutionary history. The rate of retention of replacement substitutions in three genes of the rifin*-family has been greater than the rate of retention of silent substitutions. In other words, those lineages of *P. falciparum* in which these new mutations occurred had a selective advantage as a result. Those lineages in which these substitutions did not occur were less fit and are not part of the extant genome of *P. falciparum*.

Although it is dangerous to conclude that evidence of adaptive evolution is indicative of lifestyle change, analyses of adaptive evolution may provide testable hypotheses. Subsequently, it may be possible to investigate further (using biochemical or other means) amino acids that are identified as having been under positive selection.

Frank (2002) recently reviewed the advantages of measuring selection in understanding further the immunology and evolution of infectious diseases. In the absence of structural data, detecting positive selection through gene sequence analysis has enabled the prediction of 'which site may be structurally exposed and can change and which sites are either not exposed or functionally constrained' (Frank, 2002). This has led to the recognition of epitopes, the sites within macromolecules to which a specific antibody binds, as well as other positively selected amino acid sites that may or may not constitute unidentified epitopes (Endo *et al.*, 1996; Yang and Bielawski, 2000). Sites under strong positive selection may represent candidate vaccination targets (Suzuki and Gojobori, 2001), although there is no means of knowing at present, whether such sites will continue to be primary targets of selection and thus to what extent they will either change or remain important in the host–parasite interaction. Identifying genes and gene regions under selection brings the effect of the environment on the genotype into focus. When selection within a parasite's genome can be demonstrated to be in response to the host, or indeed when host genes can be demonstrated to be under selection in response to a parasite, we can narrow our focus on the evolutionary genetic basis of the host–parasite interaction. For vaccine development there may come a time when routine genome analysis involves the assessment of whether genes appear to be under positive selection even before the function of the genes is fully understood. Phylogenetically based methods of detecting selection within and between lineages clearly have great potential in elucidating gene and protein evolution, and host–parasite interactions. Thus, once again, an accurate phylogeny is pivotal, whether founded on proteins, protein-coding genes, or species.

ACKNOWLEDGEMENTS

We thank Gayle Philip and Peter G. Foster for their help with this manuscript. J.O.M. and C.J.C. acknowledge the financial support of the Irish Research Council for Science, Engineering and Technology, Enterprise

Ireland, The Health Research Board (H.R.B) and H.E.A. PRTLI Cycle II and III. D.T.J.L. was funded by a Wellcome Trust Fellowship (043965). We thank two anonymous reviewers for their invaluable positive contribution to this manuscript.

REFERENCES

Abdel-Latif, M.S., Khattab, A., Lindenthal, C., Kremsner, P.G. and Klinkert, M.-Q. (2002). Recognition of variant rifin antigens by human antibodies induced during natural *Plasmodium falciparum* infections. *Infection and Immunity* **70**, 7013–7021.

Altschul, S.F., Madden, T.L., Schäffer, A.A., Zhang, J., Zhang, Z., Miller, W. and Lipman. (1997). Gapped BLAST and PSI-BLAST: a new generation of protein database search programs. *Nucleic Acids Research* **25**, 3389–3402.

Bielawski, J.P. and Yang, Z. (2001). Positive and negative selection in the DAZ gene family. *Molecular Biology and Evolution* **18**, 523–529.

Creevey, C. and McInerney, J.O. (2002). An algorithm for detecting directional and non-directional positive selection, neutrality and negative selection in protein coding DNA sequences. *Gene* **300**, 43–51.

Creevey, C. and McInerney, J.O. (2003). CRANN: a program for detecting positive selection in protein-coding genes. *Bioinformatics* **19**, 1726.

Creighton, T.E. and Darby, N.J. (1989). Functional evolutionary divergence of proteolytic enzymes and their inhibitors. *Trends in Biochemical Sciences* **14**, 319–324.

Endo, T., Ikeo, K. and Gojobori, T. (1996). Large-scale search for genes on which positive selection may operate. *Molecular Biology and Evolution* **13**, 685–690.

Felsenstein, J. (1981). Evolutionary trees from DNA sequences: a maximum likelihood approach. *Journal of Molecular Evolution* **17**, 368–376.

Felsenstein, J. (1996). Inferring phylogenies from protein sequences by parsimony, distance and likelihood methods. In: *Computer Methods for Macromolecular Sequence Analysis* (R.F. Doolittle, ed.), *Methods in Enzymology*, Vol. 266, pp. 418–427. Orlando: Academic Press.

Fernandez, V., Hommel, M., Chen, Q., Hagblom, P. and Wahlgren, M. (1999). Small, clonally variant antigens expressed on the surface of the *Plasmodium falciparum*-infected erythrocyte are encoded by the *rif* gene family and are the target of human immune responses. *Journal of Experimental Medicine* **190**, 1393–1403.

Forsyth, K.P., Philip, G., Smith, T., Kum, E., Southwell, B. and Brown, G.V. (1989). Diversity of antigens expressed on the surface of erythrocytes infected with mature *Plasmodium falciparum* parasites in Papua New Guinea. *American Journal of Tropical Medicine and Hygiene* **41**, 259–265.

Frank, S.A. (2002). *Immunology and Evolution of Infectious Diseases*. Princeton: Princeton University Press.

Galindo, B.E., Vacquier, V.D. and Swanson, W.J. (2003). Positive selection in the egg receptor for abalone sperm lysin. *Proceedings of the National Academy of Sciences of the USA* **100**, 4639–4643.

Gardner, M.J. and 26 others (1998). Chromosome 2 sequence of the human malaria parasite *Plasmodium falciparum*. *Science* **283**, 1126–1132.

Gaut, B.S. and Clegg, M.T. (1991). Molecular evolution of alcohol dehydrogenase 1 in members of the grass family. *Proceedings of the National Academy of Sciences of the USA* **88**, 2060–2064.

Gaut, B.S. and Clegg, M.T. (1993). Molecular evolution of the adh1 locus in the genus *Zea*. *Proceedings of the National Academy of Sciences of the USA* **90**, 5095–5099.

Goldman, N. and Yang, Z. (1994). A codon-based model of nucleotide substitution for protein-coding DNA sequences. *Molecular Biology and Evolution* **11**, 725–736.

Good, M.F. and Doolan, D.L. (1999). Immune effector mechanisms in malaria. *Current Opinions in Immunology* **11**, 412–419.

Hughes, A.L. (1992). Positive selection and interallelic recombination at the merozoite surface antigen-1 (MSA-1) locus of *Plasmodium falciparum*. *Molecular Biology and Evolution* **9**, 381–393.

Hughes, A.L. (1999). *Adaptive Evolution of Genes and Genomes*. Oxford: Oxford University Press.

Hughes, M.K. and Hughes, A.L. (1995). Natural selection on *Plasmodium* surface proteins. *Molecular and Biochemical Parasitology* **71**, 99–113.

Ina, Y. (1995). New methods for estimating the numbers of synonymous and nonsynonymous substitutions. *Journal of Molecular Evolution* **40**, 190–226.

Kimura, M. (1968). Evolutionary rate at the molecular level. *Nature* **217**, 624–626.

Kyes, S.A., Rowe, J.A., Kriek, N. and Newbold, C.I. (1999). Rifins: a second family of clonally variant proteins expressed on the surface of red cells infected with *Plasmodium falciparum*. *Proceedings of the National Academy of Sciences of the USA* **96**, 9333–9338.

Kyes, S., Pinches, R. and Newbold, C. (2000). A simple RNA analysis method shows *var* and *rif* multigene family expression patterns in *Plasmodium falciparum*. *Molecular and Biochemical Parasitology* **105**, 311–315.

Li, W.-H. (1993). Unbiased estimation of the rates of synonymous and nonsynonymous substitution. *Journal of Molecular Evolution* **36**, 96–99.

Li, W.-H., Wu, C.-I. and Luo, C.-C. (1985). A new method for estimating synonymous and nonsynonymous rates of nucleotide substitution and considering the relative likelihood of nucleotide and codon changes. *Molecular Biology and Evolution* **2**, 150–174.

Maynard-Smith, J. (1970). Population size, polymorphism, and the rate of non-Darwinian evolution. *American Naturalist* **104**, 231–236.

McDonald, J.H. and Kreitman, M. (1991). Adaptive protein evolution at the ADH locus in *Drosophila*. *Nature* **351**, 652–654.

McInerney, J.O. (1998). Replicational and transcriptional selection on codon usage in *Borrelia burgdorferi*. *Proceedings of the National Academy of Sciences of the USA* **95**, 10698–10703.

Messier, W. and Stewart, C.B. (1997). Episodic adaptive evolution of primate lysozymes. *Nature* **385**, 151–154.

Nei, M. and Gojobori, T. (1986). Simple methods for estimating the numbers of synonymous and nonsynonymous nucleotide substitutions. *Molecular Biology and Evolution* **3**, 418–426.

Nielsen, R. and Yang, Z. (1998). Likelihood models for detecting positively selected amino acid sites and applications to the HIV-1 envelope gene. *Genetics* **134**, 1271–1276.

Smith, N.H., Smith, J.M. and Spratt, B.G. (1995). Sequence evolution of the *porb* gene of *Neisseria gonorrhoeae* and *Neisseria meningitidis* – evidence of positive Darwinian selection. *Molecular Biology and Evolution* **12**, 363–370.

Suzuki, Y. and Gojobori, T. (2001). Positively selected amino acid sites in the entire coding region of hepatitis C virus subtype 1b. *Gene* **276**, 83–87.
Swanson, W.J. and Vacquier, V.D. (1995). Extraordinary divergence and positive Darwinian selection in a fusagenic protein coating the acrosomal process of abalone spermatozoa. *Proceedings of the National Academy of Sciences of the USA* **92**, 4957–4961.
Thompson, J.D., Higgins, D.G. and Gibson, T.J. (1994). CLUSTAL W: improving the sensitivity of progressive multiple sequence alignment through sequence weighting, position-specific gap penalties and weight matrix choice. *Nucleic Acids Research* **22**, 4673–4680.
Weber, J.L. (1988). Interspersed repetitive DNA from *Plasmodium falciparum*. *Molecular and Biochemical Parasitology* **29**, 117–124.
Winzeler, E.A., Liang, H., Shoemaker, D.D. and Davis, R.W. (2000). Functional analysis of the yeast genome by precise deletion and parallel phenotypic characterization. *Novartis Foundation Symposium* **229**, 105–109; discussion pp. 109–111.
World Health Organization (1997). World malaria situation in 1994. Part I. Population at risk. *Weekly Epidemiological Record* **72**, 269–274.
World Health Organization (2002). Economic costs of malaria. *Roll Back Malaria Information*, sheet 10 of 11.
Yang, Z. (1997). PAML: a program package for phylogenetic analysis by maximum likelihood. *Computer Applications in the Biosciences* **13**, 555–556.
Yang, Z. (1998). Likelihood ratio tests for detecting positive selection and application to primate lysozyme evolution. *Molecular Biology and Evolution* **15**, 568–573.
Yang, Z. (2000). Maximum likelihood estimation on large phylogenies and analysis of adaptive evolution in human influenza virus A. *Journal of Molecular Evolution* **51**, 423–432.
Yang, Z. and Bielawski, J.P. (2000). Statistical methods for detecting molecular adaptation. *Trends in Ecology and Evolution* **15**, 496–503.
Yang, Z. and Nielsen, R. (2002). Codon-substitution models for detecting molecular adaptation at individual sites along specific lineages. *Molecular Biology and Evolution* **19**, 908–917.
Yang, Z. and Swanson, W.J. (2002). Codon-substitution models to detect adaptive evolution that account for heterogeneous selective pressures among site classes. *Molecular Biology and Evolution* **19**, 49–57.
Yang, Z., Kumar, S. and Nei, M. (1995). A new method of inference of ancestral nucleotide and amino acid sequences. *Genetics* **141**, 1641–1650.
Yang, Z., Swanson, W.J. and Vacquier, V.D. (2000a). Maximum-likelihood analysis of molecular adaptation in abalone sperm lysin reveals variable selective pressures among lineages and sites. *Molecular Biology and Evolution* **17**, 1446–1455.
Yang, Z., Nielsen, R., Goldman, N. and Pederson, A.-M.K. (2000b). Codon-substitution models for heterogeneous selection pressure at amino acid sites. *Genetics* **155**, 431–449.
Zanotto, P.M., Kallas, E.G., de Souza, R.F. and Holmes, E.C. (1999). Genealogical evidence for positive selection in the nef gene of HIV-1. *Genetics* **153**, 1077–1089.

INDEX

Page numbers in *italic* indicate figures and tables

Acanthocephala *105*
Acanthocheilonema viteae 127
Acanthocolpidae *201*, 212, 217
Accacoeliidae *201*
Achatinoidea *213*
Acmaeoidea *213*
Acoela *105*
Aculeata
   nest-building 88–9
   sociality 88–9
adaptation 296
   study of 282
adaptive evolution, example of 366–73
adaptive immunity 161
adaptive molecular evolution 6, 359–79
   methodological advances 362–6
Aedes aegypti 129, 132
Aedes albopictus 127
Agrobacterium tumefaciens 33
Alaimina *109*, *130*
Alfonsiella longiscapa 87
Allocreadiidae 204, 212, 222
Allocreadioidea *201*, *208*, *209*, *210–1*, *215–6*, *218*, 219, *221*, *223*, 248
alpha-proteobacterium 14, *14*
alpha-tubulin 38
ALT proteins 163
Aloutta seniculus 294
Alveolata 20
Alysiinae 80, *81*
amino acid positions, positive selection 374
Amphiboloidea *213*
Amphilinidea 204

Ampullarioidea *213*
Anaplasma marginale 127
Anastrentha suspensa 127
ancestral diversity 334–7, *335*
ancestral genealogy, estimation of 346
Ancylostoma 130
Ancylostoma caninum 131
Ancylostoma ceylanicum 131
Ancylostoma duodenale 146
Annelida *105*, 215–6
Anopheles 266
Anopheles funestus 274
Anopheles gambiae 274
anti-bacterial factor peptides (ABF) 162
Antistrophus 86
Aotes trivirgatus 294
Apicomplexa 20, 31, 38–46, 256–8
   evolution of life cycles 257–8, *258*
   origin of plastid 256–7
apicoplast *258*
Aplacophora 210
Apocreadiata 201
Apocreadiidae *201*, 248
Apocreadioidea *201*, *208*, *210–1*, *215–6*, *218*, 219, *221*, *223*, *228*, 233
   key life cycle characteristics 248
Apoidea 75
Aquifex aeolicus 41
Arabidopsis thaliana 28, *34*, *41*
Araeolaimida *109*, *130*
Archamoebae 30
Argobacterium tumefaciens 28, *41*
Arionoidea *213*
Armadillium vulgare 127

arthropod association and vertebrate parasitism 120–2
arthropods *105*, 116, *215–6*, 233
Ascarididae 283
Ascaridomorpha *109*, 118, *130–1*
Ascaris 130
Ascaris lumbricoides 131, 132
Ascaris suum 34, *131*, 120, 132, 139, *141*, 145, *146*, 156, 158–9, 162
Aschelminthes 103–4
Aspidogastrea *201*, *204*, 206–9, *208*, *211*, 214, *218*, *221*, *223*, 225, *228*, *235*, *237*, 239, 242
Aspidogastridae *201*, 206
Astasia longa 52
Asteraceae *86*
Astomonema 114
Ateles ater 294
Atractotrematidae *201*
Aulacidea phlomica 86
Aulacidea verticillica 86
Auridistomidae *201*
Aylax 86
Azorhizobium caulinodans 34
Azygiidae *201*, 217, 224
Azygioidea *201*, *208*, *210–1*, 214, *215*, *218*, *221*, *223*, *226*, *228*, 230, 245
key life cycle characteristics 245–6

Babesia 256
Bacillus subtilis 28, 33, 41
Barbotina 86
basal metabolic rate (BMR) 291
Basommatophora *213*
Bayesian inference
  codons under positive under positive selection 373, *374*
  DEN-4 *350*
  evolutionary rates 347–50
  HIV-1 *351*
Bayesian Markov chain Monte Carlo inference (MCMC) *352*
Bayesian methods 304, 365
beta-tubulin 38
biogeography 305

Biguetius trichuroides 294
Bivesiculata *201*
Bivesiculidae *201*, 230, 245
Bivesiculoidea *201*, *208*, *210–1*, 214, *215*, *218*, 218, *221*, *223*, 224, 226, *228*
key life cycle characteristics 245
BLAST 154–5
BLASTP 367
Bolbocephalodidae 244
Bordetella pertussis 28
Borrelia burgdorferi 28, *33*
Botulisaccidae 246
Brachiopoda *105*
Brachycladiidae *201*, 212, 219
Brachycoeliidae *201*
Brachylaimidae *201*, 244
Brachylaimoidea *201*, 205, *208*, *210*, *211*, *215–6*, *218*, 220, *221*, *223*, 224–5, 234, *235*, *237*, 237–9
key life cycle characteristics 244
Brachyteles arachnoides 294
Braconidae 80
Bradyrhizobium japonicum 28
Brauninidae 244
Brevibuccidae *109*, *130*
broken-stick model 290
Brooks' Parsimony Analysis 305
Brucella ovis 33
Brugia 129, *130*
Brugia malayi 127, *131*, 132, 139, *140–1*, *146*, 144–5, 155–60, 162–3, 166
linkage and microsynteny conservation *140*
number of genes 165
Brugia pahangi 127, *131*, 160
Bryozoa *105*
Bucephalata *201*, 207, *208*, 212, 214, *216*, 222
Bucephalidae *201*, 214, 217, 220, 224, 231
Bucephaloidea *201*, *208*, *210–1*, *215–6*, *218*, *221*, *223*, 225, *228*, 230
key life cycle characteristics 246
Buchnera 111

Buchnera aphidicola 156
Bunonematomorpha *109*, *130*
Burkholderi cepacia 33

Cableia 201
Caenorhabditis 130, 132, 143
    patterns of genome evolution 137–8
Caenorhabditis briggsae 124, 128, *131*, 132, 138, *141*
Caenorhabditis elegans 4, *28*, *33*, *34*, 104, 113, 119–20, 124, 128–9, *131*, 134–45, *146*, 150–1, *153*, 155–9, 162–3, 165
    chromosome arms and chromosome centres *136*
    genes co-transcribed in operons 142
    genome 132–7, *134–5*
    HOX genes *141*
    linkage and microsynteny conservation *140*
    number of genes 149–52
    ORFeome 152–6, *153*
    orthologues 141
    proteome *147*
Caenorhabditis remanei vulgaris 113, 119
CAIC program 285, 287, 293
Callicebus moloch 294
Callicebus torquatus 294
Callimico goeldi 294
Callithrix jacchus 294
Callodistomidae *201*, 246
Calyptraeoidea *213*
Campydorina *109*, *130*
Campylobacter jejuni 41
Cantharoctonus 81
Cardiochilinae *85*
Carpediemonas membranifera 24, 38
Caulobacter crescentus 28, *33–4*
Cebus albifrons 294
Cecconia 86
central posterior density (CPD) intervals 348
Cephalaspidea *213*
Cephalobomorpha *109*, *130–1*

Cephalogonimidae *201*
Cephalopoda 210
Cephoidea 75
Ceraphronoidea 75
Ceratosolen arabicus 87
Ceratosolen bisulcatus 87
Ceratosolen capensis 87
Ceratosolen constrictus 87
Ceratosolen gailili 87
Ceratosolen notus 87
Ceratosolen pilipes 87
Ceratosolen solmsi 87
Ceratosolen vetustus 87
cercaria, behaviour during infection 224-7
cercaria, tails 223-4
Cerithioidea *211*, *213*, 213
Ceroptres 86
Chaetognatha *105*, 216
Chalcidoidea 75, 81
    aculeate Hymenoptera 88
    fig-pollinating in 87–8, *87*
    gall-forming in 84–5
chaperonin 60 (Cpn60) *28*, 29, 31, 37
Cheirogaleus major 294
Cheloninae *85*
cherry-tree picking test 325
Chiropotes chiropotes 294
Chlamydia trachomatis 33, *41*
Chlamydomonas reinhardtii 28, *41*
Chlamydophila pneumoniae 41
Chlorarachnion 16
Chlorocebus aethiops 294
Chloroflexus aurantiacus 41
Choanaflagellida *105*
Choanocotylidae *201*
Chordata *105*, 215
Chromadoria *109*, 118
Chromadorida *109*, *130*
Chromatium vinosum 28
Chrysidoidea 75
ciliates *20*, *258*
Ciona intestinalis 132
Cladorchiidae *201*
Clinocentrus 81

Clinostomidae *201*, 207, *208*, 209, *210–1*, *215*, *218*, *221*, 222, *223*, *235*, 235, *237*, 237, 245
Clinostomoidea 245
CLF/CHF ratio 288
CLG/CHG ratio 288
Clinocentrus 80–1
Clinostomidae 209
Clostridium-like bacterium 27
Clostridium perfringens 28, *33*
Clostridium thermocellum 41
CLS/CHS ratio 288
Cnidaria *105*, 215–6
coalescent methods 348
codivergence 306–7, *307*, 319, 324
codivergence events (CEs) 309, 311, 323–4, *324*, 326
codons under positive selection, Bayesian inference of 373, *374*
coevolution 324
Colobenterobius colobis 294
Colobenterobius entellus 294
Colobenterobius guerezae 294
Colobenterobius inglisis 294
Colobenterobius paraguerezae 294
Colobenterobius pesteri 294
Colobenterobius presbytis 294
Colobenterobius zakiri 294
Colobus guereza 294
combinatorial problems 312–13
comparative method 281–302
  essence of 282
  scepticism concerning 295–6
continuous time Markov process 365
Conoidea *213*
Convoluta roscoffensis 111
cophylogenetic behaviour for particular sets of model parameters 327
cophylogenetic events 306–8, *307*, 309
  distributions of numbers of *321*, 322
cophylogenetic simulation 321
cophylogeny
  confounding 324–6
  in associations with viruses 82–4
  modelling 316–22
  parallel problems *306*
  with hosts 82
cophylogeny mapping 5, 303–30
  complexity 312–18, 326
  discipline of 326
  empirical complexity testing 313
  future 327–8
  interpretability 308–9
  number of potentially optimal maps vs. max. codivergences *316*
  numbers of maps vs. degree of fit for different numbers of taxa *315*
  numbers of maps vs. number of taxa *314*
  optimality 309–10
  overview 304–6
  problem phrasing 308
  statistical inference 312
  theory involved 305
cophylogeny modelling
  phase 1, codivergence 319
  phase 2, duplication 319–22
COSPEC 0.51 program 306, 318, *321*
Cossonus 127
Cotesia 83
Cotesia congregata 85
Cotesia glomerata 85
Cotesia mariginiventris 85
Cotesia melitaearum 85
Cotesia orobenae 85
Cotesia rubecula 85
cotylocidium 206
Courtella armata 87
Courtella bekiliensis 87
CoxII 44
CRANN program 368
Cricetulus griseus 28, *33*
Crithidia fasciculata 33
Critogaster 87
cryptic organelles
  detection and identification 21–3
  in protists and fungi 9–67
  mitochondria 21–3
  plastids 23
  use of term 12

Cryptogonimidae *201*, 208, 247
cryptomonads *20*, *258*
Cryptosporidium 50, 256
Cryptosporidium parvum 257
Ctenophora *105*, *215–6*
Culex pipiens 127
Cuscuta europeae 48
Cyanidium caldarium 28
cyanobacteria 16–21, *18*, *20*
Cyanophora paradoxa 28
Cyathocotylidae 244
Cycliophora *105*
Cyclocoelidae *201*, 221, 225, 232, 247
Cynipidae
  gall-forming in 84–5
  host–plant associations in *86*
Cynipini *86*
Cynipoidea 75

Dactylogyrus 286
Daubentonia madagascariensis 294
DEN-4 333, 340–3, *341*, *345*, 349, *350*, *352*
dengue virus serotype 4 *see* DEN-4
Derogenes 201
Derogenidae *201*
Desmodorida *109*, 114, *130*
Desulfovibrio desulfuricans 41
Deuterostomia *105*, *141*
Diabrotica virgifera 127
Diastrophus 86
Dicrocoeliidae *201*, 219–20
Dictyocaulus viviparus 159
Dictyostelium discoideum 28, *34*
Didymozoidae *201*, 230, 246
Digenea 4, 198–254
  basal host group 212
  basal superfamilies 214
  character state patterns 199–200
  departures from developmental pattern 205–6
  hypothesis of evolution within Diplostomida 234–9, *235*, *237*

  hypothesis of evolution within Plagiorchiida 227–33, *228*
  individual host associations 212
  life cycle database 200–2
  life cycle evolution 197–254
    conflict between hypotheses 241–2
    questions remaining to be answered 243–4
    shortcomings of parsimony approach 240–1, *240*
  life cycle stages 205
  life cycle variation
    cercarial tail 223–4, *223*
    definitive hosts 217–20, *218*
    first intermediate hosts 209–14, *210–11*
    infection processes and cercarial behaviour *223*, 224–7
    infection processes and miracidial behaviour 220, *221*
    second intermediate hosts 214–17, *215–16*
  mapping and interpreting life cycle traits 209–39
  mapping life cycle characters 202–3
  overview 204–6
  phylogeny 199–200, *201*
  relationships and higher classification of superfamilies *208*
  relationships within Neodermata 203
dinoflagellates *20*, 42, *258*
Dioctophymatida *109*, *130*
Diphtherophorina *109*, *130*
Diplodiscidae *201*
Diplogasteromorpha *109*, *130–1*
Diplolepidini *86*
diplomonads, mitochondria in 36–8
Diplostomata *201*
Diplostomida *201*, 207, *208*, 212, *213*, 214, *216*, 218–20, *228*, 239, 242–5
  hypothesis of evolution within 234–9, *235*, *237*
  key life cycle characteristics 244–5
  major gastropod taxa as first intermediate hosts *213*

Diplostomidae *201*, 225, 244
Diplostomoidea *201*, *208*, *210–1*,
  215–6, *218*, *221*, *223*, *235*, *237*,
  237–9, 244
Diptera 80
Dirofilaria 130
Dirofilaria immitis 127, *131*
Dirofilaria repens 127
Dirrhopinae *85*
distance-based methods 337–43
divergence times, estimation of 351
diversification 287–8
diversification ratio and body size *289*
DNA-based genetic transformation of
  parasitic nematodes 165
DNA sequences 368
DNA viruses 333
Dolichoris 87
Dorylaimia *109*, 117
Dorylaimida *109*, 114, *130*
double-stranded RNA (dsRNA) 166
DR (different rates) model 346
Dracunculoidea *109*, *130*
Drosophila 72, 364
Drosophila melanogaster 28, 128–9, 132,
  137, *140*, 150–2
  number of genes 149–51
  ORFeome *153*
Drosophila orientacea 127
Drosophila recens 127
Drosophila sechellia 127
Drosophila simulans 127
Duffy blood group proteins 263
duplication, cophylogeny and 306–7,
  307, 309, 319–22

Ecdysozoa *105*, *141*
Echinodermata *105*, *215–6*, *201*
Echinostomatidae *201*, 232, 247
Echinostomatoidea *201*, *208*, *210–1*,
  214, *215–6*, *218*, 219, *221*, *223*,
  225–6, *228*, 231–2
  key life cycle characteristics 247
Echiura *105*

ectoparasitism, evolution of
  endoparasitism and koinobiosis
  79–81
EF-1α 32
EF-2 32
Ehrlichia chaffeensis 28
Ehrlichia ruminantium 28
Eimeria tenella 33
Elisabethiella baijnathi 87
Elisabethiella glumosae 87
Encephalitozoon 15, 32, 35, 49
Encephalitozoon cuniculi 32, *33–4*, 35,
  148
Encyclometridae *201*
endoparasitism, evolution from
  ectoparasitism 79–81
endosymbiosis
  mitochondria 12–16, *14*
  plastids 16–21
Enenteridae *201*, 233, 248
Enoplia *109*, 117
Enoplida *109*
Enoplina *109*, *130*
Entamoeba 15, 49
entamoebid mitosomes 30–1
Entamoeba histolytica 28, 30
Enterobius anthropopitheci 294
Enterobius bipapillatus 294
Enterobius brevicaudata 294
Enterobius buckleyi 294
Enterobius lerouxi 294
Enterobius macaci 294
Enterobius vermicularis 294
Entoprocta *105*
Eogastropoda *213*
Ephestia cautella 127
Epifagus virginiana 48
Epstein-Barr Nuclear Antigen-1
  (*EBNA-1*) 272–3
Epstein-Barr virus (EBV) 272–3
error estimation 339–40
  using MCMC 348–9
Erysipelothris rhusiopathiae 33
erythrocyte membrane protein 1
  (PfEMP-1) 366–7

Eschatocerus 33, 41, 86
Escherichia coli 166
EST analyses 162
EST-based gene discovery 157
EST datasets 154–5, 158, 160–1, 163, 165
EST matches 155
EST programme 154
EST sequencing 152, 163
EST strategies 164
Eucestoda *204*
Eucotylidae *201*, 225, 232, 248
Euglena 16–7
Euglena gracilis 28
euglenids 20
Eulemur fulvus 294
Eulemur macaco 294
eukaryotes
  and prokaryotes, difference between 10–2
  structure and function 3
Eupristina verticillata 87
Eupulmonata *211*, *213*, 213
Evanoidea 75
event costs 312
evolutionary non-independence 334–7
evolutionary rates
  Bayesian inference of 347–50
  variation through time 344
Exothecini *81*
exploitative symbioses 111–2

FabI (enoyl-ACP reductase) *41*
Fagaceae *86*
Fahrenholz's Rule 70
Fasciolidae *201*, 232, 247
fatty acid synthase (FAS) 46
Faustulidae *201*, 212, 217, 219, 222, 248
feasible map 308
Fellodistomidae *201*, 224, 230, 231, 246
Ficus 87
fig-pollinating in Chalcidoidea 87–8, *87*
Fissurelloidea *213*
Francisella tularensis 28, 33
Fugu rubripes 132
fungi, cryptic organelles in 9–67

gall-forming
  in Chalcidoidea 84–5
  in Cynipidae 84–5
Gastrotricha *105*
generalised least-squares (GLS) on a tree 339–40
generation length estimation 353–4
Geobacter metallireducens 41
Giardia 15, 37–8, 50
Giardia intestinalis 28, 36
Glaucophytes *20*
Globodera 118, *130*
Globodera pallida 145, *146*
Globodera rostochiensis 164
Globodera pallida 131, *146*
Globodera rostochiensis 131
glutathione peroxidase (GPX) 161–2
glyceraldehyde-3-phosphate dehydrogenase (GAPDH) 36–7, 43–5
Gnamptodontinae *81*, 81
Gnathostomatomorpha *109*, *130*
Gnathostomulida *105*
Gonaspis 86
Gorgocephalidae *201*, 233, 248
Gorgoderidae *201*, 212, 217, 219, 222, 224, 226, 230
Gorgoderoidea *201*, *208*, *210*, *211*, *215–6*, *218*, *221*, *223*, 225, 248
Gorilla gorilla 294
green algae *20*
green algal parasites 46–7
Green Fluorescent Protein (GFP) 40
Gryllus integer 127
Gryllus rubens 127
Gyliauchenidae *201*, 233, 248
Gymnophallidae 219, 246
Gymnophalloidea *201*, *208*, 209, *210–1*, 214, *215–6*, *218*, 218–9, *221*, *223*, 226, *228*, 230
  key life cycle characteristics 246–7
Gyrocotylidea *204*

Haemonchus 130
Haemonchus contortus 131, 142, 161

Haemophilus ducreyi 33
Haemophilus influenzae 28, *33*, 41
Haemoproteus 294
Haemoproteus kopki 294
Haemoproteus majoris 294
Haemoproteus ptyodactylii 294
Haemoproteus sylvae 294
Halicephalobus gingivalis 115, 119
Halobacterium cutirubrum 33
Halobacterium marismortui 33
Haminoeoidea *213*
Haploporidae *201*, 225–6, 232–3
Haplosplanchnata *201*
Haplosplanchnidae *201*
Haplosplanchnoidea *201*, *208*, *210–1*, *215*, *218*, 219, *221*, *223*, 225–6, *228*, 231–2
  key life cycle characteristics 247
Haploxyuris callithricis 294
Haploxyuris goeldii 294
Haploxyuris oedipi 294
Haploxyuris tamarini 294
haptophytes *20*
Hasstilesiidae 225, 234
heat shock protein *see* HSP70
Hedickiana 86
Helicobacter pylori 41
Helicoidea *213*
Helicosporidia 46–7
Helicosporidium parasiticum 46–7
Heliothis virescens 28
helminth parasite species richness across mammalian host species 290
Hemichordata *105*
Hemiperina 201
Hemiurata *201*, 207, *208*, 218, 220
Hemiuridae *201*
Hemiuroidea *201*, 205, *208*, *210–1*, 212, 214–5, *215–6*, *218*, 218, *221*, *223*, 224–6, *228*, 230
  key life cycle characteristics 246
Hepatocystis 265
Heronimata *201*
Heronimidae *201*, 220, 225, 246

Heronimoidea *201*, *208*, *210–1*, *215*, *218*, *221*, *223*, 224–5, *228*, 230
  key life cycle characteristics 246
Heronimus 246
Heterodera 118, *130*
Heterodera glycines 131, 164
Heterodera schachtii 131
heterokonts *20*, *258*
Heterophyidae *201*, 204, 208, 247
Heterorhabditis 113–4, 119, 121, 125
highest posterior density (HPD) intervals 348–50
Himalocynips 86
HIV-1 333, 341–4, 346, 349, 352, *352*, 353–4
  Bayesian inference approach *351*
  evolution rate 338
  origin of 340
  transmission history 339
HKY model 345–6
Homo sapiens 28, *33*, *34*, 132, 137–8, 151–2, *294*
  number of genes 149–51
  ORFeome *153*
host body mass and parasite species richness relationship *286*
host-driven evolution 71
host longevity and parasite body length 293
host–parasite co-adaptation
  independent contrasts method 292–3
  PER method 293–5
host–plant associations in Cynipidae *86*
host-switching 306–7, *307*, 308–9, 311–12, 318, 325–6
hosts and parasites 1–2
HOX gene cluster 143–4
HOX genes, evolution *141*, 143–144
HSP10 22
HSP70 22, 30–2, *33*, 34
human immunodeficiency virus (HIV) 332, 348
human immunodeficiency virus type 1 *see* HIV-1
hydrogenosomes 24–9

INDEX

Hymenoptera 4, 69–99
  comparative method 89–90
  comparative traits 90
  diversification 71
  DNA sequence data 77
  earliest form of parasitism found in 77
  ectoparasitoid 88
  evolution from parasitism to other
    lifestyles 84–9
  evolutionary transitions to and from
    parasitoidism 73
  host and parasitoid phylogenies 82
  non-aculeate apocritan 88
  origin of parasitism 76
  parasitoid lifestyle as evolutionary
    'dead end' 74–6
  parasitoids 71–4
  partial phylogenetic tree 78
  phylogenetic hypothesis 77
  phylogeny 74–84
  species richness 75
  summary tree 78–9
  venoms and viruses 79
hyperjungle 328

Ichneumonoidea 75, 80–1
  aculeate Hymenoptera 88
idiobionts 80
independent contrasts method 296
  host–parasite co-adaptation 292–3
Iraella 86
Ironina *109, 130*
Isoculus 86

jungles 310–12, *311*

Kardibia gestroi 87
Khoikhoiinae *85*
Kinorhyncha *105*
koinobionts 80
koinobiosis, evolution from
  ectoparasitism 79–81

Labicolidae *201*
Lagothrix lagothricha 294
Lamiaceae *86*
Lecithasteridae *201*, 212
Lecithodendriidae *201*
Legionella pneumophila 28, 41
Lemuricola baltazardi 294
Lemuricola contagiosus 294
Lemuricola daubentoniae 294
Lemuricola lemuris 294
Lemuricola microcebi 294
Lemuricola vauceli 294
Lepocreadiata *201*
Lepocreadiidae *201*
Lepocreadioidea *201, 208, 210–1*, 214,
    215–6, 218, 219, *221, 223,* 225, *228,*
    231, 233
  key life cycle characteristics 248
Leptopilina australis 127
Leptospira interrogans 28
Leucochloridiidae *201*, 225, 234, 244
Leucochloridiomorphidae 224, 238, 244
Leucocytozoon dubreuli 294
Leucocytozoon simondi 294
likelihood ratio test (LRT) 344, 366,
    370–3
Limacoidea *213*
linear regression methods 337–43,
    341–2, 375
  examples 341–3
Liolopidae 209, 245
Liposthenes kerneri 86
Lipporrhopalum tentacularis 87
Lissorchiidae *201*, 248
Listeria innocua 41
Littorinoidea *213*
Litomosoides 130
Litomosoides sigmodontis 127, *131,* 156,
    158
Lobatostoma manteri 206
Lophotrochozoa *105, 141*
Loricifera *105*
Lymnaeioidea *211,* 213, *213*
lysozyme proteins 361
lysozyme sequences 364

McDonald–Kreitman test 364
Macaca fascicularis 294
MacroCAIC program 287–8, *289*
Macroderoididae *201*, 219
malaria
  malignant human 268–74
  research 5, 255–80
  rifin family of proteins 366–73
  species base on circumsporozoite gene *259*
malaria parasites
  affiliation with arthropod vectors 266
  host affiliation 260–6
  molecular phylogenetics 261
  morphology, phylogenetics and plasmodium systematics 258–66
Malaria's Eve hypothesis 268–74
  arguments against 269
  independent information in support of 273
mammalian phylogenetic relationships 290
Markov chain Monte Carlo method *see* MCMC methods
maximum likelihood analysis of rifin data *371*
maximum likelihood estimation 343–6, *343*, *352*
maximum likelihood implementations, shortcomings of 346
maximum likelihood methods 364–5
  examples 345–6, *345*
maximum parsimony (MP) methods 364
MCMC methods 304, 346–8, 353
  error estimation using 348–9
  examples of 349–50
Megalodontoidea 75
Megalyroidea 75
Meloidogyne 118, 145, *130*
Meloidogyne hapla 131
Meloidogyne incognita 131, 164
Meloidogyne javanica 131, *141*
Mendesellinae *85*
Mermis nigrescens 158
Mermithida *109*, *130*

Mesometridae *201*
Mesorhizobium loti 34, *41*
Mesozoa 105
Messier and Stewart method 368–70
Methanosarcina mazei 33
Microcebus murinus 294
Microgastrinae *85*
Microphallidae *201*, 204, 219, 249
Microphalloidea *201*, *208*, *210–1*, *215–6*, *218*, 219, *221*, *223*, 225, 248
Microphallus fusiformis 201
Microscaphidiidae *201*
microsporidia 31
microsporidian mitosomes 31–6
migration inhibitory factor (MIF) 163
miracidium, behaviour during infection 220
Miracinae *85*
mitochondria 11–12
  case histories 24–8
  cryptic organelles 21–3
  endosymbiosis 12–16, *14*
  future directions 49–51
  origin 12–16
mitochondrial DNA (mtDNA) 22, 260, 269
mitochondria in diplomonads 36–8
molecular clock 336, 352–3
molecular evolution, neutral theory of 352–3
Mollusca *105*, 215–6
Monhysterida *109*, 114, *130*
Monogenea 204
Mononchida *109*, *130*
Monorchiata *201*
Monorchiidae *201*, 212, 222, 232, 248
Monorchioidea *201*, *208*, *210–1*, *215–6*, *218*, 219–20, *221*, *223*, *228*, 233
  key life cycle characteristics 248
Multicalycidae *201*, 206
multiple rates dated tips (MRDT) models 344, 353
Muricoidea 213
Mus musculus 33, 132

mutation rates
  and substitution rates 333–4
  differences 332
  estimation of 333
Mycobacterium leprae 33
Mycoplasma genitalium 33
Mymarommatoidea 75
Myolaimina *109, 130*
Myzostomida *105*

Naegleria gruberi 28
Naticoidea *213*
Neaylax 86
Necator 130
Necator americanus 131, *146*, 155–6, 162
Neisseria meningitides 28, 41
Nematoda 4, 101–95
  association to parasitism 112–16
  bacterial symbionts *114*
  bacterial symbiosis *127*
  body sizes 283
  Clades I, II and C&S 108
  coevolution 124–6
  definition 103
  evidence for variation in and selection on parasitism genes 160–1
  expressed sequence tag sequences *131*
  genes with parasite-specific signatures 157–9
  genome evolution 129–48
    and parasitism 148
  genomes 103–11
  genomes and parasitism 126–65
  genomics programmes *130*
  L3 rule 119–20
  limitations of current molecular phylogeny 108–11
  mitochondria 144–8
  mitochondrial genomes *146*
  molecular analyses of diversity and phylogeny 107–8
  necromenic 113
  number of genes 149–56
  number of species 107
  parasite defences 161–3
  parasites of vertebrates 116
  parasitic groups underrepresented in datasets 110
  parasitism 103–26
  patterns in parasitism of animals 119–26
  patterns of genomic evolution 138–43
  phenotype of parasitism 126
  phenotypes 104
  phylogenetic analysis of parasites 166
  phylogenetic placement of parasites 116–9
  phylogeny 106–11, *109*
  phylum status 106
  ranked as phylum 103–6
  rapid genomic change 143–4
  response to necrosis 164
  spectrum of exploitative symbioses 112–6
  tree of metazoan life *105*
Nematomorpha *105*
Nemertea *105*
Neocallimastix 25
  hydrogenase proteins 26
Neodermata
  phylogeny *204*
  relationships 203, 242
Neorickettsia risticii 28
Nephrops 207
Neritoidea *213*
Neurospora crassa 34
neutral theory 352–3, 364
Nicotiana tabacum 41
Nippostrongylus 130
Nippostrongylus brasiliensis 131, 139, 158, 163
Nitrosomonas europaea 41
non-codivergence events (NCEs) 309, 323–4, *324*, 327
non-independence problem 336
non-parametric bootstrapping of sequence data 340
nonsynonymous mutations 364

Nosema locustae 33–4
Nostoc punctiforme 41
Notocotylidae *201*
Nycticebus coucang 294
Nyctotherus ovalis 25

Oceanobacillus iheyensis 41
Odontella sinensis 28
Omphalometridae *201*
Onchocerca 130
Onchocerca flexuosa 126, *127*
Onchocerca gibsoni 127
Onchocerca gutturosa 127
Onchocerca ochengi 124, *127*, *131*, 156, 158–9
Onchocerca volvulus 124, *131*, 146
Onchocercinae 114
Oncholaimina *130*, *109*
Onychophora *105*
Opecoelidae *201*, 204, 212, 217, 249
Opiinae 80, *81*
Opisthobranchia *213*
Opistholebetidae *201*
Opisthorchiata *201*
Opisthorchiidae *201*
Opisthorchioidea *201*, 205, *208*, 208, 210, 211, 214, *215*–6, *218*, 219–20, 221, 223, 224–5, *228*, 231–3, 247
  key life cycle characteristics 247
Opisthotrematidae *201*
Orchipedidae *201*
ORFeome/proteome evolution 148–65
organellar genomes, size and characteristics *13*
Orthogastropoda *213*
Orussoidea 75–6
Oryza sativa 41
Oscheius brevesophaga 142
Ostertagia 130
Ostertagia ostertagi 131
Ostreococcus tauri 35
Oxyuridae 283
Oxyuridomorpha *109*, 110, 118, *130*
oxyuroid nematodes and primate hosts, phylogenies of 294, *294*

oxyuroids, parasite body length 295, *295*

Pachypsolidae *201*
pairwise distance linear regression 338–9, 341–3, *341–2*
pairwise distance methods 362–3, 368
pairwise linear regression *352*
PAML 345–6, 368
Pan troglodytes 294
Panagrellus silusiae 139
Panagrolaimomorpha *109*, 114, *130–1*
Panteliella 86
Papaveraceae *86*
Papio hamadryas 294
Parabasalia 29
parabasalian hydrogenosomes 26–9
parabasalids 27
Paragonimidae *201*
Paramphistomata 207, *208*, 224–6, 231
Paramphistomoidea *201*, *208*, *210*, *211*, *215*, *218*, 219, *221*, *223*, *228*, 231–2
  key life cycle characteristics 247
Paraoxyuronema atelis 294
Paraoxyuronema brachytelisis 294
Paraoxyuronema lagothricis 294
parasite body length
  and host longevity 293
  of oxyuroids and primate longevity 295, *295*
parasite evolutionary ecology 5, 281–302
parasite species richness correlations *292*
parasitic lineages, number surviving *320*
parasitic phenotype 148–65
  genes and proteins implicated in 156–61
parasitic plants 47–9
parasitism
  and phylogeny 70–1
  evolution of 1–7
  nature of 2
  origins of 112

use of term 111–12
parasitoids
  comparative method 89–90
  cophylogeny with hosts 82
  definition 71
  future prospects 89–90
  host specificity 79
  Hymenoptera 71–4
  use of term 72
parasitology 2
parasitome 156–61
Parastrongyloides 130
Parastrongyloides trichosuri 122, *123*, *131*
Parnipinae *86*
parsimony approach 240–1, *240*
Partuloidea *213*
Patellogastropoda *213*
Patelloidea *213*
Paulinella chromatophora 19
PAUP* program 317, 345–6
Pediaspis 86
Pegoscapus gemellus 87
Pegoscapus hoffmeyerii 87
Pegoscapus lopesi 87
Pelodera strongyloides 119
Periclistus 86
peroxisomal targeting signal 1 (PTS1) 26
Petunia 41
Phaeodactylum tricornutum 36, *41*
Pharmacosycea 87
Philinoidea *213*
Philocaenus warei 87
Philophthalmidae *201*, 232, 247
Photorhabdus 114, 125
photosystems I and II 16
PHYLIP program 346
phylogenetic constraints 283
phylogenetic distance matrix 290, *291*
phylogenetic effects 283, 296–7
phylogenetic eigenvector regression (PER) method 290–5, *292*
phylogenetic inertia 283
phylogenetic niche conservatism 283, 296–7

phylogenetically independent contrasts method 284–7
phylogenetically structured environmental variation 296–9, *297*
Phytopthora 44
Pisum sativum 28, *33–4*
Pithecia pithecia 294
Placozoa *105*
Plagiorchiida *201*, 207, 208, 212, *216*, 217–8, *235*, *237*, 239, 242, 245–9
  hypothesis of evolution within 227–33, *228*
  major gastropod taxa as first intermediate hosts *213*
Plagiorchiidae *201*
Plagiorchioidea *201*, *208*, *210–1*, *215–6*, *218*, *221*, *223*, 248
Plasmodium 38, 40, 44, 255–6
  definitive hosts 258
  evolution 258
  evolutionary divergence 260–1
  intermediate hosts 258
  phylogeny *265*
Plasmodium agamae 265
Plasmodium atheruri 265
Plasmodium aurophilum 259, 265
Plasmodium berghei 259, 265
Plasmodium brasilianum 259, 261–2, 264, 266
Plasmodium chabaudi 265
Plasmodium chiricuhae 265
Plasmodium cynomolgi 259, 265
Plasmodium elongatum 265
Plasmodium fairchildi 265
Plasmodium falciparum 6, *28*, 39, *41*, *259*, 259, 261, *263–4*, *265*, 266, 360, 366–7, *369*, 374–6
  ancestry 270
  "ancient" origin 269
  constraints on synonymous substitutions 272–4
  evolutionary history 267
  extant distribution 268, 270–2
  extant populations 268

genome sequence 268
malignant human malaria 266–74
mitochondrial genome 269
most recent common ancestor
   (MRCA) of extant populations
   271
parasite-vector-host system 274
polymorphism 270
rhoptry-associated protein gene 270
worldwide distribution 274
Plasmodium fieldi 265
Plasmodium floridensi 265
Plasmodium gallinaceum 265
Plasmodium giganteum 265
Plasmodium gonderi 265
Plasmodium hylobati 265
Plasmodium inui 265
Plasmodium knowlesi 259, 263, 265
Plasmodium malariae 259, *259*, 261–4, 265
Plasmodium mexicanum 265
Plasmodium ovale 259, 265
Plasmodium reichenowi 259, 262, 264, 265, 266–7, 269–70
Plasmodium relictum 265
Plasmodium simium 259, 261–2, 264, 265
Plasmodium simovale 259, 265
Plasmodium vinckeri 265
Plasmodium vivax 259, *259*, 261–4, *265*
Plasmodium yoelii 28, *41*, 259, 265
Platygastroidea 75
Platyhelminthes *105*, *216*, 242
Platyscapa soraria 87
plastids 11–12
   case histories 38–49
   cryptic organelles 23
   endosymbiosis 16–21
   endosymbiotic origins and spread *18*
   evolutionary history in
      eukaryotes *20*
   future directions 51–3
   in parasitic plants 47–9
Plectida *109*, *130*
Pleistodontes froggati 87

Pleistodontes imperialis 87
Pleurogenidae *201*, 219
Pogonophora *105*
polydnaviruses, coevolution with
   braconid parasitoid wasps 85–6
Polygyroidea *213*
Polyplacophora 210
Pongo pygmaeus 294
Porifera *105*
Porphyra purpurea 28, 33
Porphyromonas gingivalis 28
Postgaardi mariagerensis 24
Pratylenchus 130
Pratylenchus penetrans 131
Priapulida *105*
primate/lentivirus tanglegram *325*
principal coordinate analysis (PCoA)
   290, *291*, 294
prior probabilities 347
Pristionchus 130
Pristionchus lheritieri 115
Pristionchus pacificus 131, *141*, 142
probability theory 347
Prochlorococcus marinus 41
Procolobus badius 294
Proctotrupoidea 75
prokaryotes and eukaryotes, difference
   between 10–12
Pronocephalata *201*
Pronocephalidae *201*
Pronocephaloidea *201*, 205, *208*, *210–1*, *215*, *218*, 219–20, *221*, *223*, *228*, *231–2*
   key life cycle characteristics 247
Propithecus verrauxei 294
Prosthogonimidae *201*, 219
protists, cryptic organelles in 9–67
Protenterobius nycticebi 294
Proterodiplostomidae 244
Prototheca 46–7, 52
Prototheca richardsi 47
Psalteriomonas lanterna 24
Pseudomonas aeruinosa 41
Pseudoterranova decipiens 158
Psilostomidae *201*, 232, 247

Ptychogonimidae 212, 217
Pulmonata *213*
Pupilloidea *213*
Pyramidelloidea *213*
Pyrenomonas salina 28
pyridine nucleotide transhydrogenase (PNT) 30–1
pyridoxal-5'-phosphate-dependent cysteine desulfurase (IscS) 29
pyruvate:ferredoxin oxidoreductase (PFOR) 26–7, 31
pyruvate dehydrogenase (PDH) 32, *34*, 35
Pythium 44

Raja clavata 207
Ralstonia metallidurans 41
rate-per-lineage models 353
Rattus norvegicus 28, *34*
reconciled tree analysis 308
red algae *20*
rediae 220–2, *221*
relative rate ratio test (RRRT) 370, 373
Renicolidae *201*, 219
replacement-invariable (RI) sites 364
replacement-variable (RV) sites 364
Rhabdias 122
Rhabdias bufonis 123
Rhabdiopoeidae *201*
Rhabditia 119–20
Rhabditina *109*
Rhabditis coarctata 119
Rhabditis orbitalis 119
Rhabditomorpha *109*, 114, *130–1*
Rhabditophanes 123, *123*
Rhigonematomorpha *109*, 118 *130*
Rhodobacter capsulatus 28
Rhodobacter sphaeroides 41
Rhodospirillum rubrum 41
Rhodus 86
Rhoophilus 86
rhoptry-associated protein gene 270
Rhysipolis 80–1, *81*
Rhytidoidea *213*

Rickettsia 156
Rickettsia prowazekii 28, *33–4*, *41*
rif (repetitive intersperses family) gene family 367
rifin data, ML analysis of *371*
rifin family genes 375
rifin family of proteins 366–73
Rissooidea *211*, *213*, 213
RNA interference (RNAi) 166
RNA viruses 333
Rogadinae, partial phylogenetic tree *81*
Rogadini *81*
Romanomermis culcivorax 145
root-to-tip linear regression 336, *336*, 337–8, 341–3, *341–2*, 352
Rosaceae *86*
Rotifera *105*
Rugogastridae *201*, 206

Saccharomyces cerevisiae 33–4, 137, 152, *153*
Saguinus nigricollis 294
Saguinus oedipus 294
Saimiri sciureus 294
Sanguinicolidae *201*, 207, *208*, 209, 210, *211*, 214, *215*, *218*, *221*, 222, *223*, 224, 234, *235*, 235, *237*, 245
Scaphopoda 210
Schistosomatidae *201*, 207, 222, 224, 235, 245
Schistosomatoidea *201*, 207, *208*, 209, 210–1, *215–6*, *218*, *221*, *223*, 224, 235, *237*, 237–9
 key life cycle characteristics 245
Schizosaccharomyces cerevisiae 28, *33–4*
Sclerodistomidae *201*
Semnopithecus entellus 294
Seres solweziensis 87
silent-invariable (SI) sites 364
silent-variable (SV) sites 364
single rate dated tips (SRDT) model 343, *343*, 345, *345*, 346

Sinorhizobium meliloti 28, *33–4*
Siphonarioidea *213*
Sipunculida *105*
Siricoidea 75
small subunit ribosomal DNA (SSU rDNA) 124
small subunit ribosomal RNA (SSU rRNA) 13, 15, 25, 32, 38–9, 46, 107–8, 110, 121–3
software packages for comparative and statistical analyses *298*
Solanum tuberosum 33
Spinacia oleracea 33
Spirorchiidae *201*, 207, 222, 224, *235*, 235, *237*, 245
Spiruria 118, 121–2
Spirurida 121
Spirurina *109*, 119, 121, 167
Spiruromorpha *109*, 114, *130*, *131*
sporocysts 220–2, *221*
Steinernema 113–4, 121, 125, *130*
Steinernema carpocapsae 123
Steinernema feltiae 131
Steinernema kari 123
Stephanoidea 75
Stephanostomum 201
Stichocotyle nephropis 207
Stichocotylidae 206
Stichorchis 221
Stilbonematidae 114
Streptomyces coelicolor 28, *33*, 137
Strigeidae *201*, 244
Striropius 81
Strongylida 119
Strongyloidea 167
Strongyloides 122, *130*
Strongyloides cebus 123
Strongyloides fuelleborni 123, *123*
Strongyloides kelleyi 124
Strongyloides papillosus 123
Strongyloides ratti 120, *123*, *131*, *141*
Strongyloides stercoralis 123, *131*
Strongyloides suis 123
Strongyloides venezuelensis 123
Strongyloides westeri 123

Strongyloidoidea 118–9, *123*
life cycle evolution 122–3, *122*
substitution process 365
substitution rates
and mutation rates 333–4
estimates of 337
Succineoidea *213*
superoxide dismutase (SOD) 162
Sus scrofa 34
Sycidium 87
Sycocarpus 87
Sycomorus 87
symbiosis 111
spectrum of exploitative 112–16
Syncoeliidae *201*
Synechococcus 28, *41*
Synechocystis 28, *33*, *41*
Synergus 86
Syngamus trachea 159
synonymous mutations 364
Synophromorpha 86

Tandanicolidae *201*, 224, 230, 246
tanglegrams 294, 308–10, *311*, 325
Tardigrada *105*
Teladorsagia 130
Teladorsagia circumcincta 131
Telorchiidae *201*
Tenthredinoidea 75
Teratocephalidae *109*, *130*
Tetrapus americanus 87
Tetrapus costaricanus 87
Thelazia lachrymalis 127
Thermosynechoccus elongatus 41
Thermus thermophilus 28
thioredoxin peroxidase (TPX) 162
Thermoanerobacter brockii 28
Thrasorinae 86
TICs and TOCs (Translocation Inner Chloroplast membrane and Translocation Outer Chloroplast membranes) 19
TIMs and TOMs (Tranlocation Inner Mitochondrial membrane and

Translocation Outer
  Mitochondrial membranes) 15
TipDate 345–6
tissue-dwelling parasites 116
Tobrilina *109*, *130*
Toxocara 130
Toxocara canis 131
Toxoplasma 40, 44, 256
Toxoplasma gondii 28, 39
Trachipleistophora hominis 33, 34–5
Trachypithecus phayrei 294
Transversotremata *201*
Transversotrematidae *201*, 229, 239, 242, 245
Transversotrematoidea *201*, *208*, *210–1*, 214, *215*, *218*, 218, *221*, *223*, 224, *228*
  key life cycle characteristics 245
TREEMAP 318, 325
TREEMAP2 306, 323
Trefusilina *109*, *130*
Trematoda 198, 203, 206, 210, 242
  phylogeny 207–9
Treponema pallidum 28
Trichinella 116, *130*
Trichinella spiralis 112, 117, 120, *131*, *141*, 145, *146*
Trichinellida *109*, *130–1*
Trichodesmium 41
Trichodorus 117
Trichomonas 15, 27–9
Trichomonas vaginalis 28
Trichuris 130
Trichuris muris 131
Trimastix pyriformis 24
triosephosphate isomerase (TPI) 37
Triplonchida *109*
Tripylina *109*, *130*
Tripyloidina *109*, *130*
Triticum aestivum 28
Tritrichomonas foetus 24
Trochoidea *213*
Troglotrematidae *201*
Trypanosoma 17
Trypanosoma brucei 28

Trypanosoma cruzi 28, 33
Trypanoxyuris callicebi 294
Trypanoxyuris clementinae 294
Trypanoxyuris croizati 294
Trypanoxyuris microon 294
Trypanoxyuris minutus 294
Trypanoxyuris satanas 294
Trypanoxyuris sceleratus 294
Trypanoxyuris trypanuris 294
Ts/Tv (transition/transversion) ratio 317
'Turbellaria' *204*
Tylenchia 118
Tylenchina *109*
Tylenchomorpha *109*, 114, *130*, *131*
type I errors 282
type II errors 282

UPGMA 336
Urostigma 87

Valvatoidea *213*
variation partitioning in phylogenetic context 297, *297*
Varimorpha necatrix 33
Velutinoidea *213*
Vermetoidea *213*
Vertebrata *216*
vertebrate immune system 161
vertebrate parasitism and arthropod association 120–2
Vespoidea 75
Vetustia 86
viral evolutionary rates 331–58
viruses
  cophylogeny in associations with 82–4
  evolutionary divergence 6

Waterstoniella 87
Wiebesia pumilae 87
Wiebesia puntatae 87

Wolbachia 114, 125–6, *127*
Wucheria 130
Wucheria bancrofti 127, *131*, 160

Xenopus laevis 34
Xenorhabdus 114
Xestophanes 86
Xenorhabdus 125
Xiphidiata *201*, 207, *208*, 214–5, *216*, 217, 219, 220, 224, 225, *228*, 231, 233, 248
    key life cycle characteristics 248

Xiphinema 114, 125, *130*
Xyeloidea 75

Yule–Markov model 318

Zea mays 28, *34*
Zeldia 130
Zeldia punctata 131
Zoogonidae *201*, 217, 219
Zymomonas mobilis 34

Contents of Volumes in This Series

Volume 40

Part 1 *Cryptosporidium parvum* and related genera

Natural History and Biology of *Cryptosporidium parvum* 5
    S. TZIPORI AND J.K. GRIFFITHS
Human Cryptosporidiosis: Epidemiology, Transmission,
    Clinical Disease, Treatment and Diagnosis 37
    J.K. GRIFFITHS
Innate and Cell-mediated Immune Responses to
    Cryptosporidium parvum... 87
    C.M. THEODOS
Antibody-based Immunotherapy of Cryptosporidiosis 121
    J.H. CRABB
Cryptosporidium: Molecular Basis of Host–Parasite
    Interaction 151
    H. WARD AND A.M. CEVALLOS
Cryptosporidiosis: Laboratory Investigations and
    Chemotherapy 187
    S. TZIPORI
Genetic Heterogeneity and PCR Detection of
    Cryptosporidium parvum.. 223
    G. WINDMER
Water-borne Cryptosporidiosis: Detection Methods and
    Treatment Options. 241
    C.R. FRICKER AND J.H. CRABB

Part 2 *Enterocytozoon bieneusi* and Other Microsporidia

Biology of Microsporidian Species Infecting Mammals 283
    E.S. DIDIER, K.F. SNOWDEN AND J.A. SHADDUCK
Clinical Syndromes Associated with Microsporidiosis 321
    D.P. KOTLER AND J.M. ORENSTEIN
Microsporidiosis: Molecular and Diagnostic Aspects....................... 351
    L.M. WEISS AND C.R. VOSSBRINCK

Part 3 *Cyclospora cayetanensis* **and related species**

Cyclospora cayetanensis 399
  Y.R. ORTEGA, C.R. STERLING AND R.H. GILMAN

Volume 41

Drug Resistance in Malaria Parasites of Animals and Man 1
  W. PETERS
Molecular Pathobiology and Antigenic Variation of
  Pneumocystis carinii 63
  Y. NAKAMURA AND M. WADA
Ascariasis in China 109
  PENG WEIDONG, ZHOU XIANMIN AND D.W.T. CROMPTON
The Generation and Expression of Immunity to *Trichinella spiralis*
  in Laboratory Rodents.. 149
  R.G. BELL
Population Biology of Parasitic Nematodes: Applications of Genetic Markers 219
  T.J.C. ANDERSON, M.S. BLOUIN AND R.M. BEECH
Schistosomiasis in Cattle 285
  J. DE BONT AND J. VERCRUYSSE

Volume 42

The Southern Cone Initiative Against Chagas Disease 1
  C.J. SCHOFIELD AND J.C.P. DIAS
Phytomonas and Other Trypanosomatid Parasites of Plants and Fruit 31
  E.P. CAMARGO
Paragonimiasis and the Genus *Paragonimus* 113
  D. BLAIR, Z.-B. XU AND T. AGATSUMA
Immunology and Biochemistry of *Hymenolepis diminuta* 223
  J. ANREASSEN, E.M. BENNET-JENKINS AND C. BRYANT
Control Strategies for Human Intestinal Nematode Infections 277
  M. ALBONICO, D.W.T. CROMPTON AND L. SAVIOLI
DNA Vaccines: Technology and Applications as Anti-parasite and
  Anti-microbial Agents 343
  J.B. ALARCON, G.W. WAINE AND D.P. MCMANUS

Volume 43

Genetic Exchange in the *Trypanosomatidae* 1
  W. GIBSON AND J. STEVENS
The Host–Parasite Relationship in Neosporosis............................. 47
  A. HEMPHILL
Proteases of Protozoan Parasites 105
  P.J. ROSENTHAL

Proteinases and Associated Genes of Parasitic Helminths 161
  J. TORT, P.J. BRINDLEY, D. KNOX, K.H. WOLFE AND J.P. DALTON
Parasitic Fungi and their Interactions with the Insect Immune System 267
  A. VILCINSKAS AND P. GÖTZ

Volume 44

Cell Biology of *Leishmania* 1
  E. HANDMAN
Immunity and Vaccine Development in the Bovine Theilerioses 41
  N. BOULTER AND R. HALL
The Distribution of *Schistosoma bovis* Sonsino, 1876 in Relation to Intermediate
  Host Mollusc–Parasite Relationships 98
  H. MONÉ, G. MOUAHID AND S. MORAND
The Larvae of Monogenea (Platyhelminthes) 138
  I.D. WHITTINGTON, L.A. CHISHOLM AND K. ROHDE
Sealice on Salmonids: Their Biology and Control 233
  A.W. PIKE AND S.L. WADSWORTH

Volume 45

The Biology of some Intraerythrocytic Parasites of Fishes, Amphibia and
  Reptiles 1
  A.J. DAVIES AND M.R.L. JOHNSTON
The Range and Biological Activity of FMRFamide-related Peptides and
  Classical Neurotransmitters in Nematodes 109
  D. BROWNLEE, L. HOLDEN-DYE AND R. WALKER
The Immunobiology of Gastrointestinal Nematode Infections in
  Ruminants 181
  A. BALIC, V.M. BOWLES AND E.N.T. MEEUSEN

Volume 46

Host–Parasite Interactions in Acanthocephala: a Morphological Approach ... 1
  H. TARASCHEWSKI
Eicosanoids in Parasites and Parasitic Infections 181
  A. DAUGSCHIES AND A. JOACHIM

Volume 47

An Overview of Remote Sensing and Geodesy for Epidemiology and Public
  Health Application 1
  S.I. HAY
Linking Remote Sensing, Land Cover and Disease 37
  P.J. CURRAN, P.M. ATKINSON, G.M. FOODY AND E.J. MILTON

Spatial Statistics and Geographic Information Systems in
    Epidemiology and Public Health 83
    T.P. ROBINSON
Satellites, Space, Time and the African Trypanosomiases 133
    D.J. ROGERS
Earth Observation, Geographic Information
    Systems and Plasmodium falciparum Malaria
    in Sub-Saharan Africa 175
    S.I. HAY, J. OMUMBO, M. CRAIG AND R.W. SNOW
Ticks and Tick-borne Disease Systems in Space and from Space 219
    S.E. RANDOLPH
The Potential of Geographical Information Systems (GIS)
    and Remote Sensing in the Epidemiology and Control of
    Human Helminth Infections.. 247
    S. BROOKER AND E. MICHAEL
Advances in Satellite Remote Sensing of Environmental Variables for
    Epidemiological Applications 293
    S.J. GOETZ, S.D. PRINCE AND J. SMALL
Forecasting Disease Risk for Increased Epidemic Preparedness in
    Public Health 313
    M.F. MYERS, D.J. ROGERS, J. COX, A. FLAUHALT AND S.I. HAY
Education, Outreach and the Future of Remote
    Sensing in Human Health 335
    B.L. WOODS, L.R. BECK, B.M. LOBITZ AND M.R. BOBO

Volume 48

The Molecular Evolution of Trypanosomatidae 1
    J.R. STEVENS, H.A. NOYES, C.J. SCHOFIELD AND W. GIBSON
Transovarial Transmission in the Microsporidia 57
    A.M. DUNN, R.S. TERRY AND J.E. SMITH
Adhesive Secretions in the Platyhelminthes 101
    I.D. WHITTINGTON AND B.W. CRIBB
The Use of Ultrasound in Schistosomiasis 225
    C.F.R. HATZ
Ascaris and ascariasis 285
    D.W.T. CROMPTON

Volume 49

Antigenic Variation in Trypanosomes: Enhanced Phenotypic Variation in a
    Eukaryotic Parasite 1
    J.D. BARRY AND R. MCCULLOCH
The Epidemiology and Control of Human African Trypanosomiasis 71
    J. PÉPIN AND H.A. MÉDA

Apoptosis and Parasitism: from the Parasite to the Host Immune
  Response 132
  G.A. DosReis and M.A. Barcinski
Biology of Echinostomes except *Echinostoma* 163
  B. Fried

Volume 50

The Malaria-Infected Red Blood Cell: Structural and Functional Changes ...   1
  B.M. Cooke, N. Mohandas and R.L. Coppel
Schistosomiasis in the Mekong Region: Epidemiology and
  Phylogeography 88
  S.W. Attwood
Molecular Aspects of Sexual Development and Reproduction in Nematodes
  and Schistosomes 153
  P.R. Boag, S.E. Newton and R.B. Gasser
Antiparasitic Properties of Medicinal Plants and Other Naturally Occurring
  Products 200
  S. Tagboto and S. Townson

Volume 51

Aspects of Human Parasites in which Surgical Intervention
  May Be Important   1
  D.A. Meyer and B. Fried
Electron-transfer Complexes in *Ascaris* Mitochondria 95
  K. Kita and S. Takamiya
Cestode Parasites: Application of *In Vivo* and *In Vitro* Models for
  Studies on the Host–Parasite Relationship 133
  M. Siles-Lucas and A. Hemphill

Volume 52

The Ecology of Fish Parasites with Particular Reference to Helminth Parasites
  and their Salmonid Fish Hosts in Welsh Rivers: A Review of Some of the
  Central Questions   1
  J.D. Thomas
Biology of the Schistosome Genus *Trichobilharzia* 155
  P. Horák, L. Kolářová and C.M. Adema
The Consequences of Reducing Transmission of *Plasmodium falciparum* in
  Africa 234
  R.W. Snow and K. Marsh
Cytokine-Mediated Host Responses during Schistosome Infections; Walking
  the Fine Line between Immunological Control and Immunopathology 265
  K.F. Hoffmann, T.A. Wynn and D.W. Dunne

Volume 53

Interactions between Tsetse and Trypanosomes with Implications for the
Control of Trypanosomiasis 1
   S. Aksoy, W. C. Gibson and M. J. Lehane
Enzymes Involved in the Biogenesis of the Nematode Cuticle 85
   A. P. Page and A. D. Winter
Diagnosis of Human Filariases (Except Onchocerciasis) 149
   M. Walther and R. Muller